ILL-STARRED
CAPTAINS

ILL-STARRED CAPTAINS

CAPTAINS

FLINDERS and BAUDIN

Anthony J. Brown

With a foreword by
Tim Flannery

STACKPOLE
BOOKS

For Malcolm and Jackie,
for Michael,
and for Andrew and Eve-Marie

Published in North America by
STACKPOLE BOOKS
5067 Ritter Road
Mechanicsburg, PA 17055
www.stackpolebooks.com

Originally produced and published in Australia by
Crawford House Publishing Pty Ltd

Published in association with the
Royal Geographical Society of South Australia

Cover design by Travis Crawford

ISBN 0-8117-0849-7

Printed in Malaysia

10 9 8 7 6 5 4 3 2 1

FIRST EDITION

Contents

Foreword

The early 19th-century voyages of Captains Matthew Flinders and Nicholas Baudin arguably represent the high-tide mark of Australian maritime exploration. The expeditions were sent out – Flinders by England and Baudin by France – with the same general purpose in mind: to explore and map the largely uncharted continent soon to be known as Australia. Despite their parallel origins the expeditions were very different. Matthew Flinders was dispatched by the Admiralty in the *Investigator*, a vessel similar in construction to Cook's bark *Endeavour*. It was a rather poorly supplied enterprise – the best the British Admiralty could do in a time of war – yet through Flinders's sheer determination and brilliance it resulted in such detailed and accurate mapping of much of Australia's coastline that Flinders's maps were drawn on for nearly two centuries. The French sailed in the corvettes *Géographe* and *Naturaliste*. They were equipped with the best scientific materials and *savants* that France could assemble. The expedition was sent out by Napoleon himself at the height of his power; its opulence reflecting both his personal interest in the great south land and the achievements of French enlightenment science. Unlike that of Flinders, the Baudin expedition's greatest achievement was not maps, but tens of thousands of biological and ethnographic specimens, including priceless observations of Tasmania's Aborigines.

At the beginning of the 19th century, England and France were jealous rivals, more frequently at war than at peace. One might have predicted that Flinders and Baudin would have carried the war with them into the Southern Hemisphere, but these were gentlemen of the Enlightenment, and war was not the absolute horror it was to become the following century. Instead the English and French met as men of science with a common interest, although rivalry was never far from the surface. What strikes me most about this story of the ill-starred captains, so ably told by historian Anthony Brown, is the way that science overcomes, even if only tentatively, the impulse to war.

Not all, however, is lightness and peace in the story of Flinders and Baudin, for it reveals the most base as well as the most noble of human instincts as they were played out during the geographic, scientific and anthropological discovery of Australia by Europe. As the book's title indicates, it also tells of the hard fate that can await even the best of men. Brown's historical research is impeccable, bringing to light much that was hitherto unknown, while his ingenious interweaving of the tales of these two very different expeditions brings the story of Australia's exploration to life in a riveting and insightful new narrative.

Tim Flannery
Director, South Australian Museum

Preface

'The only history worth reading is that written at the time of which it treats; the history of what is done and seen, out of the mouths of the men who did and saw it.' Ruskin's sweeping generalisation is perhaps anathema to the professional historian, whose business it is to interpret the past through systematic and rigorous examination of the surviving evidence, but I suspect many general readers share his view. It seems especially applicable to the narratives of the maritime explorers and discoverers, from the 15th century onwards, whose voyages enlarged the boundaries of the known world and challenged man's understanding of the natural order.

Ill-Starred Captains is the story, as distinct from a history, of two such voyages – the French and British expeditions of discovery to the *Terres Australes* or New Holland (yet to become Australia) from 1801 to 1803. They took place amid the first world war, the struggle between Revolutionary and Napoleonic France and Great Britain, which involved all the European powers and, briefly, the fledgling United States of America, and lasted, with just two short breaks, from 1792 to 1815. The war is no more than a distant thunder, but its impact on the two expeditions – and on Matthew Flinders personally – plays an integral part in the story.

The narrative is based, in the main, on the words of 'the men who did and saw it'. I have used the printed and manuscript records held in libraries, museums and maritime archives to reconstruct the 'lived experience' of both voyages. They supply the voices not only of the captains, Nicolas Baudin and Matthew Flinders, but also those of the naturalists Francois Péron and Robert Brown, and many others, from Seaman Smith on the *Investigator* to Captain Milius who sailed the *Géographe* home to France after Baudin's death. Governor King of New South Wales and Governor Decaen of Isle de France (Mauritius) both have their say, as does Flinders' patron, Sir Joseph Banks. Ann Chappelle,

Flinders' bride of three months separated from her husband for nine long years, and Mary Beckwith, the convict girl who sailed with Baudin, are also part of the story.

The structure of the book developed, by trial and error and with many false starts, over several years. It did not come easily, for the subject matter involved two major expeditions, each absent from Europe for three or more years, six vessels, the circumnavigation of Australia, the exploration of vast stretches of the continent's south, west, north-west and north coasts, the systematic investigation of its flora and fauna, shipwreck, the death of one captain and the captivity for six and a half years of the other.

The idea for the present framework – not least the interweaving of the present tense with the factual prose of historical narrative to add immediacy to the story – came from reading Penelope Lively's complex novel *City of the Mind*. I am comfortable with it, and it has led to the sort of book I aimed to write – one that might fit within the age-long tradition of 'sea voyage narratives', from the *Odyssey* to the present, embracing 'both literature and history, imagination and expectation' (to quote Robert Foulke in his recent study of the genre). Readers will decide for themselves whether I have succeeded.

During my research a correspondent in Perth, Dr Tom Cullity, wrote to say, 'I can't really imagine how these astonishing brave men felt, behaved and lived.' No more can I. They are far more removed from our experience than are the American and Russian astronauts circling Earth, in instantaneous communication with their tracking stations in the Global Village. Once they left port Baudin's and Flinders' men 'vanished trackless into blue immensity', their whole world the hundred or so feet from stem to stern.

Yet their words, set down in their journals, in reports to government officials and letters to colleagues, family and friends back home, and in later published narratives, survive to give us a glimmer of understanding of what their lives were like, crammed together on their stinking, leaking, overcrowded ships. For today's reader these documents throw random shards of light on particular actions and events, revealing something of the relationships and tensions on board, of their individual hopes and fears, and of the rivalries between the English and French. I have tried to remain impartial, according to each narrator the right to his (and occasionally her) viewpoint, and leaving readers the space to draw their own conclusions.

Searching for a fitting epitaph for the two expeditions and their commanders, I came across these lines by W.H. Auden: 'No, no one was to blame. Blame if you like the human situation.'

Acknowledgements

During the six years of research and writing that have gone into *Ill-Starred Captains*, I have received generous support and encouragement from many people, in Australia and overseas. My particular thanks go to the following fellow-travellers on a voyage of discovery:

To David Moore, of Aldershot, UK, for so warmly welcoming me to his family home on a visit to England in 1996, for providing privileged access to the *Investigator* diary of Robert Brown (in press), and not least for his untiring efforts to visit every significant Flinders site in Lincolnshire; to Huguette and Madeleine Ly-Tio-Fane of Beau Bassin, Mauritius, for unreservedly sharing their research into, respectively, Flinders's detention on Mauritius and Baudin's earlier career as a botanical voyager; to Jacqueline Bonnemains, Conservateur of the Lesueur Collection, Muséum d'Histoire Naturelle, at Le Havre, France, for introducing me to the art treasures in the collection, and for her kindness in selecting and making available so many of the illustrations in this book; and to Dr Tom Cullity of Perth, for the gift of his insightful monograph on the disappearance of Helmsman Vasse off Wonnerup Beach, near Busselton, Western Australia. I owe a special debt of gratitude to my long-time friend Jack Ward, of Melbourne, whose offer of hospitality in Paris first opened a window of opportunity for research in France.

My thanks are also due to Dr Frank Horner, author of *The French Reconnaissance*, for his advice and comments on an early section of my manuscript, and to Professor Alan Atkinson for information on Judge Advocate Richard Atkins. Jenny Treloar and Christine Cooper (née Cornell) have made significant contributions, Jenny by translating many passages from the original documents that tested my undergraduate French, and Christine by permitting me to quote extensively from her masterly translations of Baudin's journal and volume II of Péron and Freycinet's *Voyage de découvertes aux terres australes* (unpublished).

The unfailing help of Valerie Sitters, Librarian of the Royal Geographical Society of SA, is also gratefully acknowledged.

I appreciate the ready assistance I have received from staff of the following institutions: the Mortlock Library, State Library of South Australia; the Mitchell Library, State Library of New South Wales; the La Trobe Library, State Library of Victoria; the Manuscript and Pictorial Reference Collections of the National Library of Australia; Flinders University Library; the South Australian Museum; Thomas Ramsay Collection, Scotch College, Melbourne (Mr R. Briggs, Archivist); and the Kerry Stokes Collection, Perth (Mr J. Stringer, Curator). The Natural History Museum, London, the National Maritime Museum, Greenwich, UK, and the United Kingdom Hydrographic Office, Taunton, have also been most helpful.

I am likewise indebted to the governing bodies of these institutions – especially the Libraries Board of South Australia, the Trustees of the State Library of New South Wales, the Council of the National Library, Scotch College, the Muséum d'Histoire Naturelle at Le Havre, and the Trustees of the National Maritime Museum, Greenwich – and to Mr Kerry Stokes for permission to quote from the historical records in their possession, and/or to reproduce in the book black-and-white and colour illustrations from the originals in their collections.

My family and friends in Adelaide have been supportive in a score of ways. My son Malcolm, son-in-law Craig, and friend and colleague Alan Wallace all came to the rescue at various times when my ageing computer experienced partial memory loss and other malfunctions. John Emery, Tom Nelson, Greg Mackie, Garth Morgan and Professor Alex Castles have provided advice and encouragement over the years. Bev and Hartley Willson twice hosted me at their seafront home at Penneshaw, Kangaroo Island, overlooking Frenchmans Rock and Backstairs Passage. I am grateful to them all.

The support and cooperation of my colleagues on the Council of the Royal Geographical Society of South Australia has been much appreciated. Access to the rich historical collections in the society's York Gate Library has greatly aided my research, as has the council's generosity in contributing to my travel expenses within Australia, and in meeting the costs of illustrations for the book. It is particularly pleasing that the society is associated with my publisher, Tony Crawford, in the book's publication. I owe a great deal to Tony and his editorial director, David Barrett, whose interest in the subject matter has helped shape the book.

Finally, I would like to acknowledge with thanks the two writer's grants I received from the SA Department for the Arts in the project's early years. For many new writers the psychological boost received from such grants may be as important as their monetary value. In my case, they provided the impetus to continue with what at times seemed an almost impossible task.

Pictorial acknowledgements

The author and publishers wish to thank the following institutions and individuals for the use of charts, maps, sketches, paintings and portraits that illustrate this volume:

- Dixson Galleries and Mitchell Library, State Library of NSW;
- Flinders University Library, Adelaide, SA;
- Kerry Stokes Collection, Perth, WA;
- Collection Lesueur, Muséum d'Histoire Naturelle, Le Havre, France;
- National Library of Australia, Canberra, ACT;
- Royal Geographical Society of South Australia;
- Scotch College, Hawthorn, Victoria;
- UK Hydrographic Office, Taunton;
- John Ford, Port Adelaide, SA;
- Colin Gill, Port Lincoln, SA;
- Jill Gloyne, Kangaroo Island, SA;
- M. Hobbs, Kangaroo Island, SA;
- Madeleine Ly-Tio-Fane, Mauritius;
- Robert Marshall, Norwood, SA;
- David Moore, Aldershot, UK;
- R.T. Sexton, Adelaide, SA;
- Jack Ward, North Melbourne, Victoria;
- Neville Weston, Perth, WA; and
- Mr T. Young, Honorary Consul for the Republic of Mauritius, Port Adelaide, SA.

A Note on Sources

In reconstructing the voyages of Baudin and Flinders, I have drawn from a wealth of contemporary records, French and British. Both captains left detailed personal accounts, Baudin in his *Journal de Mer* and Flinders in *A Voyage to Terra Australis*. Though the former has only been available in a microfilm copy,[1] an acclaimed English translation by Christine Cornell, published by the Libraries Board of South Australia in 1974, makes a splendid substitute. The journal, in effect Baudin's private diary, was written up at the time, and provides an almost daily record of events on board, from the expedition's departure in October 1800 to 5 August 1803, six weeks before Baudin's death. There are, unfortunately, some major gaps in the narrative, most notably from June to November 1802, covering the period of the French stay in Port Jackson.

The two volumes of Péron and Freycinet's official history, *Voyage de découvertes aux terres australes* (Paris 1807-1816), offer a contrasting perspective on Baudin's voyage. An inadequate English translation of the first volume, *A Voyage of Discovery to the Southern Hemisphere*, appeared in London in 1809 (reprinted in Melbourne, 1975); and translation of volume II by Christine Cornell remains unpublished. Further records of the voyage can be found in the Hélouis transcripts, a selection of extracts from the journals (in the French archives) of many of the expedition's officers, made for Professor Ernest Scott of Melbourne by Mme Robert Hélouis before World War I. Copies are held in the National, Mitchell and La Trobe libraries, and on microfilm. Another important source is the journal of *Capitaine de frégate* Pierre-Bernard Milius, who succeeded to the command after Baudin's death; the original is in the possession of the Kerry Stokes Collection, Perth, but a transcription by J. Bonnemains and P. Haughel has been published by the Muséum d'Histoire Naturelle at Le Havre (1987).

For a French view of the British colony of New South Wales, the reader can turn to Péron's published narrative (volume I), or to Ernest Scott's *The Life of Captain Matthew Flinders* (1914), which includes a translation of the naturalist's subsequent report on the settlement to General Decaen at Isle de France. The *Historical Records of Australia*, volumes III to V, and *Historical Records of New South Wales*, volumes IV and V, reprint Baudin's correspondence with Governor King, and numerous other documents relating to his visit (see in particular *HRNSW*, vol. IV, Appendix: 'Baudin Papers).

Flinders's *Voyage* was first published in July 1814, the day before his death. Its republication in facsimile by the Libraries Board of South Australia in 1966 has made it more readily available to a general readership. Compared to Baudin's journal, the *Voyage* has the advantage (for its narrator) of being fashioned as a unified narrative, linking historical data with literary structure, and allowing scope for hindsight. More recently (1986) Genesis Publications and Geoffrey C. Ingleton have produced a facsimile edition of Flinders's *Private Journal 1803-1814*, from the original in the possession of the Mitchell Library, State Library of New South Wales. The journal covers the six-and-a-half years Flinders spent in detention on Isle de France, and his final years in England, and is vital reading for any student of his life and achievements. Though its price puts it beyond the reach of most readers, it can be accessed through state or university libraries. I believe the Mitchell Library plans to put it on the Internet.

Thanks to David Moore of Aldershot, I have also been able to draw on the (as yet) unpublished *Investigator* diary of the botanist Robert Brown. David and his coeditors have laboured for many years to transcribe Brown's jottings (held by the Natural History Museum, London) into a coherent narrative that will surely rank as a seminal document in the history of Australian science. The diary of Brown's assistant, the gardener Peter Good, published in the museum's *Bulletin (Historical Series)*, is a more personal, straightforward daily account by a likeable, modest man.

The seamen on such voyages are generally faceless and anonymous. 'When you look down from the quarter-deck to the space below,' Dr Johnson observed to Boswell, 'you see the utmost extremity of human misery – such crowding, such filth, such stench!' Seaman Samuel Smith, one of Flinders's crew, shatters the stereotype with his journal, kept throughout the voyage. My grateful thanks go to the Mitchell Library for supplying a copy of the manuscript in their possession.

The definitive biographies of Flinders and Baudin – Geoffrey C. Ingleton's *Matthew Flinders: Navigator and Chart-Maker* (1986) and Frank Horner's *The French Reconnaissance: Baudin in Australia 1801-1803* (1987) – have been my constant companions for several years. One might say of them, as La Pérouse observed of Captain Cook, 'they have done so much, they have left nothing for their successors to do'. Other informative studies of Flinders are *The Voyage of the* Investigator, *1801-1803*, by K.A. Austin (1964), Sidney J. Baker's *My Own Destroyer: A Biography of Matthew Flinders, Explorer and Navigator* (1962), and *Matthew Flinders 1774-1814*, by James D. Mack (1966). To these may be added two more specialised works: *In the Grips of the Eagle: Matthew Flinders at Ile de France 1803-1810* (1988), by Huguette Ly-Tio-Fane Pineo, an analysis of the navigator's detention by General Decaen; and more recently, *Letters to Ann: The Love Story of Matthew Flinders and Ann Chappelle* (1999), told by Catharine Retter and Shirley Sinclair, largely through Flinders's letters to his wife.

On the French side, Leslie Marchant's *France Australe: A Study of French Explorations and Attempts to Found a Penal Colony and Strategic Base in South Western Australia 1503-1826* (1982) reviews the successive French voyages to the western coasts of New Holland prior to British annexation in 1826. *The Baudin Expedition and the Tasmanian Aborigines 1802* (1983), by N.J.B. Plomley, is a fascinating study of the anthropological observations made by Péron, Baudin, and other members of the expedition, and includes many original manuscript reports translated into English for the first time.

Both expeditions left rich pictorial records, which complement the narrative accounts. The Muséum d'Histoire Naturelle at Le Havre is the most important resource for the drawings and paintings of Baudin's two talented artists, Charles-Alexandre Lesueur and Nicolas-Martin Petit. An outstanding selection of their work can be found in *Baudin in Australian Waters: The Artwork of the French Voyage of Discovery to the South Lands*, edited by Jacqueline Bonnemains, Elliott Forsyth and Bernard Smith (1988). A beautifully presented exhibition, *Terre Napoléon: Australia through French Eyes 1800-1804*, displayed a collection of these works at the Museum of Sydney – later repeated at the National Library of Australia – in 1998-99.

The artwork of Flinders's voyage is no less absorbing. Many of the sketches and paintings of William Westall, the landscape artist, are now held by the National Library of Australia, and recorded by Elisabeth Findlay in *Arcadian Quest: William Westall's Australian Sketches* (NLA, 1998). His colleague Ferdinant Bauer is now recognised as one of the greatest botanical artists of the 19th century; his beautifully detailed botanical and zoological drawings were exhibited at the Museum of Sydney in 1997-98, under the title *An Exquisite Eye: The Australian Flora and Fauna Drawings 1801-1820 of Ferdinand Bauer.*

Note

1. A transcription by Jacqueline Bonnemains under the title *Mon voyage aux Terres Australes: Journal personnnel du commandant Baudin* is to be published in Paris in October 2000.

Weights, Measures and Currency

For narrative authenticity, this book has been written using the imperial system of weights and measures, which was used in Britain at the time of the expeditions. Metric equivalents for most of the measures used in the book are given below.

Currency is more problematic; to say £1 equals $2 is to distort matters considerably, for various reasons. Currencies mentioned in the text are therefore described below without reference to modern values.

Conversions of weights and measures (imperial to metric); currency

	Imperial	Metric
Length and distance	1 inch	2.54 centimetres
	1 foot	30.48 centimetres
	1 yard	0.91 metres
	1 mile	1.61 kilometres
	1 league	3.18 nautical miles* (5.89 kilometres)
	1 fathom	1.83 metres (6 feet)
Volume	1 gallon	4.55 litres
	1 bushel†	36.37 litres (8 gallons)
Weight (or mass)	1 pound	0.45 kilograms
	1 ton	1.02 tonnes
Temperature	°F	$\frac{2}{5}$°C + 32
Currency	pound (£): British unit of currency, formerly valued at 240 pence (240d) or 20 shillings (20s) farthing: British coin worth a quarter of an old penny piastre: former Spanish peso or dollar franc: former French silver coin, worth 100 centimes sou: former French bronze coin worth 5 or 10 centimes	

Notes

* For practical purposes, a league was often taken as 3 nautical miles.

† A bushel could also be used to measure the weight of specified materials; for example, 60 pounds (60 lb) of wheat, 40 lb of oats, and so on.

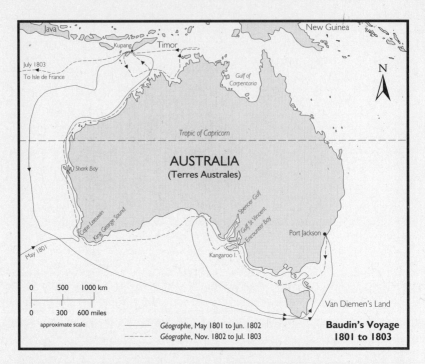

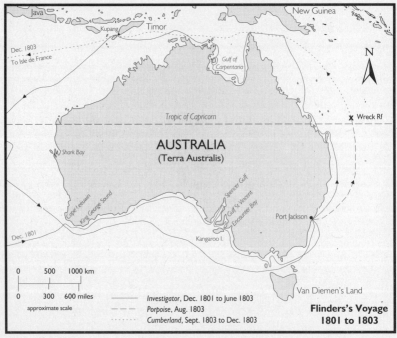

⚓

CHAPTER ONE

Introduction: Finding Australia

Terra Australis nondum Cognita

The southern zone has a warm climate, but is unknown to the sons of Adam. It has no links with our race. No human eye has seen it yet ... Access is barred to man, and the sun makes it impossible to enter this region. In the opinion of the philosophers it is supposed to be inhabited by the antipodes who live through seasons which are the exact opposite of ours. For when we are dried up by heat they are freezing in the cold. [Legend on medieval map]

The historian Herodotus, writing in the 5th century BC, recorded that a Phoenician ship had sailed from the Red Sea into the southern ocean, and after a voyage lasting two years had returned through the Pillars of Hercules (the Straits of Gibraltar) and docked at Alexandria:

> These men made a statement which I do not myself believe, though others may ... that as they sailed on a westerly course round the southern edge of Libya [Africa] they had the sun on the right to the northward of them. This is how Libya was first discovered to be surrounded by sea.[1]

Two millenia passed before a ship from western Europe equalled the Phoenicians' feat, although ships of the Roman Empire are known to have sailed from Red Sea ports to India's Malabar Coast – some may have voyaged as far as southern China.

The Greeks observed the curvature of the earth, and deduced from it that the earth was spherical, like the great globes that circled it by day and night. It followed logically that the Southern Hemisphere must contain a vast land mass roughly equal in size to the three known continents, Europe, Asia, and Africa, or the world would surely overbalance. The two hemispheres were thought to be separated by a middle band 'scorched by an incessant belt of heat', uninhabited and impassable.

With the decline and fall of Rome, classical science went into a long hibernation, the idea of a spherical earth kept alive by a handful of scholars in Islam and the Christian West. It re-emerged in the 14th and 15th centuries, spurred by the translation into Latin of the surviving manuscripts of Claudius Ptolemy, a geographer and cartographer of genius who had lived in the city of Alexandria in the 2nd century.

The logic and symmetry of the Ptolemaic world view persuaded the medieval cartographers to include in their maps of the world a vast southern land linking Africa with Asia, and closing off the Indian Ocean to the south. Described as *Terra Incognita secundum Ptolemeum* (Unknown Land according to Ptolemy) or *Terra Australis nondum Cognita* (Southern Land not yet known), it spread across the southern hemisphere, mostly below 40°S. Maps such as these encouraged Prince Henry the Navigator of Portugal in the mid-15th century to send his captains ever further south down the Atlantic coast of Africa, and led the Genoese-born mariner Christopher Columbus to believe he could reach Asia by sailing west across the Atlantic.

Following the example of Portugal and Spain, England, France and the Netherlands dispatched their ships and seamen on unimaginable voyages of discovery and trade. The ships were tiny – at 100 tons, Columbus's flagship, the *Santa Maria*, was larger than most – and their captains lacked maps and any accurate means to judge the distance travelled. Yet by the beginning of the 17th century, southern Africa, the Americas, the East Indies and the Philippines had been discovered, European settlements made, and the process of exploitation was well advanced. As the captains and crews who survived returned home with reports of their discoveries, the cartographers inscribed the outlines of new continents on their maps, and added islands where before had been trackless ocean. One area remained virtually untouched – across the bottom of each map *Terra Australis* was still *incognita*, apart from Tierra del Fuego in the west and a land mass south of Java and New Guinea.

The Portuguese settled in Timor about 1516, and the northern coast of New Guinea was partly known by 1545; possibly Portuguese vessels had also sighted land to the south. A series of maps, reputedly based on charts stolen from the Portuguese, and drawn by French cartographers at Dieppe between 1536 and 1550, showed a large projection named Java la Grande, later identified by some writers with Australia.

Joseph Banks owned a copy, which he may have carried on board the *Endeavour* when he sailed with Cook.

From the late 1590s the Dutch set about establishing a trading empire in the East Indies. Unlike the Portuguese, wrote Manning Clark, 'they were prepared to grant toleration to the Muslim, because holy wars were not compatible with uncommonly large profit'. The first Dutch fleet reached Java in 1596, and within ten years they were trading from the Moluccas to Brunei, Banda and Ambon. In 1606 they accidentally stumbled on Australia.

The discovery of New Holland

Dutch discoveries in Australian waters began in 1606. Captain Willem Jansz was on a voyage eastwards from the Banda Sea, on the lookout for new trading opportunities, in the 60-ton *Duyfken* (*Little Dove*), when shoal water forced him south. He made landfall on a flat, dull, and distinctly unpromising coast; his men, sent ashore to barter, as often as not found themselves facing hostile, spear-waving savages. In one clash a seaman was killed; elsewhere (it is not clear whether this was on Cape York, where he had unknowingly landed, or New Guinea) nine of the crew 'were killed by Heathens, which are man-eaters; so they were constrained to returne, finding no good to be done there'. During the voyage, Jansz without realising it had charted some 300 kilometres of the coast of a new continent. He thought it was part of New Guinea's south coast.

For the next century and a half, seamen of the Dutch East India Company (Vereenigde Oost-Indische Compagnie, or VOC), followed the northern and western coasts of the continent from the Gulf of Carpentaria in the north to the Great Australian Bight in the south, tried unsuccessfully to trade with the natives, dotted the map with Dutch names, and called the barren land they had found New Holland. As traders they had little interest in claiming possession   most VOC captains agreed with Jan Carstensz, returning from the Gulf of Carpentaria in 1623, that the country seemed 'the driest, poorest area to be found in the world, [with] not one fruitful tree nor anything that could be of use to mankind'; its inhabitants were 'the poorest and most wretched creatures ever seen'.

Yet there were some in the VOC's headquarters at Batavia (Jakarta) who believed in the *Terra Australis* of the ancient maps. In 1642

Governor-General Van Diemen sent Abel Janszoon Tasman, a tough and experienced navigator, south to search for the undiscovered continent, there being

> good reasons [to suppose that] many attractive and fruitful lands are located therein, as being in the cold, temperate and hot zones, where necessarily there must be many inhabited places in the pleasant climate and attractive sky.[2]

Tasman was away for ten months. In the south he discovered a mountainous and well-wooded country, which he named after the governor-general, and next came upon the coast of an even more mountainous land, to which he gave the name New Zealand. On Van Diemen's Land he left a marker inscribed

> *Those who come after us may become aware that we have been here and have taken possession of the said land as our lawful property.*

The ships returned to a lacklustre welcome – the VOC was not impressed by the results of the voyage. Tasman had lost four men killed by natives in New Zealand; more to the point, he had found none of the expected fertile regions or well-populated towns, and there seemed little prospect of future profits from his discoveries. A few hoped New Zealand might yet prove to be the westernmost part of Terra Australis.

William Dampier, the first Englishman to land on New Holland, was no more enthusiastic about the country than the Dutch. Part buccaneer and part scientist, he first visited the north-west coast in 1688, in the piratical vessel *Cygnet*. He returned eleven years later as commander of HMS *Roebuck*, on a voyage of discovery for the British Admiralty. He landed at Shark Bay and spent five weeks on the coast before a shortage of water forced him to sail for Timor. His writings provided for the first time a naturalist's impressions of the country's flora and fauna, and also of its inhabitants, 'the miserablest People in the world'. Though lacking none of the ethnocentric prejudices of his day, Dampier nonetheless wrote perceptively of their concerns for the weaker members of the group:

> They do live in Companies of 20 or 30 Men, Women, and Children together ... the Old People that are not able to stir abroad by reason of their Age, and the Tender Infants, await [the food gatherers] return; and what Providence has bestowed on them, they presently broil on the Coals, and eat it in common. But whether they get little or much, every one has

his part, the young and tender as well as the old and feeble, who are not able to go abroad as can the strong and lusty.[3]

Scientists in Europe were stirred by his observations of trees, plants and flowers 'unlike any I had seen elsewhere', and of strange creatures such as 'a Sort of Raccoons different from those of the West Indies', which had very short forelegs, but still went 'jumping upon them as the others do, and like them [made] very good Meat!' His impact on English literature was no less significant. The hero of *Gulliver's Travels* is said to be Dampier's cousin, and Swift locates Lilliput off the shores of southern Australia. In his *A Tale of a Tub* (1704), the antisocial Dean also anticipated the use to which the continent would be put eighty-four years later:

> This work contains exact accounts of all the provinces, colonies, and mansions of that spacious country, where, by a general doom, all transgressors of the law are to be transported.

Rivals in the South Seas

Britain's next serious attempt to locate the 'missing' southern continent came six decades later, following the Seven Years' War (1756-1763). The Treaty of Paris left France humiliated and stripped of her prized North American territories, and Britain unquestionably the world's leading maritime power. Looking to consolidate this supremacy, the Admiralty dispatched two expeditions in rapid succesion, in 1764 and 1766, to search for Terra Australis in the South Pacific.

Both voyages managed to avoid any grand discoveries, although the second, under the command of Captain Samuel Wallis, discovered King George III Island (Tahiti) in June 1767. His report influenced the Royal Society of London to recommend the island as one of the scientific sites for observing the transit of Venus across the sun, due in 1769. The Admiralty agreed, but took advantage of the voyage to instruct the commander of the ship carrying the scientists to Tahiti, Lieutenant James Cook, that after the completion of their observations, he was

> to proceed to the southward in order to make discovery of the Continent [that may be found there] until you arrive in the Latitude of 40°, unless you sooner fall in with it.[4]

Cook sailed in the bark *Endeavour* in August 1768, and was away for almost three years. Tahiti was reached on 10 April 1769 with not a single man suffering from scurvy, the result of Cook's dogged attention to the crew's diet (including 'sour krout', portable (dried) soup, malt and citrus juice, plus 'scurvy grass' gathered ashore whenever possible). An observatory was set up ashore at the appropriately named Fort Venus, and the transit observed in good conditions on 3 June. During the next two years Cook circumnavigated the southern oceans, but without much success, informing the Admiralty on his return that he had not discovered 'the so much talked of Southern Continent (which perhaps do not exist)'. However, he continued, 'I flatter myself that the discoveries we have made tho' not great will apologise for the length of the voyage'. In His Majesty's name he had 'taken possession of the whole Eastern coast [of New Holland] by the name of New South Wales'. Although the country did not produce any 'Articles in trade' that might invite Europeans to settle there, there were many fine harbours along the coast; a possible gateway had opened for European settlement.

What Cook had failed to discover seemed of greater import than what he had actually seen. Within a year he was dispatched again – this time with two ships, *Resolution* and *Adventure* – to solve the mystery of Terra Australis once and for all. In three years (1772 to 1775) he covered some 70000 miles of ocean, crisscrossing the South Pacific, twice crossing the Antarctic Circle, revisiting Tahiti and New Zealand (but avoiding New Holland), calling at Easter Island, New Caledonia and other island groups, and on the homeward voyage discovering South Georgia in the South Atlantic. He was satisfied that if a southern continent did exist, it must lie 'within the Polar Circle where the Sea is so pestered with Ice, that the Land is thereby inaccessible'.

Meanwhile, the defeated French had not been idle. Stirred by the publication in 1756 of the *Histoire des Navigations aux Terres Australes,* by Charles de Brosses – who claimed that three unknown continents (*Magellanie* in the South Atlantic, *Australasie* in the southern Indian Ocean, and *Polynesie* in the South Pacific) lay in the South Seas – Louis XV and his advisers dispatched Louis-Antoine de Bougainville, a veteran of the campaign that had lost French Canada to the British, to the Pacific in 1766 to restore French pride by discovering a new France in the south.

After a lengthy stay in South America and the Falklands, and a terrible passage of more than seven weeks through the Straits of Magellan, Bougainville reached Tahiti in April 1768. His two ships were at once surrounded by a hundred canoes 'filled with women whose beauty was the equal of European women', most of them nude. 'I ask you, given such a spectacle, how could one keep at work 400 Frenchmen?' wrote their commander. He claimed possession for France, despite obvious signs that other Europeans had visited there recently, and named the island New Cythera, after the Greek island where the goddess Venus had her temple.

From Tahiti the ships headed west, touching at Samoa and Vanuatu, until on 4 June the lookouts sighted a line of reefs dead ahead:

> We could see the tops of rocks over which seas kept breaking with the utmost violence. In this discovery we saw a warning from God that we could not disregard.[5]

With no passage visible through the reefs, food and water running low, and his men weakened by scurvy and disease, Bougainville put about and headed north. He returned to France in 1769, the first French captain to circumnavigate the world and the first European explorer known to have seen the Great Barrier Reef.

Bougainville's exploits inspired further French attempts to discover the southern continent (or continents). In 1771 two rival expeditions were at Isle de France (Mauritius)[6] preparing to sail south in search of de Brosses's *Australasie* and/or *Polynesie*. The first, led by Captain Marion Dufresne, a colonist, left in October 'to reconnoitre the southern lands and find out if they exist, as I think they do'. He discovered and named the Crozet Islands (after his navigator), before dropping anchor in Blackman's Bay in Van Diemen's Land in March 1772 – the first European since Tasman to set foot there, and the first to make contact with the island's Aborigines. He then sailed for New Zealand, where he and several of his crew were massacred by Maori warriors.[7] The survivors withdrew to Mauritius.

The second expedition, with two ships under the command of Kerguelen de Trémarec, had come from France, and its sailing was delayed until January 1772. Separated from his consort in a fierce gale, Kerguelen turned back, but his second-in-command, Louis François de St Allouarn, continued eastwards in his ship, the *Gros Ventre*, towards New Holland. Cape Leeuwin was sighted in March, and St Allouarn turned north, following the line of the coast to Shark Bay. Here he landed on Dirk Hartog Island, and on 30 March took possession of the land in the name of the French king – the third claim, after Tasman's in Van Diemen's Land in 1642 and Cook's on the east coast in 1770.

Before the *Gros Ventre* left the area, the body of a French seaman who had died on board was ferried ashore for burial. St Allouarn died shortly after his return to Mauritius in September.

Botany Bay, 1788

London, 18 August 1786. The Home Secretary, Lord Sydney, informs the Lords Commissioners of the Treasury:

> My Lords, the several gaols and places for the confinement of felons in this kingdom being in so crowded a state that the greatest danger is to be apprehended, not only from their escape, but from the infectious distempers, which may hourly be expected to break out amongst them, His Majesty, desirous of preventing by every possible means the ill consequences which might happen from either of these causes, has been pleased to signify to me his royal commands that measures should immediately be pursued for sending out of this kingdom such of the convicts as are under sentence of transportation. [Historical Records of New South Wales, *vol. I, part 2, 1783-1792, pp. 14-19]*

The outbreak of the United States War of Independence in 1776 had caught the British government by surprise. Faced with a land campaign in North America and a resurgent French Navy at sea, it called a halt to further exploration – excepting only the 'immortal' James Cook, who sailed on his third and last great voyage (1776-1779) in July to search for the Pacific outlet to the supposed North-West Passage, bearing safe-conducts from the American rebels and the French monarchy which supported them.

France openly joined the struggle in 1777. General Rochambeau's troops and Admiral de Grasse's fleet were key factors in the colonists' final victory in 1783. This time it was Britain's turn to lose her colonies, while autocratic France emerged as a champion of liberty. In fact, in helping to ensure the rebels' victory the monarchy had committed a long-drawn-out suicide – the country was made bankrupt, and Frenchmen had seen for themselves how free men could 'snatch the sceptre from the hands of tyrants'.

The loss of her American colonies left Britain with a festering social problem – what to do with the felons crowding the prisons, filled to overflowing by a harsh penal code. An estimated 30 000 to 40 000 convicts were sent across the Atlantic between 1715 and 1776, to the mutual benefit of the government and the private contractors who transported them.[8] The victorious rebels refused to continue this satisfactory arrangement, which had relieved the homeland of unwanted rogues and vagabonds while providing Treasury with needed funds and the colonies with a regular source of cheap labour.

During the war the pressure on the crowded gaols was relieved for a time by housing the felons aboard rotting hulks moored in the Thames and other rivers. These, however, soon became overcrowded in their turn as the crime rate soared, and epidemics cut a swathe through the inmates. On some hulks the mortality rate rose above one in four, and by the mid-1780s the situation was at crisis point. Attempts to establish convict settlements in Central America and West Africa came to nothing, and in desperation the government turned to Cook's previously unwanted legacy – the distant New South Wales – as a possible place of exile for the surplus prisoners. It was suitably far removed, a British possession, and *terra nullius* (unowned if not unoccupied).

Called before the House of Commons Committee on Transportation as an expert witness, Cook's botanist in the *Endeavour*, Joseph Banks (now Sir Joseph, baronet, president of the Royal Society, friend and adviser to the king, and a man of influence in government circles), told the Honorable Members 'that the Soil of many Parts [of New South Wales] is sufficiently fertile to support a Considerable Number of Europeans who would cultivate it in the Ordinary Modes used in England'. It was 'in every respect adapted to the purpose', and from the fertility of the soil, 'the timid Disposition of the Inhabitants' and the climate he gave 'this place the preference to all that I have seen'.

His Majesty's government agreed, and lost little time fixing upon Botany Bay 'as a place likely to answer its purposes'. These extended beyond relieving the crisis in the prisons. The war had led to a shift in the balance of power, and Britain needed time and opportunity to restore her economy and rebuild her strategic position, not least in India and the East, the new focus of empire. A settlement at Botany Bay could provide an alternative trade route to China, serve as a base for British commerce in the Pacific, and counter French influence in the area. It could also, as Banks was quick to point out, prove an important source for such vital naval supplies as flax for ropes and sails, and Norfolk Island pines for ships' masts. Thus when Captain Arthur Phillip put to sea in May 1787 with eleven ships, some 780 convicts, 210 officers and marines, and forty or so dependents, his instructions went beyond founding a penal colony and ensuring its self-sufficiency.

Phillip, as captain general and governor-in-chief, was directed to take possession of the territory extending from Cape York in the north to South Cape in Van Diemen's Land, and all the country inland as far west as 135°E longitude – the meridian neatly bisected the continent into two roughly equal parts, with the western half tacitly remaining New Holland. In doing so he also assumed responsibility for perhaps 200 000 of its inhabitants, whose 'affections [he was] to conciliate, enjoining all our subjects to live in amity and kindness with them'.

It quickly became clear that Botany Bay could not support a settlement. The lush meadows described by Cook and Banks were not to be found – only coarse grass, sandy and indifferent soil, little fresh water, and an unsheltered anchorage. Luckily Phillip discovered a magnificent harbour a few miles to the north, 'in which a thousand Sail of the line may ride in the most perfect security'. Cook had passed it by, but had noted the two heads guarding the entrance and mapped it as Port Jackson. Phillip ordered the fleet to sail for the new harbour, where he selected a sheltered cove as the site for his settlement; he named it in honour of Lord Sydney.

The transfer was under way when, to the 'infinite surprise' of Phillip and his officers, two large ships appeared far out to sea. Some supposed they were English, bringing more convicts and supplies, others that they were Dutch 'sent to oppose our landing'. Others again thought they must be the two French ships 'so long out upon Discoveries in the South Seas'. The wind blew strong from the north-west, and the strangers were lost to view. It was not until two days later, 26 January, as HMS *Sirius* and the last of the transports were about to sail, that they reappeared at the entrance to the bay. With French colours flying they sailed in and dropped anchor.

Captain Hunter in the *Sirius* at once sent two officers across to present his compliments and offer assistance if need be to the visiting commander, and with orders to find out all they could about his intentions. The ships were indeed the *Boussole* and the *Astrolabe*, they reported, on a voyage of discovery under the Comte de La Pérouse. Leaving Botany Bay to the French, Hunter rejoined the governor at Sydney Cove and informed him of their identity. That evening Phillip assembled his officers and the marines at the point where they had landed; 'an union jack [was raised], when the marines fired several vollies', wrote

Captain David Collins, 'between which the Governor and the officers
… drank the healths of His Majesty and the Royal Family and success
to the new Colony'.

*The colony's survival is uppermost in the governor's mind. He had asked
for able-bodied convicts, preferably with skills in a trade or in husbandry
– the government has sent the aged, the diseased, duffers and footpads,
the riffraff of the London streets, strumpets and petty thieves. There are
a dozen carpenters – where three times that number would not be
enough – and even fewer with experience in agriculture. The building
of the hospital begins at once, but it is filled to overflowing before it can
be completed – 'chiefly of Dysenteries', writes Surgeon Bowes Smyth of
the transport* Lady Penrhyn*. There are few blankets for the sick.*

*The female convicts are landed on 6 February. In general they are
dressed 'very clean', but, says Bowes Smyth, the male convicts 'got to
them very soon after they landed, & it is beyond my abilities to give a
just discription of the Scene of Debauchery & Riot that ensued during
the night'. Marines and sailors from the transports join in, bringing grog
with them. The saturnalia ends in a violent storm of thunder, lightning
and heavy rain.*

*Officers and marines alike refuse to take any responsibility for the
convicts' conduct. The governor has no option but to appoint some of
the better-behaved as overseers:*

> *Here are only convicts to attend to convicts, and who in general fear to
> exert any authority and very little labour is drawn from them in a coun-
> try which requires the greatest exertions.*

*Winter is not far off. The late-planted vegetables have withered in the
soil. Survival promises to be a close-run thing.*

The French return – La Pérouse and d'Entrecasteaux

*Paris, 1785. King Louis XVI presents his personal instructions to the
Comte de La Pérouse:*

> *The Sieur de La Pérouse, on all òccasions, will treat the different peoples
> visited during his voyage with much gentleness and humanity. He will
> zealously and interestedly employ all means capable of improving their
> condition by procuring for their countries useful European vegetables,*

fruits, and trees, which he will teach them to plant and cultivate ... [Quoted in Brosse, 1983, p. 79]

As Phillip suspected, the arrival of the French ships was not a mere coincidence. La Pérouse had sailed from Brest in August 1785 on a voyage inspired by Louis XVI himself; an admirer of Captain Cook, the king was determined that France should match the Englishman's discoveries and complete the mapping of the globe. His orders, drawn up by Charles-Pierre Claret de Fleurieu, a noted navigator and director-general of Ports and Arsenals, concentrated on the Pacific, where major discoveries might yet be made, and also singled out for investigation the south coast of New Holland, 'the greater part of which has not been visited'.

In September 1787, after some eighteen months circling the North Pacific, the expedition called at the tiny Russian outpost of Petropavlovsk, on the remote Kamchatka Peninsula, more than 5000 miles from St Petersburg. It was a prearranged visit to a friendly port, and while the voyagers were in the midst of a continuous round of feasts and celebrations laid on by their Russian hosts, letters and dispatches arrived from Paris. Among them were fresh sailing orders, directing La Pérouse (promoted *chef d'escadre*, or rear admiral) to proceed at once to inspect and report on the new English settlement at Botany Bay. It was a clear indication that Paris was as much concerned by British ambitions in the Pacific as was London by the French – the dispatches would have taken a year or more to cross the Siberian wilderness, and must have been sent as soon as Louis's ministers heard of plans to settle the colony.

La Pérouse farewelled the Russians and headed south, calling first at the island of Samoa. Here a landing party led by Fleuriot de Langle, his second-in-command, was ambushed by natives, and de Langle and eleven of his men were killed, and others wounded. Shocked by his friend's death, La Pérouse wrote bitterly of the *philosophes* and their fanciful concept of the 'noble savage':

I am a hundred times angrier at the philosophers who praise them ... than at the savages themselves. Lamanon, whom they massacred, was saying the day before that these men were worth more than we are ...[9]

The French found Botany Bay without difficulty, thanks to the copies of Cook's charts carried on board. Happy to find themselves once more among friends, they set up camp ashore, protected by a stockade to deter an attack by the local natives. It proved more useful as a deterrent to convicts and deserters, La Pérouse complaining that he had to dismiss them 'with threats, giving them a day's provisions to carry them back to the [British] settlement'. Perhaps not all convicts were turned away, for it was rumoured in the British camp that 'after the French sailed, two women were missing'.

Before departing, La Pérouse forwarded his dispatches, charts and private correspondence to the French ambassador in London on an English ship. He wrote to a friend: 'When I return you will take me for a centenarian. I have no teeth and no hair left and I think it will not be long before I become senile' (he was forty-six). He sailed on 10 March, after a stay marked by 'visits and other interchanges of friendship and esteem' between the officers of both nations. The two ships disappeared over the horizon, and into mystery; nothing was heard of them for thirty-nine years.[10] Phillip knew only that the French commodore intended to make for the Friendly Islands (Tonga).

By 1791, French hopes of La Pérouse's safe return had been abandoned. In Paris, now in the throes of revolution, the Society of Natural History petitioned the National Assembly to investigate his fate. Strongly backed by Louis XVI, still on the throne in the unwonted role of constitutional monarch, the Assembly commissioned two ships to sail in search of the missing explorer, and to make investigations 'advantageous to navigation, geography, commerce, and the arts and sciences'. Fleurieu, now Minister of Marine, drew up the instructions for the search expedition, again pointing to the unknown southern coast of New Holland:

> No navigator has penetrated in that part of the sea; the reconnaissances and discoveries of the Dutch, the English and the French commenced at the south of Van Diemen's Land.

The expedition, commanded by Rear Admiral Bruny d'Entrecasteaux, sailed in September 1791. During the next two years the ships twice circled the southern continent, on both occasions failing to explore the southern coast. Discipline fell apart as relations festered between the royalist and republican factions on board. Dysentery, scurvy and fever ravaged the crews. Huon de Kermadec, the second-in-command, died in May 1793, shortly followed by d'Entrecasteaux. The new commander, Captain d'Auribeau, took the ships to Java before dying in his turn – depriving the revolutionary delegates on their way to the island of the chance to ask the Dutch authorities for his head.

Within the space of twenty-two years, Terra Australis had claimed, directly or indirectly, the lives of seven French captains, from St Allouarn to d'Auribeau. The king who had so keenly followed their voyages paid the price for neglecting his subjects at home. Tried and sentenced to death for conspiring against the nation's freedom, on 21 January 1793 he was taken from his prison to the Place de la Révolution and guillotined. On the night before his execution Louis is said to have asked his gaolers: '*A t'on des nouvelles de M. de La Pérouse?*' ('Is there any news of La Pérouse?')

The first Australians

Captain Willem Jansz and his Dutch, French and British successors could scarcely have conceived that the land they found had been settled for 40 000 years or more. For them *Terra Incognita* had become *terra nullius*, a no-man's-land there for the taking by any European nation with sufficient interest in its inhospitable shores to seize possession. Its inhabitants appeared no more than treacherous and ignorant savages, living in a state of raw nature – the 'miserablest and most wretched' people on earth – lacking any cultural identity or rights to the land on which they lived.

An estimated 300 000 to 1 million of these first Australians occupied the continent when the First Fleet sailed into Botany Bay. They were divided into some 500 to 600 tribal or dialect groupings, speaking perhaps 250 distinct languages. Each grouping believed itself descended from the same ancestors, mythical beings half spirit and half human. Ownership of territory in the European sense was unknown, but members of each tribe felt an intense kinship with the land – it had given them birth, nourished their lives, and held the spirits of their ancestors.

Their social organisation was complicated and cohesive. Authority within the tribe lay with the elders, men steeped in tribal lore and custom. In the absence of individual leaders, lust for power or self-aggrandisement was rare. Small-scale intertribal warfare, triggered perhaps by conflict over hunting rights or by young men stealing wives, took place, but seldom lasted for long and casualties were light. Traditional enmity lasting for generations seems to have been unknown, in contrast to more 'advanced' societies. 'These historically conditioned characteristics,' says the historian Russel Ward, 'handicapped them against the highly competitive, bellicose and industrious European invaders of their land.'

The king's instructions to Governor Phillip suggest that the British government intended to deal fairly with Aboriginal Australians. What followed – the decimation of the population, the destruction of their culture and identity – may not have been meant, but in the circumstances was probably inevitable. In 1837 a select committee of the House of Commons inquiring into the impact of settlement reported that

> These people, unoffending as they were towards us, have ... suffered in an aggravated degree from the planting among them of our penal settlements. In the formation of these settlements it does not appear that the territorial rights of the natives were considered, and very little care has since been taken to protect them from the violence or the contamination of the dregs of our countrymen. The consequences have been dreadful beyond example, both in the diminution of their numbers and in their demoralization.[11]

Port Lincoln, Thursday, 4 March 1802. Matthew Flinders writes of a small party of 'Australians' heard calling in the bush:

> *No attempt was made to follow them, for I had always found the natives of this country to avoid those who seemed anxious for communication; whereas, when left entirely alone, they would usually come down after having watched us for a few days. Nor does this conduct seem to be unnatural; for what, in such case, would be the conduct of any people, ourselves for instance, were we living in a state of nature, frequently at war with our neighbours, and ignorant of the existence of any other nation? On the arrival of strangers, so different in complexion and appearance to ourselves, having power to transplant themselves over, and even living upon an element which to us was impassable; the first sensation would probably be terror, and the first movement flight.*
>
> *We should watch these extraordinary people from our retreats in the woods and rocks, and if we found ourselves sought and pursued by them, should conclude their designs to be inimical; but if, on the contrary, we saw them quietly employed in occupations which had no reference to us, curiosity would get the better of fear, and after observing them more closely, we should ourselves seek a communication. Such seemed to have been the conduct of these Australians ...[12]*

King Island, Thursday, 23 December 1802. Captain Nicolas Baudin pens a personal letter to his friend Governor P.G. King at Port Jackson:

I have never been able to conceive that there was justice and equity on the part of Europeans in seizing, in the name of their Governments, a land seen for the first time, when it is inhabited by men who have not always deserved the title of savages or cannibals which has been given them, whilst they were but the children of nature and just as little civilised as are your Scotch Highlanders or our peasants in Brittany, who, if they do not eat their fellow men, are nevertheless just as objectionable. From this it appears to me that it would be infinitely more glorious for your nation, as for mine, to mould for society the inhabitants of the respective countries over which they have rights, instead of wishing to occupy themselves in improving those who are so far removed by seizing the soil which they own and which has given them birth.

These remarks are no doubt impolitic, but at least reasonable from the facts; and had this principle been generally adopted you would not have been obliged to have formed a colony by means of men branded by the law, and who have become criminals through the fault of Government which has neglected and abandoned them to themselves. It follows therefore that not only have you to reproach yourselves with an injustice in seizing their land, but also in transporting on a soil where the crimes and the diseases of Europeans were unknown all that could retard the progress of civilisation, which has served as a pretext to your Government, &c.

If you will reflect upon the conduct of the natives since the beginning of your establishment upon their territory, you will perceive that their aversion to you, and also for your customs, has been occasioned by the idea which they have formed of those who wished to live amongst them. Notwithstanding your precautions and the punishments undergone by those among your people who have ill-treated them, they have been enabled to see through your projects for the future; but being too feeble to resist you, the fear of your arms has made them emigrate, so that the hope of seeing them mix with you is lost, and you will presently remain the peaceful possessors of their heritage, as the small number of those surrounding you will not long exist ...[13]

Notes

1. Herotodus, 1996, pp. 228-229.
2. Kenny, 1995, p. 32.

3. Dampier, 1998, p. 219.
4. Grenfell Price, 1971, pp. 18-20.
5. Lacour-Gayet, 1976, p. 39.
6. The island of Mauritius lies in the Indian Ocean, some 700 km east of Madagascar, just north of the Tropic of Capricorn. First discovered by the Portuguese in the early 16th century, it was claimed by the Netherlands in 1598, and named after the Stadthouder of the Netherlands, Prince Mauritius van Nassau.

   In 1715 Guillaume Dufresne d'Arsel took possession of the island for France, renaming it Isle de France so that its ownership should not be in doubt. By the close of the century, the *s* in Isle, though often elided, survived in many documents of the time. Contemporary British documents refer to it variously as the Mauritius, Isle of France, and Isle (or Ile) de France.

   Following the British invasion in 1810, and subsequent annexation, the island reverted to its former Dutch name. In 1968 Mauritius became an independent country within the Commonwealth.

   For consistency, in the narrative I have standardised on the spellings Isle de France and Mauritius, depending on the context.
7. Crozet returned to Paris, where it seems he told Rousseau of his misgivings about the 'noble savages' who had killed and perhaps eaten his commander. The philosopher was chagrined: 'What,' he cried, 'is it possible that the goodness of Nature's children should be compatible with a certain perversity?'
8. 'Outsourcing' has a long history. Government sold the convicts at £5 a head to the contractors, who then resold them to colonial employers for a tidy profit.
9. Shelton, 1987, p. 91.
10. Relics of the two ships were found on the island of Vanikoro (Santa Cruz Islands, north of Vanuatu) in 1827 by Captain Peter Dillon.
11. *Historical Records of Victoria*, vol. 24.
12. Flinders, 1814, vol. 1, p. 146.
13. *Historical Records of New South Wales*, vol. 5, 1803-1805, p. 826.

CHAPTER TWO

⚓

The Captains

NICOLAS BAUDIN AND THE FRENCH VOYAGE OF DISCOVERY, 1800 TO 1803

Science was not at war

Paris, Friday, 16 May 1800. Professor Antoine-Laurent de Jussieu and his colleagues of the Institut National write to Sir Joseph Banks, president of the Royal Society of London.

> *The Institut National of France is desirous that several distant voyages useful to the progress of human knowledge should begin without delay. Its wishes have been endorsed by our Government which has just issued orders for the preparation as soon as possible of expeditions led by skilful navigators as well as enlightened men of science, and will approach the Government of your country for the necessary passports or safe-conducts for our vessels.*
>
> *The Institut National considers that it is precisely at the moment when war still burdens the world that the friends of humanity should work for it, by advancing the limits of science and of useful arts by means of enterprises similar to those which have immortalised the great navigators of our two nations and the illustrious men of science who have scoured sea and land to study nature, Sir, where they could do so with the greatest success.*
>
> *We hasten to beg you, as one of the most distinguished members of the commonwealth of learning, to use your good offices with your Government with that zeal which has always inspired you to work in the interest of humanity, to renew those marks of respect for science which our two nations have more than once given, and therefore to secure the prompt despatch of the passports which will be requested ... [Quoted in de Beer, 1960, p. 238]*

Britain and France had been at war for eight years when, in June 1800, the Republic's resident commissioner in London, Citizen Louis-Guillaume Otto, lodged his government's application. It sought passports for two ships under the command of Captain Nicolas Baudin, 'ready to sail to continue the useful discoveries which your navigators made in their voyages round the world'. Though Otto's official duty was to arrange the exchange of prisoners of war, his office also served as a useful channel for informal contacts between the two governments. Through tact and diplomacy he had won the esteem of Sir Joseph Banks and other influential men in London.

Prime Minister William Pitt's government referred the French request to the First Lord of the Admiralty, Earl Spencer, for a decision; Spencer in turn called on his close friend Sir Joseph for advice. Banks recalled that Cook had sailed on his third voyage with safe-conducts from America and France, and could not ignore Jussieu's appeal to 'the commonwealth of learning'. He was himself a corresponding member of the Institut National, and remained on friendly terms with Jussieu and other French men of science despite the war. In 1796 he had not hesitated to recommend the provision of a passport for a French scientific voyage to the West Indies, commanded by the same Captain Baudin. On that occasion he had written to Jussieu:

> Whatever the fortune of War may be, Science and those who possess the liberal views of it which you have ever done will be the nearest to the heart of one who is with distinguished consideration and unvaried esteem – Your most h[um]ble Servant, J. Banks.[1]

There could be little doubt of Banks's response to this latest request. Notwithstanding any personal reservations – that perhaps *this* French voyage was not intended solely for discovery and scientific research – he recommended that the passports be granted. Signed by Spencer and his fellow commissioners, they provided 'protection against His Majesty's cruisers for the Ships *Géographe* and *Naturaliste* ... intended to proceed from Havre under the direction and command of Captain Nicolas Baudin on a voyage round the world' (see appendix A).

The French Republic professed the same enlightened principles. Citizen Forfait, Minister for Marine and Colonies, stated them clearly in his instructions to Baudin:

> Since you are sailing under the flag of truce, and since the sole aim of your labour is the perfecting of the sciences, you must observe the most

complete neutrality and not give rise to a single doubt as to your exactitude in confining yourself to the object of your mission, such as it is announced in the passports obtained for you.

It was, he added, Baudin's responsibility to

make France's name honoured in all the countries that you visit, and, especially, make it beloved of those uncivilised peoples to whom you are taking nothing but benefits; for if the principal purpose of your relations with them is to enrich France with the products of their countries, you should also invite them to introduce into their homelands those whose usefulness you are charged with demonstrating to them …

I invite you to seize every opportunity … to send me details on the events of your voyage. They will be received with all the interest that is aroused by an expedition whose aim is to increase the scientific field, to add (if it be possible) to what Nature has done for the nations that live in another hemisphere, and to form men destined some day to augment the numbers of celebrated mariners and naturalists.[2]

A botanical voyager

Paris, 21 July 1798. Professor Jussieu writes to the Minister of Marine, Admiral Bruix, in praise of Captain Nicolas Baudin:

This excellent Mariner can render further service to his country; he can combine geographical researches with those which more particularly interest us; the experience of the past and the knowledge of his former achievements make us believe that he follows worthily in the steps of Bougainville, La Pérouse, and d'Entrecasteaux, and that he will be more fortunate than the last two. [Quoted in Horner, 1987, p. 36]

Nicolas Thomas Baudin had little in common with his three great predecessors – he was a commoner by birth, he had not made his career in the French Navy but had been a merchant captain sailing under the Spanish and Austrian flags. His reputation was made, not in naval exploits, but by safely transporting living plants and animals across the oceans back to Europe. As the historian Jean-Paul Faivre suggests (in his introduction to Christine Cornell's translation of Baudin's journal), his officers and scientists, coming as many did from good families and endowed with some social standing, 'found it possible to think themselves outraged at being under the orders of an officer risen from the ranks'.

Baudin was born at the seaport town of St Martin on the Ile-de-Ré, the 'bright island', off the west coast of France near La Rochelle, on 17 February 1754. His parents were merchants, and as their fifth child Nicolas could expect little help from them in the way of advancement. He went to sea at the age of fifteen as a cabin boy on coastal vessels – a harsh apprenticeship, for the 'boy' was generally the ship's drudge, the crew's scapegoat, beaten and abused by all until he grew big enough to defend himself. At twenty he enlisted as a naval cadet at St Martin, and the following year embarked as quartermaster on a troop transport bound for India, hoping to make a career in the French East India Company. Disillusioned, he returned home in 1777.

France's entry into the United States War of Independence offered the chance of a naval career. Obtaining a commission as an *officier bleu* (not of noble birth) he served in the Caribbean theatre from 1779 to 1780, before returning to France to command the sloop *Apollon* on convoy duty in the English Channel. Without warning he was relieved of his command by the Comte d'Hector, commanding officer at Brest, and replaced by an *officier rouge* (a noble). Baudin blamed this 're-volting injustice' on an intrigue of aristocratic officers against a commoner, though perhaps his abrasive personality may also have played a part. Embittered, he resigned his commission and went abroad, determined to mend his fortunes in the merchant service.

He is next heard of in October 1785 as captain of the *Caroline*, a 200-ton ship carrying emigrants from France to New Orleans, Louisiana; after taking on a load of timber, he returned to Nantes. Early in 1787 he arrived at Port Louis, capital of the French colony of Isle de France (Mauritius), in command of the *Pepita*, a Spanish vessel under charter to French merchants in New Orleans. On the voyage he had called at the Cape of Good Hope, where he embarked as a passenger Franz Boos, the Austrian emperor's head gardener and botanist. Their chance meeting signalled the start of a new career for Baudin; from 1787 to 1794 he made four successive voyages on botanical expeditions for the Austrians.

Boos had been at the Cape collecting for Emperor Joseph II, and was on his way to visit the botanical gardens at Pamplemousses, near Port Louis, whose director, Jean-Nicolas Céré, had offered his help. While Céré and Boos collected specimens on Isle de France, Baudin sailed to Mozambique to fetch a cargo of slaves for the colony's plantations. On his return he agreed on terms to carry the Austrian's collections back to

Europe; this captain, Céré wrote approvingly, 'is as distinguished for his talents as for his honesty'.

The *Pepita* sailed in December, carrying the botanical treasures Boos had gathered with Céré's help. In January 1788 she called at the Cape to take on board the flora and fauna collections that the Austrian had left in the care of his assistant Georg Scholl. These had now expanded so much that only the birds and mammals (among them ostriches and zebras) and the rarer plants could be taken on board. Scholl, with the remaining plants and seedlings, was left behind, to continue his collecting until another ship could be sent. By now a floating herbarium and ark, the *Pepita* continued on to the Austrian port of Trieste, whence the collections were transported overland to Vienna. During the voyage, writes Frank Horner, Baudin learned a great deal from Boos about botany and keeping plants and animals alive at sea. The experience sparked a genuine interest in natural history, which lasted until Baudin's death.

His success in repatriating Boos's collections made Baudin's name with the Austrians; in the following five years he made three further voyages – to the Indian Ocean and the Pacific – under the imperial flag. All three were essentially joint ventures between the government and mercantile interests, in which the latter underwrote most of the costs; all three were on vessels named the *Jardinière*; and all ended in shipwreck – the first in the Pacific, the second in Port Louis harbour during a hurricane, and the third as his ship came in sight of Cape Town.

This last voyage, the most ambitious, appeared doomed from the start. Baudin, with a temporary commission as a post captain in the Imperial Navy, sailed from Genoa on a voyage to the Pacific in April 1792 on the very day that Revolutionary France declared war on Imperial Austria. At his first port of call, Malaga in Spain, he contacted the French consul offering to rejoin the French Navy 'in the same grade that he held in the Austrian'. The offer was refused, but some of Baudin's officers got wind of it and informed the imperial ambassador in Madrid; the Spanish authorities seized the ship and briefly imprisoned Baudin. Though he was soon released and allowed to proceed, the officers resigned in disgust; meanwhile many of the crew had deserted.

Calling at the Cape of Good Hope, Baudin arranged with the long-suffering Scholl to embark him and his collections on the homeward voyage. He then headed east for New Holland, but was driven north by successive hurricanes and forced into Bombay for repairs, unaware that France was now also at war with Britain. Here 'British intrigue and money' lured more seamen to desert, and he had to replace them with an Indian crew. Abandoning plans for a cruise in the Far East, he turned west and coasted the shores of the Persian Gulf, the Red Sea, and East Africa. Finally, the *Jardinière* III foundered like her earlier namesakes, driven ashore in a storm as she came into Table Bay.

Scholl, now in his eighth year of exile at the Cape, his hopes of returning home dashed yet again, accused Baudin of deliberately driving his ship onto the sands so that he could sell the slaves he had taken on board at East African ports. There is no firm evidence, but the Mauritian historian Madeleine Ly-Tio-Fane,[3] who has painstakingly researched Baudin's Austrian links, writes that he is deliberately vague about his visits to these ports, and adds 'one can guess he called at Mozambique', a slaving centre.

Citizen Baudin and the Belle Angélique

At sea, October 1796. Citizen André-Pierre Ledru describes his captain's skill in saving the vessel Belle Angélique *from shipwreck:*

> *Nothing could equal the zeal shown by [Captain] Baudin. Amidst the greatest dangers his example inspired the company, his coolness spread confidence, and the orders he gave, with the greatest precision, were always those demanded by the critical urgency of the moment. Although badly wounded in the head, although deprived of food and sleep, he was heedless of his own needs, and concerned himself only with our own. His experience and his skill rescued us from the horrors of shipwreck. [Quoted in Horner, 1987, p. 32]*

The aftermath of the *Jardinière*'s voyage left Baudin with many enemies among the court at Vienna, the merchant houses of Genoa, and – as later events were to show – the colonists of Isle de France. His movements after her shipwreck are no less obscure than the course of the voyage and the manner of its ending. At some stage he succeeded in transporting to the Spanish Caribbean island of Trinidad a rich botanical collection of living plants and trees, apparently saved from the wreck, and left it in the care of a friend, the botanist Labarrère. He then made his way to the United States, obtained a passport from the French ambassador and returned to France on an American ship. Back in Paris at the end of 1795, after an absence of more than a decade, Nicolas Baudin again offered his services to his country.

The Minister of Marine, Admiral Truguet, listened attentively to his plan to harass and destroy English East Indiamen homeward-bound from the Indies and China at the most vulnerable point on their route – the island of St Helena, where they assembled to form convoys for the final stage home. The Minister considered that conditions did not favour the scheme's execution and regretted that at present he could not offer its author a commission.

Judging that his wartime service under the Austrian flag marked him in his countrymen's eyes as a turncoat and possible traitor, Baudin set about fashioning an alternative persona – one based on his credentials as a botanical voyager. He called on Professor Antoine-Laurent Jussieu at the Muséum National d'Histoire Naturelle in Paris to inform him of the rich collection of exotic plants left behind on Trinidad; prudently he had with him specimens of seeds and plants to whet the museum's appetite. It would be a simple matter, he suggested, to retrieve the collection for the Republic: 'I do not propose to make this voyage a speculative venture, my sole purpose is to contribute by all appropriate means to the progress of the sciences and to the public good'.

Jussieu and his colleagues urged the Directory, the Republic's executive government, to accept the offer without delay: 'Citizen Baudin [asks] … only that the government pays for the charter of a small ship which a friend will lend to him for this purpose.' Eager for kudos from any source, the cash-strapped Directory agreed, and within six months, on 30 September 1796, the expedition sailed from Le Havre in the *Belle-Angélique*, 350 tons, protected by a British safe-conduct issued on Sir Joseph Banks's recommendation. On board were four *savants* (scientists) nominated by the museum – René Maugé, Anselm Riedlé, André-Pierre Ledru and Stanislas Levillain.

Three weeks out, a howling gale blew up from the west, raging seas broke across the deck, and after four days the ship wallowed helplessly, almost wholly dismasted, without a rudder, holed fore and aft, and barely seaworthy. Night and day Baudin battled the sea, certain in his own mind that but for him the ship would surely sink. Injured and short of sleep, it took him three weeks to bring the battered vessel into port at Tenerife. Ledru and his friends were united in admiration of their captain's courage.

The *Belle-Angélique* was beyond repair, but Baudin replaced her with a smaller vessel and sailed for Trinidad. He found the island under British occupation, and the authorities refused him permission to land

and recover the collection. 'All Europe will hear of this unjust treatment,' he promised as he left. From Trinidad the expedition moved on to the Danish colony of St Thomas (one of the Virgin Islands), where they remained for ten weeks. Was it coincidence that Baudin anchored in the deepwater harbour of Charlotte Amalie, the largest slaving port in the Americas? The naturalists found many new treasures in the rich volcanic soil, while Baudin, importantly, was able to purchase a larger ship. He renamed her the *Belle-Angélique*.

The next port of call was San Juan, on neighbouring Puerto Rico. Here they stayed nine months, gathering, in Baudin's words, 'the most beautiful collection of plants it is possible to see'. Ledru wrote to Jussieu:

> As for our captain, as eager as any of us and more tireless, he puts his own hands to the task of pulling out, carrying, and planting our living trees and shrubs, and sets us an example by his ceaseless activity.[4]

On the homeward passage, Baudin worked with the naturalists to protect the plants and trees from the hazards of the ocean voyage; despite strong winds and seas in mid-Atlantic, few were lost or damaged. Entering the Channel he faced a challenge of another sort – the Royal Navy blockade of French ports. Taken aboard an English warship, the indignant Baudin was interrogated for hours by Commodore Strachan in his cabin; as a parting insult the French ship was forbidden to make port at Le Havre. Baudin determined to have the last word: 'Commodore, it would have reflected more glory on you to show favour to an expedition undertaken for the progress of science, than to bombard our ports which you will never destroy!' He anchored at Fécamp, in Normandy, on 7 June 1798, after an absence of twenty months.

Unloading of the *Belle-Angélique*'s precious cargo began at once under Baudin's supervision – 450 stuffed birds, 4000 butterflies and other insects, 200 shells, seven cases of corals, crabs, sea urchins and starfish, 200 specimens of wood, four cases of seeds of 400 different species, 8000 dried plants from some 900 species, 800 living plants and shrubs from 350 species. He took it upon himself to organise their transport by wagon train to Paris – nothing must interrupt their safe delivery to the museum.

Fortune smiles on Nicolas Baudin. The collections arrive in the capital in time for the fête de la liberté *on 10 Thermidor (28 July, the fourth anniversary of Robespierre's execution). This year the grand parade features the spoils of General Bonaparte's Italian victories – the bronze horses of St Mark's; the Capitoline Venus, Apollo and the Muses, the Dying Gladiator, and other glories of the past; splendid paintings by Titian, Raphael, Paul Veronese and others – vivid reminders of the invincibility of the Republic's armies.*

In intermittent rain the parade passes the great golden dome of the Hotel des Invalides, the École Militaire – the tricolores flapping proudly in the breeze – and moves on to the Champ de Mars. At its head the cheering crowds see some curious trophies – wagons carrying banana and coconut palms, pawpaw trees, and other exotic plants. Word spreads that they have been landed by a French mariner, Citizen Baudin, after a voyage to the West Indies.

A grand venture for a great nation

Paris, 1800. 'Discoveries in the sciences have been with reason placed amongst the chief records of the glory and prosperity of nations …'

> *… the honour of the nation and the progress of science amongst us combined together to require an expedition of discovery to the Southern Hemisphere, and the Institute of France thought it a duty to lay the proposition before the government. War at this epoch seemed to rage with redoubled fury; the political existence of France was in danger; her territory was usurped; but Bonaparte was first consul; he received, and was interested in the proposition of the Institute, which, some years before, had been gratified in nominating him one of its members; and even at the time when the army of reserve was on the move to cross the Alps, he gave the order to hasten the execution of this great undertaking. [Péron, 1809, p. 10]*

In Paris the enthusiasm for Baudin's achievements came near to obsession. The *Moniteur* described his treasure-trove as 'the richest and most beautiful collection of living plants ever brought to Europe … an infinitely precious cargo', while the gazettes – the *Magasin encyclopédique* and the *Décade philosophique* – lauded him as the greatest navigator and naturalist of all time. Jussieu, gripped by the same fever, moved to capitalise on his protégé's success, proposing to the Minister of Marine, Admiral Bruix, that he should 'make further use of Captain Baudin's ability by sending him … on another voyage in which the interests of natural history would not be forgotten'.

On 30 July 1798 Jussieu wrote again, recommending a detailed plan for a voyage of discovery *autour du monde* that Baudin had submitted to the museum on his return. The Ministry moved with incredible speed – within two weeks Baudin had been reinstated in the French Navy, promoted *capitaine de vaisseau* (post captain), and nominated as leader of the proposed expedition; the ships for the expedition had been selected and a budget of 150 000 francs prepared. The Directory, however, vacillated. France was at war, funds were short, and the government's political survival was in peril; the voyage was supported in principle, but in practice was postponed indefinitely. Baudin was appointed to the fleet at Brest as chief of staff to Admiral Bruix, now returned to active service.

In late April 1799 Bruix slipped out of Brest with twenty-five ships of the line and sailed through the Straits of Gibraltar to Toulon, took on supplies for the Army of Italy, besieged in Genoa, and landed them without interference from Admiral Lord Keith's pursuing squadron, always in the wrong place at the wrong time. On the return voyage he was joined by a large Spanish force, and the combined fleet arrived safely at Brest in mid-September. It was a rare French success in the war at sea.

Baudin took leave, and in the New Year of 1800 was back in Paris, again lobbying for command of a scientific expedition. The Republic now had a dynamic young leader at its head – General Bonaparte, evading Nelson's patrols, had returned from his Egyptian adventure, overthrown the fading Directory in the coup of 18 Brumaire (9 November 1799), and been proclaimed first consul. Baudin appealed directly to Bonaparte for justice:

> Little accustomed to solicit favours or advancement, I have seen, without jealousy, several officers whom I have trained rapidly obtain distinguished appointments; but the apparent forgetfulness regarding my own services affects me, and I count on your Justice to obtain an activity suited to my character.[5]

His letter remained unanswered, and Baudin turned once more to the *savants* for support, presenting a more ambitious version of his earlier plan to a session of the Institut National in March. As before, he proposed a voyage round the world, taking in the Americas, the Pacific Islands, and southern Africa, but also focusing on New Holland, 'to discover in a clear and precise manner whether or not … these lands form part of a single island'. He appealed to members to use their influence with 'the Hero, who holds today between his hands the high destinies of France' to initiate the project.

A commission of eminent members, including Jussieu, Bougainville, and Fleurieu, reviewed the plan, and within three weeks produced a scaled-down version that omitted the Americas and the Pacific, and concentrated instead on the coasts of New Holland and southern New Guinea. The commissioners and Baudin presented the revised proposal to General Bonaparte on 25 March, and a month later Minister Forfait was able to report that

> the First Consul was ... disposed to order a voyage which will have for its main object the exploration of the south-west coast of New Holland where Europeans have not yet penetrated.[6]

Baudin's appointment as its commander was confirmed on 12 June; three days later he was at Le Havre.

The expedition's ships

Le Havre, 1800:

> *Two ships in the port of Havre had been prepared for this expedition; the* Géographe *a fine corvette of 30 guns, drawing from 15 to 16 feet water, an excellent sailer, but rather too slightly built for such service; and the* Naturaliste, *a large and strong built store-ship, drawing much about the same waters as the* Géographe, *not so good a sailer, but more seaworthy, and on that account much superior to the corvette. [Péron, 1809, p. 13]*

At Le Havre, Baudin inspected the two ships selected by the Ministry for the voyage – two corvettes built for the planned invasion of England in 1794. Not satisfied with their condition, he chose instead a thirty-gun corvette, the *Galatée*, 350 tons, and a twenty-gun store-ship, the *Ménacante*, of the same tonnage. In keeping with the expedition's scientific aims, they were renamed the *Géographe* and the *Naturaliste*, and the armaments reduced to eight carriage guns on each ship.

Though both were good ships of their type, the marked difference in sailing speeds proved an endless source of irritation and discord, and led to two lengthy separations during the voyage. The two ships, Baudin wrote to the Minister from Tenerife, their first port of call, 'behaved perfectly at sea in the various circumstances in which we found ourselves. I regret only that the speed of the *Naturaliste* should be vastly inferior to that of the *Géographe*'. He feared, correctly, 'that it will cost us a few more weeks at sea than if she had sailed better'. He was not to know that it was also to cost him his reputation as a navigator.

Later, at Port Jackson, he decided to send the *Naturaliste* back to France with the natural-history collections gathered on the west coast of New Holland and on Van Diemen's Land. To replace her he purchased a small colonial-built schooner of about 30 tons, suitable for surveying close inshore. He named her the *Casuarina*, after the timber from which she was built (she-oak, or *Casuarina stricta*). Unfortunately she too proved a slow sailer, providing cause for still more conflict with his officers.

Baudin's officers

Le Havre, 1800.

> *The officers of this expedition were chosen with the greatest care; those who aspired to the distinction submitted to the most strict examinations to obtain admission among us, and all were worthy of the preference. Not only among the officers was this regulation observed, but the most inferior ranks of our company were thus selected, and many young men of respectable families in Normandy joined our crews ... [Péron, 1809, p. 15]*

Baudin had proposed a complement of eight officers and ninety-two crew, plus eight *savants*, for each ship, but when they sailed the *Géographe* carried 118 men in all and the *Naturaliste* 120; as well, eleven seamen were found stowed away, and there were also two passengers for Isle de France. The total of 251 included fifteen midshipman (one had deserted at Le Havre), whereas the commander had requested none. Officers, midshipmen, and scientific staff together numbered fifty-four – twenty-two more than his estimates – and there can be little doubt that much of the discontent on board, as Frank Horner suggests, had its origin in this overcrowding. Disputes over the limited cabin space available must have been rife from the outset.

Baudin seems to have had little say in the selection of his officers, with the probable exception of his second-in-command, *Capitaine de frégate* Emmanuel Hamelin, commanding the *Naturaliste*. Born in 1768, with eight years' service in the Navy, Hamelin had won his promotions under the Republic. He too had served under Admiral Bruix at

Brest, and in 1800 was second in the *Formidable*, a first-rate of eighty guns. Baudin wrote of him: 'he is, without doubt, an officer of merit who will shed lustre on the Navy'.

Capitaine de frégate Sainte Croix Le Bas, second captain in the *Géographe*, also seemed a good appointment – 'a good officer, understanding very well the duties with which he is charged'. But Le Bas was wounded in a duel at Timor, and was left behind to be repatriated to France. Other deck officers on the flagship included *lieutenants de vaisseau* Pierre-Guillame Gicquel, who had sailed with d'Entrecasteaux in search of La Pérouse, and François-André Baudin, no relation to the commandant. Gicquel and François-André Baudin clashed with their leader on the outward voyage and left the ship at Isle de France.

Hamelin's second captain in the *Naturaliste* was *Lieutenant de vaisseau* Pierre-Bernard Milius, an experienced officer who had served with his captain on previous campaigns. He was promoted *capitaine de frégate* at Timor, but ill health forced him to leave the ship at Port Jackson. From there he made his way to China, and thence to Isle de France; he was still at Port Louis when the *Géographe* arrived in port in August 1803 on her homeward voyage, and after Baudin's death was appointed his successor.

The Freycinet brothers, Henri (twenty-three in 1800) and Louis (twenty-one), had joined the Navy in 1794, but thanks to the British blockade of French ports had limited seagoing experience. Each held the rank of *enseigne* (sublieutenant), and as was customary they sailed on separate ships – Henri in the flagship and Louis in the *Naturaliste*. Their relations with Baudin were at best cool, at times openly defiant; yet despite the mutual hostility he had little option but to promote them as more senior officers left the expedition. In October 1801 Henri became senior deck-officer in the *Géographe*, while Louis was given command of the sloop *Casuarina* on leaving Port Jackson. Both proved able officers when given the responsibility, though too casual for Baudin's liking.

The engineer officer, François-Michel Ronsard, was a last-minute appointment by Minister Forfait (a marine engineer himself). His principal duties were the maintenance and repair of both vessels, but Forfait gave orders he should also act as a junior deck-officer – producing the strange anomaly that while his pay and rank as an engineer were superior to those of an *enseigne,* when on deck duty he ranked beneath them.

The expedition carried three surgeons, François Lharidon de Crémenec and Hubert Taillefer in the *Géographe*, and Jerome Bellefin in her consort. By reputation Lharidon was 'a distinguished medical officer with a taste for the sea', but proved a quarrelsome shipmate and on occasion came close to blows with other officers. Bellefin, by contrast, emerged as a proficient naval physician who earned praise for his treatment of scurvy victims.

Forfait, a firm believer in the formative role of long voyages, had dismissed Baudin's objections to taking *aspirants* (midshipmen) on the expedition:

> I expect from you, Citizen, that you will see to it that the midshipmen aboard study with care all aspects of the art of sailing, and obtain practice in the different manoeuvres … it is by accustoming them early to work and fatigue that you will be able to [form good officers].[7]

Forfait's office had been flooded with applications from influential families seeking places for their sons, nephews and protégés. The venerable Admiral Louis-Antoine de Bougainville, hero of Pacific exploration, now seventy-one but with a young wife, sought a place for his son Hyacinthe (eighteen), as did the widow of General Ransonnet for her son Joseph. Another applicant with a special claim to selection was Charles Baudin des Ardennes, whose father, a member of the Convention, had (it was said) died from joy when he heard of General Bonaparte's safe return from Egypt.

Despite the high-level patronage, Baudin had no regrets when four of the 'young gentlemen' joined the exodus of officers and *savants* from the two ships after their arrival at Isle de France. 'That is no loss to the Navy,' he commented sourly. Many of those remaining with the expedition he sent back in the *Naturaliste* from Port Jackson in 1802, along with the invalids and other 'useless men'. Charles Baudin was almost alone in earning the commandant's confidence on the voyage. Bougainville, on the contrary, he considered perhaps the least useful of all, and was more than pleased to see him depart with Hamelin.

In the end Forfait's views proved to be justified. The 'harsh school' of the voyage produced a number of officers who went on to build outstanding careers in the French Navy and Colonial Service. No less than five (Hamelin, Milius, Henri de Freycinet, Charles Baudin and Bougainville) reached flag rank. Two (Louis de Freycinet and Bougainville) led their own scientific expeditions around the world. And two (Milius and Henri de Freycinet) became colonial governors.

'An assemblage of talent'

Paris, 29 September 1800. The Minister of Marine and Colonies to Citizen Baudin, commandant-en-chef *of the corvettes* Géographe *and* Naturaliste*:*

> *I have nothing to say to you about the conduct to maintain towards the people who, at the proposal of the members of the Institut National, are specially charged, under your command, with looking after astronomical research and all that pertains to Natural History. But that conduct and those considerations that are due to men unaccustomed to the sea must not go so far as to become a condescension that would favour pretensions or be injurious to the general subordination. Maintain this vigorously, and, before your departure from Isle-de-France, notify these persons on my behalf, likewise the officers and midshipmen, that the Government's hope is that all alike will fulfil their duties. [Baudin, 1974; pp.7-8]*

The first consul, advised by the commission of the Institut National, had appointed the largest and best-qualified scientific team ever to leave Europe on a voyage of discovery. Never, wrote the naturalist Péron, 'had there been such an assemblage of talent, never had there been such preparations to ensure success'. The *savants* included astronomers, landscape and portrait artists, geographers, mineralogists, naturalists (botanists and zoologists), gardeners, and a pharmacist; there were also the three naval surgeons. As a precaution, each discipline had two or three representatives, prudently split between the two ships.

Most, writes the historian Jacques Brosse, were

> enthusiastic and inquisitive young men ... remarkable for the extent of their learning; [but] they were little able to accept shipboard discipline and quickly found themselves in conflict with a hard and authoritarian captain.[8]

Baudin had first proposed a team of eight for each ship; faced with a total of twenty-three, he complained to the Ministry: 'The number of individuals seemed to me far too many and I made some objections to that effect, but these were without result ...'.

His friends from the West Indian expedition – Maugé, Riedlé, Ledru and Levillain – all volunteered for the voyage, and were selected. Ledru, however, withdrew at the last moment 'unable to resist the tears of his elderly and infirm mother'. None of the others would return to France.

Ten *savants*, seven of them from Hamelin's ship, defected from the expedition on arrival at Isle de France (Mauritius) in April 1801, disenchanted with the voyage and the commander; most, however, claimed 'ill health' as the reason, to safeguard themselves against any government moves to recover the cost of their passage. Among them were the zoologists Bory de St Vincent and Dumont, the botanists Michaux and Delisse, and the three 'official' artists, Milbert, Lebrun and Garnier. Apart from the veteran Michaux, whose loss he regretted, Baudin was not sorry to see them go, writing to Forfait: 'I hope that the result of the expedition will confirm an observation I have often made to you, that they were not very necessary'.

He replaced the artists with two 'assistant gunners', Charles-Alexandre Lesueur and Nicolas-Martin Petit, both of them talented illustrators whom he had recruited at Le Havre, and whose skills he utilised from the start of the voyage. He informed Jussieu

These two young men ... will be more useful to the expedition and will be more deserving of national recognition than all the well-known artists who were chosen and who were so over-praised that they looked on work which would have done them honour as beneath their dignity.[9]

Of the scientists who remained, only seven survived the voyage. In addition to Maugé, Riedlé and Levillain, Pierre-François Bernier (astronomer), Louis Depuch (mineralogist), and Antoine Sautier (assistant gardener) fell victim to dysentery or fever. The survivors included the geographers Charles-Pierre Boullanger and Pierre Faure, the mineralogist Charles Bailly, Antoine Guichenot, assistant gardener, and the pharmacist Collas. Louis Leschenault de la Tour, the *Géographe*'s botanist, was left ill on Timor in 1803, and spent the next three years investigating the natural history of Java before returning to France in 1807. And there was the naturalist François Péron.

Péron, a last-minute addition to the scientific staff as a 'trainee zoologist charged with comparative anatomy', joined on the recommendation of Jussieu and the anatomist Georges Cuvier, whose pupil he had been; by the beginning of 1802, as a result of defections and death, he was the sole remaining naturalist apart from Leschenault. His studies on the voyage covered most of the natural sciences, including anatomy, anthropology, biology, botany and zoology – together with meteorology, oceanography, naval hygiene, and some unofficial espionage at Port Jackson.

'A voyage of observation and research'

Paris, September 1800. 'Plan of Itinerary for Citizen Baudin ... on the voyage of observation and research relating to Geography and Natural History, the control and direction of which has been entrusted to him':

> *In order to carry out the Government's design, Citizen Baudin will em-*
> *ploy assiduously and with all the zeal of which he has given proof, the*
> *scientists, engineers, artists and means placed at his disposal, as much to*
> *determine precisely the geographical position of the principal points along*
> *the coasts that he will visit and to chart them exactly, as to study the in-*
> *habitants, animals and natural products of the countries in which he will*
> *land. With regard to the products, he will give his attention to the collect-*
> *ing of those which appear capable of being preserved, and he will apply*
> *himself principally to the procuring of the useful animals and plants which,*
> *unknown in our climate, could be introduced here. [Baudin, 1974, p. 1]*

The choice of Charles-Pierre Claret de Fleurieu to prepare the expedition's instructions and plan of itinerary was to be expected. He had drawn up the instructions for La Pérouse and d'Entrecasteaux, he was the most knowledgeable man in France about the problems of navigation in the Pacific and the South Seas, and moreover he was a member of the new *Conseil d'État* – a council of twenty-nine eminent members appointed to advise the first consul in such key areas as finance, legislation, war and the navy.

In all probability Fleurieu would have briefed Bonaparte in advance on the advantages (scientific and strategic) to be gained from a fresh voyage of exploration. Not that the general would have needed much persuading; like Louis XVI he was a keen student of Cook's voyages (he had taken a copy with him on the Egyptian adventure), and he also shared Louis' concern for the uncivilised nations living on the shores to which the French were bound. It was his wish that 'we should appear among them as friends and benefactors'.

> By his orders [says Péron] the most useful animals were embarked in our
> vessels, a number of interesting trees and shrubs were collected in our
> ships, with quantities of such seeds as were most congenial to the tem-
> perature of the climates. The most useful tools, clothing, and ornaments
> of every sort were provided for them, even the most particular inven-
> tions in optics, chemistry, and natural philosophy, were contributed for
> their advantage, or to promote their pleasure.[10]

Fleurieu drew up the instructions in little more than a month – it was largely a matter of paring down what he had written in earlier years for La Pérouse (and for d'Entrecasteaux in his search for the missing navigator). He concentrated on the 'south-west, west, north-west and north coasts of New Holland, some of which are still unknown, while others are known only imperfectly'. His purpose, he wrote, was 'to trace [for Citizen Baudin] the route he will have to take, and to determine, on a rough estimate, the amount of time he will be able to give to the reconnaissance and examination of the different sections of coast that his navigation must take in'.

Baudin must surely have read the plan of itinerary with some foreboding. Fleurieu's timetable was tied to predictable seasonal winds and currents, and to the onset of the monsoons, and made little allowance for what Cook had called 'the Vicissitudes attending this kind of service'. Should 'circumstances of weather ... or unforeseeable events oblige Citizen Baudin to depart from the route mapped out', his orders were clear:

> he will return to it as soon as the reason for his departure from it has disappeared. The Government's intentions have been explained to him so precisely that he will have no difficulty in grasping them. And if, during any one of the periods set down in the plan ... it should not be possible for him to conform exactly to what has been prescribed, he must make every effort to revert to it as nearly and as soon as he can.[11]

Unlike La Pérouse, Baudin was given little discretion to vary his itinerary as circumstances required. Certainly he lacked the stature of his predecessor, and perhaps the rigidity of his timetable reflected doubts about his abilities as a navigator – his reputation, after all, was made as a collector-voyager, and his early record in the Navy, if consulted, gave some grounds for disquiet. The instructions hint at a further reason. Fleurieu writes:

> It is probable that the English have already made, or soon will make, a detailed reconaissance of the large strait that they have discovered and named Bass Strait ... It is too close to their settlement at Port Jackson for them not to have attended to the completion of its discovery.[12]

This suggests French naval intelligence, as well-organised as the British, may have reported the departure from Portsmouth in March 1800 of HM Surveying Brig *Lady Nelson*, under the command of Lieutenant

James Grant, bound for Port Jackson and with orders to survey Bass Strait on the way.

In any event, Fleurieu – eager that France should claim the discovery of the unknown southern coast of New Holland – must have realised time was running out. His anxiety for French ships to win the race would help to explain the inflexible itinerary and the emphasis placed on the south coast – which, he stressed,

> has not yet been discovered; no navigator has seen it, and Citizen Baudin must apply himself to establishing the geographical positions of the points that will be noticed along it, and to drawing up an accurate chart of the whole.[13]

As Fleurieu well knew, the prize would already have gone to France had either La Pérouse or d'Entrecasteaux been able to fulfil their instructions, and he had no wish to see it slip away a third time.

In addition to the itinerary, Baudin was handed 'special notes' describing in detail 'what kind of observations and research should occupy the scientists ... employed under his orders'. These included a series of reports produced by experts and learned societies at the invitation of the Institut National – it seemed no branch of science had been overlooked, from anthropology, biology and botany through geography to mineralogy, naval medicine and zoology.

The chief naval medical officer, Dr Coulomb, provided health instructions for the voyage; these contained a list of approved antiscorbutic remedies – among them powdered lemon juice and Parisian mineral water – and directions for maintaining hygiene at sea, based on the best contemporary practice. The novelist Bernardin de St Pierre, a former director of the Jardin des Plantes in Paris, offered a more innovative approach to preventing scurvy; he suggested using the Breton bagpipes to encourage the men to dance on board, thus maintaining morale and more importantly counteracting the deadly lethargy that contributed to the onset of the disease: 'Musical instruments, games, dancing will spread movement and life throughout the ship!'

The newly formed Societé des Observateurs de l'Homme (Society of the Observers of Man) contributed several papers of lasting interest. Founded the previous year by Louis-François Jauffret, the society's aim was to collect, collate, and analyse observations bearing on mankind 'in all its domains' – physical, moral, spiritual, and emotional. Both Baudin and Hamelin seem to have been corresponding members. Two papers in particular broke fresh ground in the emerging science of anthropology. The first, from the anatomist and zoologist Georges Cuvier, dealt with physical anthropology; in the second the young philosopher Joseph-Marie Degérando outlined a methodology for observing the way of life of native peoples. Now considered a classic of social anthropology, it suggested methods that were unsuited for an exploratory expedition covering vast areas and with neither the time nor opportunity for extended contact with indigenous groups.

Baudin's views on his instructions are not recorded, but it can be assumed they would be similar to Hamelin's, written when he first saw them on leaving Isle de France for New Holland:

> I am not frightened when I see before me so much to do; on the contrary my active temperament and my self-esteem are stung by the prospect and I am content, but I cannot deceive myself that the government has demanded too much of the expedition … May the winds and the seasons pose no obstacles to our zeal, I hereby swear to neglect nothing and to spare myself no pains, or care, or even repose, in order to fulfil the intentions of the government which appears to honour us with a great trust in prescribing so much work.[14]

MATTHEW FLINDERS AND THE BRITISH VOYAGE OF DISCOVERY, 1801 TO 1803

A navigator's apprenticeship

Donington, Lincolnshire, May 1791. Surgeon-apothecary Dr Matthew Flinders notes a visit by his son in his diary:

> *My son Matthew having leave of absence to visit his Friends, previous to the Commencement of his great Voyage, came here on Tuesday May 11, and left us again this day Friday May 20. He is going with Capt. Bligh in the Providence … to Circumnavigate the Globe. They expect to sail about 1st June and will be near 3 years performing this great undertaking. He had expressed a desire to go on a Voyage of this sort, and Commodore Pasley got him this situation with Capt. Bligh … I have desired him to keep an exact Journal, as if it please God we live, and he returns safe, I have some Idea a publication may be advantageous. He has made much Improvement in his knowledge of Navigation. [Quoted in Ingleton, 1986, p. 3]*

He was 'Induced to go to sea against the wishes of friends from reading *Robinson Crusoe*', Matthew Flinders wrote towards the end of his life. Only a fortnight before his death he subscribed to a new edition of Defoe's novel, requesting that 'the volume on delivery should have a neat, common binding, and be lettered'.

The writer, linguist and psychoanalyst Sidney Baker[15] has pointed to the many parallels between the lives of the fictional Crusoe and the young navigator. Crusoe

> would be satisfied with nothing but going to sea; and my inclination to this led me so strongly against the will – nay, the commands – of my father, and against all the entreaties and persuasions of my mother and other friends that there seemed to be something fatal in that propension of nature, tending directly to the life of misery which was to befall me.

Flinders's father was no less strongly opposed, having planned for his son to follow him in the family practice. His opposition was doubtless strengthened by the example of Matthew's older cousin John, still a midshipman after nine years' service in the Royal Navy.

Crusoe's father, like Dr Flinders, came from the comfortable 'middle station of life' – bidding his son take note that it had the fewest disasters, and 'was calculated for all kinds of virtues, and all kinds of enjoyments; that place and plenty were the handmaids of a middle fortune; that temperance, moderation, quiet health, society, all agreeable diversions and all desirable pleasures were the blessings attending the middle station …', and so on. Matthew was no more disposed to listen to such well-meant advice than the young Crusoe. As late as 1804, during his captivity at Isle de France, he wrote to his patron Sir Joseph Banks: 'I have too much ambition to rest in the unnoticed middle order of mankind.'

Matthew Flinders was born at Donington, in the Lincolnshire fens, on 16 March 1774, the eldest son of the town's surgeon. He first read *Robinson Crusoe* in his fifteenth year, and seems to have decided almost at once to make the Navy his career. He began to study navigation, geometry and trigonometry (probably with the help of his cousin John), and in 1789, in defiance of his father's wishes, joined the Navy as a 'lieutenant's servant' – a meaningless role involving little useful experience. The following year, thanks to a family connection with Post Captain (later Admiral Sir Thomas) Pasley, he transferred to Pasley's new command, HMS *Bellerophon*, as a junior midshipman. Known throughout the service as the *'Billy Ruff'n'*, the ship was newly com-missioned, a fine third-rate of seventy-four guns, and Pasley made sure his young protégé received a thorough grounding in his profession.

Pasley liked what he saw of Matthew's work, particularly his skill in using the new nautical timekeepers, or chronometers,[16] and recom-mended him to Captain William Bligh for appointment to HMS *Provi-dence*, then fitting out in Deptford Dockyard. Sir Joseph Banks had persuaded the Admiralty to appoint Bligh, honourably acquitted by court-martial of any blame for the loss of the *Bounty*, to command a second voyage to the South Seas to transplant the breadfruit plant from Tahiti to the West Indies (where it was seen as a cheap solution to the problem of feeding the plantation slaves). At one stroke, fate (or sim-ple good fortune) brought Midshipman Flinders into the orbit of the influential Banks, and also gave him the chance to learn the practical skills of navigation, surveying and charting at the hands of a master.

Two ships sailed on this second expedition – the *Providence*, 420 tons, and a small brig, the *Assistant*, 100 tons, commanded by Lieuten-ant Portlock, who had sailed with Bligh on Cook's last voyage. Both vessels carried a full complement of officers and marines – there was to be no chance for a second mutiny. They left Spithead in August 1791, calling at the Cape of Good Hope and Adventure Bay in Van Diemen's Land on their way to Tahiti, and dropped anchor in Matavai Bay on 10 April 1792. Here Bligh learnt to his 'great satisfaction' that Captain Edwards of HMS *Pandora* had taken prisoner fourteen of the *Bounty* mutineers, who had remained on the island rather than sail with Fletcher Christian. Meetings with several of the local women who had borne children by the mutineers satisfied him that the 'voluptu-ous gratifications of Otaheite' had led to the mutiny.

The midshipmen shared in the painstaking work of gathering the breadfruit plants, surveying the island's bays and reefs, and taking observations at the observatory set up on Cook's Point Venus. The name proved only too apt. After eight months at sea the crews of both ships, including several of the officers and 'young gentlemen', 'chased the girls only to find the girls were not chaste' – leaving many 'with warm tokens of their affection'. The *Providence* sailed with more than a quarter of her complement receiving treatment for venereal disease.

Bligh set sail in late July on the long run to Jamaica by way of the Tonga and Fiji groups and Torres Strait. Part way through the strait, four war canoes attacked the *Providence*'s cutter in an attempt to cut it off from the ship. Watching from the quarterdeck, the eighteen-year-old Flinders had to admire the islanders' seamanship:

> No boats could have been manoeuvred better in working to windward, than were these canoes of the naked savages. Had the four been able to reach the cutter, it is difficult to say whether the superiority of our arms would have been equal to the great difference of numbers, considering the ferocity of these people and the skill with which they seemed to manage their weapons.[17]

Five days later a fleet of war canoes launched a full-scale assault on both ships. Musket fire failed to drive off the attackers, and Bligh was forced to fire one of the great guns, loaded with round and grape shot, straight at the leading canoe. The survivors dived overboard and swam towards their companions, plunging continuously to avoid the musket balls showering about their heads. By the time the natives withdrew, three of the brig's crew had been wounded, one mortally; and 'the depth to which the arrows penetrated into her decks and sides ... was truly astonishing'.

The superb seamanship of Bligh and Portlock carried them through the uncharted and reef-studded waters of the strait in only nineteen days. The captains worked well together, Bligh respecting the younger man's skills as little inferior to his own. 'I left the rest to be done by Lieutenant Portlock,' he wrote in his log on one occasion, 'whose alertness to duty makes me at all times think of him with regard and esteem.' Relations with his own officers were less harmonious, though this time there was no hint of mutiny; the discontent was confined to mutterings about Bligh's abusive language and 'violent Tornados of temper'.

Flinders had his own difficulties with his irascible captain, writing to Banks many years later (from Isle de France) that he would not care to serve under Bligh's orders – as the new governor of New South Wales – 'since the credit, if any be due to my labours, would be in danger of being monopolised'. It is an ungenerous comment, for Flinders owed much to Captain Bligh. For all his defects of character the latter was a strong leader, a navigator of genius, and an accomplished chart-maker who had honed his skills as master of the *Resolution* under Cook. There is no question that Flinders's experience in the *Providence* shaped his career; he learned the essentials of navigation from Bligh, just as Bligh had learnt them from Cook. One imagines, too, that with Bligh's example daily before him, he took to heart the lesson that Cook had been unable to impress upon his stubborn sailing master: 'You do not make friends of men by insulting them.'

In August 1793 the *Providence* returned to an England once more at war with France. After a brief leave, Flinders rejoined Captain Pasley in the *Bellerophon*; with the latter's promotion to rear admiral he was appointed an aide-de-camp, seeing action in the series of skirmishes culminating in the battle of the Glorious First of June, 1794. The *Billy Ruff'n* was heavily engaged by two enemy warships when a French shot, smashing through her sides, took off the admiral's leg. Before being carried below, Pasley is reported to have said to the seamen who commiserated with him on his wound: 'Thank you, but never mind my leg, take care of my flag!'

Pasley did not go to sea again, leaving his young aide-de-camp free to look elsewhere for his next appointment. Flinders had no more than two months to wait for another chance to follow his 'favourite pastime' of navigation and exploration.

Amateur explorers: Bass and Flinders

Donington, Lincolnshire, August 1794. Dr Flinders writes that his son Matthew is about to depart on another voyage:

> Government having fitted out 2 Vessels for the New Settlement of New Holland – to carry out a New Governor [Hunter] and Captain Waterhouse, late a Lieut. in the Bellerophon, being appointed Commander – Matthew wished to go with him – the Station will be for 4 or 5 years – wch. is a long time – but he thinks he has a better Chance of Promotion than staying in the home Service as he will be the oldest Petty Officer – My son Samuel having for some time expressed a desire for the Sea and Mattw. wishing to take him also, I have advanced £30 to Mattw. to Fit him out ... he is very young (12 in Nov.) but if we missed this opportunity several years must have elapsed before so good an opportunity might again occur ... [Quoted in Ingleton, 1986, pp. 21-22]

The two ships were HMS *Reliance* (twenty guns) and the store-ship *Supply*; both had been purchased for service in New South Wales. Captain Waterhouse took on Matthew as senior master's mate, and accepted Samuel as a 'volunteer' (entitling him to a wage of £6 a year plus his keep). The passage to Port Jackson brought Flinders into frequent contact with Governor Hunter – formerly captain of HMS *Sirius*, flagship of the First Fleet, and an accomplished surveyor himself – and with the officers of the *Reliance*. In the ship's surgeon, George Bass, he found a companion after his own heart:

> I had the happiness to find a man whose ardour for discovery was not to be repressed by any obstacle, nor deterred by any danger, and with this friend a determination was formed of completing the examination of the East Coast of New South Wales by all such opportunities as the duty of the ship and procurable means would admit.[18]

Bass, three years his friend's senior and also a Lincolnshire man, by all accounts was a remarkable personality – Ernest Scott describes him as 'six feet in height, dark-complexioned, handsome in countenance, keen in expression, vigorous, strong and enterprising'. In a surviving letter to Bass, penned not long before the latter's disappearance at sea, Flinders writes:

> You have been my touchstone ... no conversation but yours could give me a degree of pleasure; your footsteps upon the quarterdeck over my head, took me from my books, and brought me upon deck to walk with you.[19]

The *Reliance* and her consort entered Port Jackson Heads on 7 September 1795, and dropped anchor in Sydney Cove. They were the first ships from England to dock in more than a year, and the crowd on the shore, a ragged mixture of convict and free, loudly cheered the governor as he was rowed ashore. The cheers turned to jeers when the news spread that these were not supply ships, and there were none on the way. The colony, with a population of about 3500, was now in its eighth year, and for most of that time had lived on short rations; scurvy and dysentery were rife, and virtually the only commodities in plentiful supply were rum and spirits, under the control of the officers of the New South Wales Corps.

Seaborne supplies were its lifeblood, a fact obvious enough to governors Phillip and Hunter, but seemingly not fully grasped in distant London. The *Reliance* now became the colony's lifeline, and during the next two years Flinders made two voyages in the ageing ship to Norfolk Island (where some 800 convicts and their guards eked out an insecure living) carrying supplies for the outpost, and a hazardous round-the-world passage to Cape Town to bring back much-needed livestock. In between, with Bass, he began his life's work in exploration.

Bass had brought out with him from England a small rowing boat, about 8 feet long and fitted with a mast and sail. Seven weeks after their arrival, he and Flinders, with his young servant William Martin as crew, sailed through the heads in the tiny craft – christened *Tom Thumb* for the occasion – and made for Botany Bay. Here they pushed up the George's River to its navigable source, 20 miles or so beyond the previous limit of survey, before returning to Sydney. So favourable was their report to Hunter that the governor set up a depot at a place he called Bankstown. The following year he approved a second expedition by the two friends, and they set sail, again with the boy Martin as crew, in a new and slightly larger *Tom Thumb* to explore the coast to the south of Botany Bay.

Tom Thumb *II's ocean-going qualities are as pitiful as her predecessor's. The second day out the boat is swamped in heavy surf and beached, leaving crew, guns, powder and provisions thoroughly drenched. Hours of hard work are needed before they are ready to put out to sea. Next day they take aboard two 'Indians', who offer in sign languauge to guide them to fresh water. All goes well until, rowing up a small creek (the estuary of Lake Illawarra, near Port Kembla), they find themselves amid a large and growing group of natives – 'reputed at Port Jackson of being exceedingly ferocious, if not cannibals'. Clearly they must extricate themselves before the group has grown large enough to mount an attack.*

Bass encourages some of the men to help him mend an oar, while Flinders diverts the others, offering to clip their coarse, untrimmed locks with a large pair of scissors. No great nicety is required, and a dozen are soon sheared:

> *Some of the more timid were alarmed at a formidable instrument coming so near to their noses, and would scarcely be persuaded by their shaven friends, to allow the operation to be finished. But when their chins were held up a second time, their fear of the instrument – the wild stare of their eyes – and the smile which they forced, formed a compound upon the rough savage countenance, not unworthy the pencil of a Hogarth.[20]*

Departing without incident, they moor at the entrance to the creek, where the extra depth of water offers security. Two nights later, on their way back to Port Jackson, they drop anchor in the shelter of a range of cliffs. The wind has been unsettled and driving electric clouds in all directions, and at ten it burst into a gale at south. Hauling up the anchor, they run north before the storm; the waves rise higher, and the blackness of the night, with the uncertainty of finding a place to shelter, adds to the danger. The only guide to their course is the shadow of the cliffs against the sky to larboard, and the roar of the surf. Bass holds the sheet of the sail in his hands, drawing in a few inches when he sees an exceptionally heavy sea following. Flinders steers with the sweep, fully conscious that the slightest wrong move or a moment's inattention will send them to the bottom. Martin bails unceasingly, as the boiling seas throw more water into the boat.

They have run for nearly an hour when Flinders makes out in the darkness ahead the spume of high breakers, with no shade of cliffs beyond. In these seas the boat will probably not live another ten minutes. At what seems to be the extremity of the reef he brings the boat's head round, mast and sail are hauled down, and pulling hard on the oars they make for the gap. White breakers close in, wind-flung spray engulfs the boat, a long rolling wave thrusts her forward past an outcrop of black rock. Minutes later they are in smooth water and entering a well-sheltered cove where they anchor for the night. Providence Cove, they agree, will be an apt name for this place.

Discoveries in the south

Port Jackson, September 1798. On board the sloop Norfolk:

> *His Excellency Governor Hunter had the goodness to give me the* Norfolk, *a colonial sloop of 25 tons, with authority to penetrate behind Furneaux's Islands; and should a strait be found, to pass through it and return by the south end of Van Diemen's Land; making such examinations and surveys on the way as circumstances might permit …*
>
> *I had the happiness to associate my friend Bass in this new expedition, and to form an excellent crew of eight volunteers from the king's ships; but a time keeper, that essential instrument to accuracy in nautical surveys, it was still impossible to obtain. [Flinders, 1814, vol. I, p. cxxxviii]*

In September 1796 the *Reliance* and the *Supply* had sailed by way of Cape Horn for Cape Town, there to take aboard cattle and sheep to replenish the colony's dwindling livestock. The *Reliance* embarked 109 head of cattle and more than a hundred sheep; 'I believe no ship ever went to sea so much lumbered,' wrote Captain Waterhouse, though

*Nicolas Thomas Baudin;
19th-century print. [Author's
collection]*

*Miniature of Matthew Flinders.
[Mitchell Library, State Library
of NSW]*

Left: *Portrait believed to be that of Pierre-Bernard Milius. [Kerry Stokes Collection]*

Below: *Self-portrait of Charles-Alexandre Lesueur on the* Géographe. *[Collection Lesueur, Muséum d'Histoire Naturelle, Le Havre]*

François **PÉRON**.

Né à Cérilly, département de l'Allier, le 22 août 1775.

Mort le 10 décembre 1810.

Right: *Portrait of François Péron drawn by Lesueur in 1810, about a fortnight before Péron's death. [Collection Lesueur, Muséum d'Histoire Naturelle, Le Havre]*

Below: *The cowshed at Cérilly, France, in which Péron spent his last days. C.-A. Lesueur, 1810. [Collection Lesueur, Muséum d'Histoire Naturelle, Le Havre]*

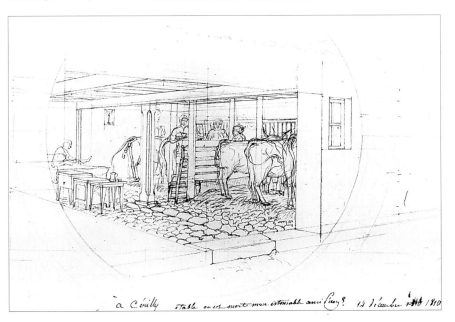

à Cérilly   étable où est mort mon estimable ami Péron?   14 Décembre 1810

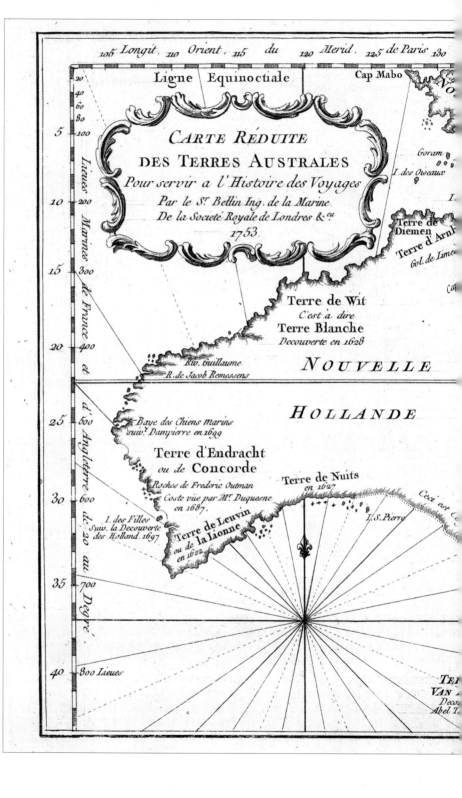

Longit. Orient. du Merid. de Paris
105 110 115 120 125 130

Ligne Equinoctiale

Cap Mabo

CARTE RÉDUITE
DES TERRES AUSTRALES
Pour servir a l'Histoire des Voyages
Par le Sr. Bellin Ing. de la Marine.
De la Societé Royale de Londres &ca
1753.

Goram
I. des Oiseaux

Terre de
Diemen
Terre d'Arn
Gol. de Lime

Terre de Wit
C'est a dire
Terre Blanche
Decouverte en 1628

NOUVELLE

Riv. Guillaume
R. de Jacob Remessens

HOLLANDE

Baye des Chiens marins
suiv. Dampierre en 1699

Terre d'Endracht
ou de Concorde

Roches de Frederic Outman
Coste vüe par Mr. Duquesne
en 1687.

Terre de Nuits
en 1627

Cea

I. S. Pierre

I. des Filles
Suiv. la Decouverte
des Holland. 1697

Terre de Leuvin
ou de la Lionne
en 1622

TE
VAN
Deco
Abel T.

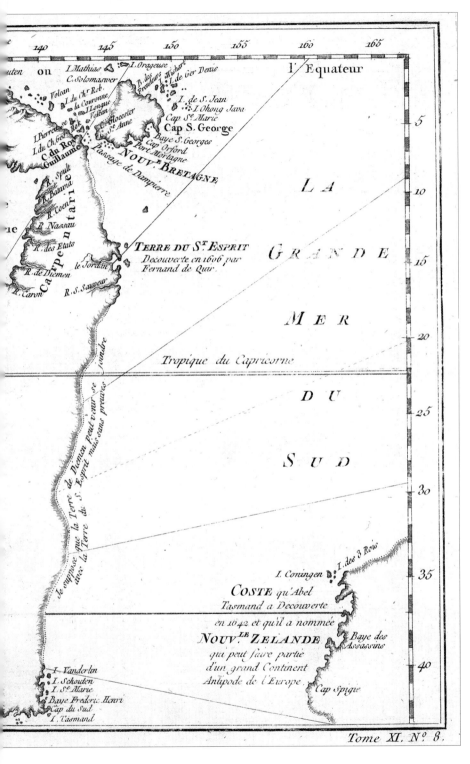

Map of the 'Terres Australes' produced by the French cartographer Jacques Bellin in 1753 – one of the very few pre-Cook maps showing only Australia.

GENERAL DECAEN.

Left: *General Decaen, Captain General of Isle de France. [Scott,* The Life of Matthew Flinders, *1914; author's collection]*

Below: *The letterhead of the Baudin expedition. [Collection Lesueur, Muséum d'Histoire Naturelle, Le Havre]*

Liberté

Egalité

BONAPARTE PREMIER CONSUL

Voyage de découvertes.

A Bord de la Corvette le Géographe.

le _____ an ____ de la République française, Une et Indivisible 9.

Matthew Flinders, painted in 1807 by his friend Toussaint de Chazal on Isle de France. [Photo courtesy of Flinders University Library]

'The Right Hon. Sir Joseph Banks', by Antoine Cardon. [Rex Nan Kivell Collection; by permission of the National Library of Australia]

*Captain Philip Gidley King,
governor of New South Wales
from 1800 to 1806. [Mitchell
Library, State Library of NSW]*

*Self-portrait of William
Westall, circa 1820. [By
permission of the National
Library of Australia]*

he personally had purchased enough stock to start a small farm. Not surprisingly it was 'one of the longest and most disagreeable passages I ever made'.

At the Cape, Flinders qualified for promotion to lieutenant, but had to wait a year for Their Lordships' approval – his seniority was dated 21 January 1798. The ship returned to Port Jackson with her pumps going, and was 'so extremely weak in her whole frame', according to Hunter, that 'it is in our situation a difficult matter to do what is necessary'. Flinders stayed aboard to assist with the refit, while Bass, 'less confined by his duty' as surgeon, made several excursions, including a bold but unsuccessful attempt to find a way through the Blue Mountains. In late 1797 he persuaded Hunter to give him a whaleboat and six volunteers from the *Reliance* to examine the coast southwards 'as far as he could with safety and convenience go'. One of the volunteers was Able Seaman John Thistle.

Bass was out for twelve weeks, from early December to the last week of February 1798, during which he explored some 600 miles of mostly unknown coast. He rounded Wilson's Promontory (where he found a small band of escaped convicts marooned by their comrades, who had sailed on westwards – the first, though anonymous and unacknowledged, discoverers of the strait), and discovered Western Port, thereby putting almost beyond doubt the existence of a great strait separating Van Diemen's Land and the mainland. Flinders wrote of his friend's voyage:

> He sailed with only six weeks' provisions; but with the assistance of occasional supplies of petrels, fish, seal's flesh, and a few geese and black swans, and by abstinence, he had been enabled to prolong his voyage beyond eleven weeks. His ardour and perseverance were crowned, in despite of the foul winds which so much opposed him, with a degree of success not to have been anticipated with such feeble means.[21]

Hunter meanwhile gave Flinders the chance to do some exploring on his own account, sending him as a passenger in the schooner *Francis* to the Furneaux Islands (off Tasmania's north-east coast) to make 'what observations he could' among them; he discovered the Kent Group (named after Captain Kent of the *Supply*) on this voyage. Noting the strength of the tides setting westwards, he deduced this could only be due to a passage through to the Indian Ocean. Comparing notes in Sydney, he and Bass concluded that the only proof now needed of

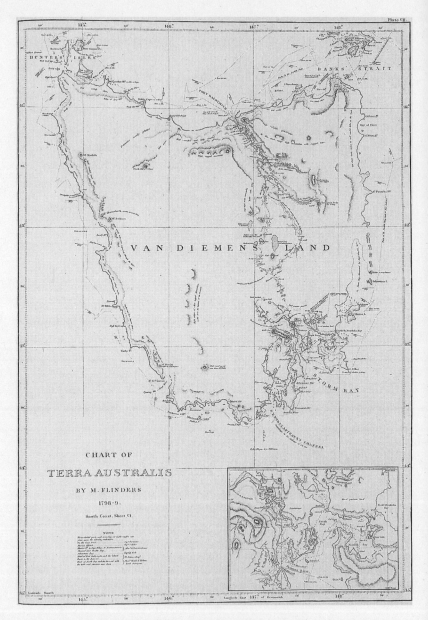

Van Diemen's Land, 1798-99. Chart by Flinders.
[By courtesy of the Royal Geographical Society of South Australia]

such a strait was to sail through it. Hunter agreed, and on Flinders's return from another supply trip to Norfolk Island authorised them to make the attempt in the *Francis*. With an 'excellent crew' of eight volunteers – John Thistle among them – they sailed on 7 October 1798.

In fourteen weeks Flinders and Bass discovered the harbour of Port Dalrymple (Launceston) in the north of Van Diemen's Land, named mounts Heemskirk and Zeehan on the west coast after Tasman's two ships, and entered the present River Derwent before heading northeast for Port Jackson. The little *Norfolk* sailed into harbour on 12 January and anchored alongside the *Reliance*. 'To the strait which had been the great object of research,' wrote Flinders, 'and whose discovery was now completed, Governor Hunter gave, at my recommendation, the name of BASS' STRAIT.'

In July 1799 Hunter sent Flinders northwards in the *Norfolk* 'to explore Glass-House [Moreton] and Hervey's Bays, two large openings ... of which the entrances only were known'. He was away for six weeks, returning to report that 'no river of importance intersected the East Coast between the 24th and 39th degrees of south latitude'. He had missed the mouths of the Clarence and the Brisbane rivers, despite running the length of Moreton Bay in the sloop.

Patron and protégé: Banks and Flinders

Portsmouth, Saturday, 6 September 1800. Lieutenant Flinders addresses a proposal to Sir Joseph Banks, PRS:

> *The detention of the* Reliance *at Spithead, prevents me from paying my respects in person to Sir Joseph Banks, and delivering the letters intrusted to me by His Excellency Governor Hunter and lieutenant-colonel Paterson, for which reason they are now inclosed, together with a small parcel of the seeds of flowers, shrubs and trees of New South Wales.*
>
> *By the first opportunity I shall do myself the honour of calling in Soho Square to give every further information within my knowledge; as also of laying the charts before the Admiralty; but previously thought it proper to make this private communication to Sir Joseph Banks, pleading on behalf of any informality there may be in thus addressing him, that almost constant employment abroad, and an education amongst the unpolished inhabitants of the Lincolnshire fens, have prevented me from learning better; but not from imbibing the respect and consideration with which the*

Right Honourable president of the most learned society in the world will always be held by his most devoted and obedient servant,

Mattw. Flinders
[Quoted in Ingleton, 1986, pp. 49-51]

The *Reliance* sailed from Port Jackson on 6 March 1800, bound for England. The ship was in such a decayed state that Hunter 'judged it proper to order her home while she may be capable of performing the voyage'. Before leaving the colony, Flinders toyed briefly with the idea of following Bass's example and running his own trading ship (his friend had returned to England the previous year and become part-owner of a trading brig, the *Venus*).[22] 'I am tired of serving for a pittance,' he wrote to an acquaintance, 'and as it were living from hand to mouth, while others with no better claim are making hundreds and thousands.'

The long passage home allowed Flinders time for reflection, and before arriving in England his decision was made – thanks to Hunter, his surveys in New South Wales and Van Diemen's Land were known to the Admiralty and to Sir Joseph Banks, and he would offer his services to complete the exploration 'of this only remaining considerable part of the globe'. He also began seriously to consider the possibility of marriage.

On the voyage he worked on his charts, preparing them for publication, and drafted an ambitious plan for surveying the remaining coasts of New South Wales and New Holland – the unclaimed western portion of the continent:

> Probably it will be found, that an extensive strait seperates [*sic*] New South Wales from New Holland by way of the Gulph of Carpentaria, or perhaps a southern gulph may only peninsulate New South Wales. The commander of an American ship, by name Williamson, reported his having sailed from the latitude 45° to 10°15′ south, in nearly a north direction, without seeing any land: his longitude being somewhat to the west of the south-west cape of Van Diemen's Land.[23]

On arrival at Portsmouth, Flinders was disconcerted to learn that Captain Philip Gidley King, RN, the colony's new governor, had sailed some months previously with orders for Lieutenant Matthew Flinders to assume command of HM Surveying Brig *Lady Nelson*, at present on her way to Port Jackson to engage on coastal survey work. (King had in fact arrived at the port some six weeks after the *Reliance* had left). The news galvanised him into action. He rapidly revised his plan, expanding it to include two survey vessels – a larger ship with the *Lady Nelson* acting in support for inshore work – and dispatched it post-haste, not to the Admiralty, but to Sir Joseph Banks.

It was a bold move for a young lieutenant of less than three years' seniority, but Flinders, possibly advised by Hunter, had chosen his target well. Banks was then at the height of his power – a Knight of the Order of the Bath, Privy Councillor, president of the Royal Society, and a confidant of the king, he was also a close friend of Earl Spencer, First Lord of the Admiralty. His influence within the government generally and with the Admiralty was immense. With Banks's support the chances of Flinders's plan being approved would be vastly improved.

Flinders was as fortunate in his timing as Baudin had been two years before. In June, Banks had recommended that the British government issue a safe-conduct to the French expedition, though not without misgivings as to the real motives behind it. Well aware of the vulnerability of the New South Wales colony, he shared the government's concern at the possibility of a rival settlement in the region. No less worried were the directors of the East India Company. French privateers from Isle de France were wreaking havoc with their Indian Ocean trade, and the prospect of another enemy base on the west coast of New Holland to complement the hornet's nest at Port Louis could not be tolerated. Political, commercial and strategic interests alike dictated that Baudin's explorations should be closely monitored.

Banks was at Revesby, his Lincolnshire estate, when Flinders's letter was delivered to his London home in Soho Square. There can be no doubt from the events that followed that he grasped the plan's potential immediately it came to his notice. His passion for science, his continuing involvement in the affairs of the colony, his close ties to the Admiralty, and his high opinion of Flinders's past surveys, no less than his concern for the national interest, convinced him that the young officer's proposals should be implemented as soon as practicable – including his offer that if his

> late discoveries ... should so far meet approbation as to induce the execution of it [completing the discovery of New Holland] to be committed to me, I should enter upon it with that zeal which I hope has hitherto characterized my services.[24]

Once discharged from the *Reliance*, Flinders travelled up to London, found lodgings in Soho a short walk from the square, and delivered his charts to the Admiralty. He visited the printer John Nichols in Little Newport Street to discuss the publication of his *Observations on the Coasts of Van Diemen's Land*, and wrote to the Court of Directors of the East India Company, enclosing a plan (based on that already sent to Banks) in which he set out

> some advantages which a Strait lately Discovered separating New Holland from Van Diemiens [*sic*] Land seems to offer the Company's Ships making an Eastern Passage to China and to Country Ships bound to Port Jackson &c.[25]

Learning that Banks was at his country estate, Flinders took a short leave in Donington, then returned to his Soho lodgings in early November. On the 16th a note was delivered to his rooms:

> Sir, – Jos. Banks presents his Compts to Mr Flinders he is sorry indeed to have been prevented by bad health from answering a Letter he Received some time ago from Mr Flinders will be happy to see him in Soho Square at any time when he will be so good as to Call upon him.[26]

Will he be so good as to call? Next morning Lieutenant Flinders presents himself early at number 32 at the corner of Soho Square, his new lieutenant's rig of blue coat and white knee-breeches freshly pressed, gold buttons adding a touch of brightness in the grey drizzle seeping down on the roofs of the capital. Sir Joseph's elegant and imposing house is daunting enough, but he controls his nerves, pulls at the bell-rope, and gives his name to the footman; he is expected.

The house is already busy, a steady stream of scientific visitors browsing in the magnificent library, studying the botanical specimens in the vast and ever-growing herbarium, or admiring the museum, with its collections of minerals, animals, fish, birds and shells, and the artefacts brought back from Banks's extensive travels or purchased from other voyagers. The young man, cocked hat in hand, is conducted through the reception rooms and ushered into the light and airy study, books lining one wall from floor to ceiling and portraits another. He notes the charts spread upon the polished tables, the specimens in the showcase. Coals glow invitingly in the grate.

Sir Joseph, grown portly in middle age, steps forward to greet his visitor, clasps his hand warmly, and seats him beside the fire. Tea, coffee and rolls are brought in by a servant. Flinders is soon at his ease, as his host praises his surveys of the east coast, in Bass's Strait, and around Van Diemen's Land – clearly Banks is well-informed about his discoveries to date. Then, leaning forward, the older man interrogates him on his proposals for completing the survey of New Holland, his views on the existence of a vast strait separating the continent into two or more parts, his proposals for examining its natural resources …

Without doubt the meeting was a great succeess for Flinders. Probably Banks had already discussed his visitor's plan with his friend Earl Spencer, and won his support; perhaps, Ingleton suggests, 'it was left to [him] to interview Flinders and judge his capability for command of the voyage'. In the event, matters moved with incredible speed – within a week the Admiralty directed the Navy Board to slip the Armed Ship *Xenophon* at Sheerness for foreign service; in January she was renamed *Investigator*.

Miss Chappelle of Partney

On board the Reliance *at the Nore (Thames Estuary), Thursday, 25 September 1800. Lieutenant Flinders writes to a friend, Miss Ann Chappelle of Partney, in the Lincolnshire Wolds:*

> *My dear friend – … the last letter which I have received from you is dated September 1797! If you think that I esteem you and value your friendship, it will be in your power to form a judgement of the uneasiness I have suffered on your account … My imagination has flown after you many a time, but the lords of the admiralty still keep me in confinement at the Nore. You must know, and your tender feelings have often anticipated for me, the rapturous pleasure I promised myself on returning from this antipodean voyage, and an absence of six years; and if I mistake not your feeling heart will well picture my disappointment and distress on finding my best beloved sister [Elizabeth] and the friend of my bosom [Mary Frank lin] both torn from my arms by that scythe bearing villain. It is a shock to my spirits … [Quoted in Mack, 1966, pp. 46-47]*

Matthew Flinders had met Ann Chappelle, the daughter of a master mariner who had died at sea, on his last home leave in 1794. She was now living with her mother and stepfather, the Reverend Mr Tyler, in the village of Partney. His last three letters to her from Sydney had gone unanswered, but now, stricken by news at Portsmouth of the deaths of his sister Betsy and a close friend, Mary Franklin, he tried again:

> As you are one of those friends whom I consider it indispensibly [sic] necessary to see, I should be glad to have some little account of your movements ... that my motions may be regulated accordingly.[27]

Ann's reply was encouraging, and they met again during his leave in October. Marriage, though not formally proposed, seems to have been in the air when he returned to London. Then, a few days before Christmas, Ann received a devastating shock – a letter from Matthew saying he had been given command of a ship going out to New South Wales, but without promotion. It would not be possible for her to accompany him, but, he suggested – with what seems a pathetic insensitivity – on his next visit they could 'meet as lovers, but part as friends!' He vowed to bury himself in his work, and hoped she would study, learn music and French, 'write a great deal, work with thy kneedle [sic] and read every book that comes thy way save trifling novels'. Somehow the relationship survived; Matthew spent the New Year in Lincolnshire and managed to soothe Ann's hurt feelings.

HM **Sloop** *Investigator*

HM Sloop Investigator *at Sheerness, January 1801:*

> On the 19th of January 1801, a commission was signed at the Admiralty appointing me lieutenant of His Majesty's sloop Investigator, to which the name of the ship, heretofore known as the Xenophon, was changed by this commission; and captain John Henry Martin having received orders to consider himself to be superseded, I took the command at Sheerness on the 25th of the same month. [Flinders, 1814, vol. I., p. 3]

Lieutenant Flinders was pleased with his ship. 'In form', he wrote, she 'nearly resembled the description of vessel recommended by Captain Cook as best calculated for voyages of discovery'. She was a north-country collier of 334 tons, built in 1795, and brought into the Navy in 1798 to serve as an armed ship on convoy duties. Her gundeck was just over 100 feet long, her maximum width 28 feet. Stout and roomy, she could hold a year's stores on a long voyage, while her shallow draught and flat bottom enabled her to sail in close and sit on the seabed if need be, with less risk of damage to the hull. It may have been Banks who suggested the name change, to signify her principal task was to investigate the unknown coasts of New Holland, particularly those in the south, and to discover whether or not a great north-south strait divided the continent.

For the voyage her twenty heavy carronades were removed and replaced with twelve long guns. Before departure Flinders asked that the armament be further reduced to eight carronades, two long six-pounders, and two swivels – 'which guns … I consider to be sufficient to repel the attack of any Indians'. The reduced weight allowed an extra 10 tons of water to be carried. Other alterations included the provision of cabins for the scientists, and extension of the copper sheathing by another two strakes.

The ship also had various defects which went unnoticed in the haste with which she was selected and refitted. Her young commander lacked Cook's experience in dealing with dockyard contractors; he had no master to check on the smallest details; and before they sailed he was distracted by the presence of a new wife on board. Most probably the Sheerness yard skimped on the essential work of caulking, replacing defective masts and spars, and checking the timbers for signs of rot.

From the beginning of the voyage the sloop leaked excessively, and at Madeira, the first port of call, the carpenter patched and caulked the leaky seams in a vain attempt to limit the mischief. More seriously, unsuspected by captain and crew, rot was silently eating away at the frame timbers behind her outer planks and copper sheathing.

The ship's company

Flinders waited anxiously for the Admiralty's decision on the ship's establishment. It came on 5 February, with advice that the approved strength was eighty-three officers and men, an increase of eight on the *Xenophon*'s complement. It allowed two lieutenants, a surgeon and his assistant, a master, two master's mates, four midshipmen, and three warrant officers, and included a contingent of fifteen marines under a sergeant. Missing from the list was the post of ship's cat, already filled by Trim, born at sea in 1797, a veteran of the *Reliance*, and the captain's trusted friend.

On the 18th, Flinders received notice of his promotion to the rank of commander. He wrote at once to his patron:

I have the satisfaction to inform you, that my commission as commander of HM sloop *Investigator*, came down here this morning, and for which, Sir Joseph, I felt myself entirely indebted to your influence and kindness. Panegyric, or a long train of sentences of gratitude, would be unpleasant to a mind like that of Sir Joseph Banks. I will therefore only add, that it shall be my endeavour to shew by my conduct and exertions that your good opinion has not been misplaced.[28]

He appreciated the promotion all the more for knowing that both Cook and Bligh had sailed on their first voyage of discovery with the rank of lieutenant. Banks replied the following day:

I give you sincere joy at the attainment of your wish … I have long known that it was certain, but I am glad it is now placed beyond the reach of accident, or the change of administration.[29]

Flinders, in contrast to Baudin, had the authority to choose his officers. Robert Fowler, the first lieutenant, virtually selected himself; formerly the *Xenophon*'s lieutenant, he petitioned the Admiralty to remain with the ship, and Flinders was happy to accept him. Flinders's younger brother Samuel, at his request, was promoted and appointed second lieutenant. His young cousin, John Franklin, and William Taylor were appointed midshipmen. Four young 'volunteers' took turns to occupy the other two posts of midshipman.

The key position of master – requiring an expert seaman to sail the ship and assist the captain with navigation and surveying – remained vacant. At the last moment, while *Investigator* lay at Spithead awaiting her sailing orders, Flinders had an unexpected stroke of luck. His old shipmate John Thistle, just returned from Port Jackson as master's mate in the *Buffalo*, came aboard and offered his services. He was recruited on the spot.

For the crew Flinders wanted experienced sailors, preferably men used to long voyages, not the sullen, resentful landsmen delivered by the press-gang, the quota system and the gaols. He was allowed to discharge all those of the *Xenophon*'s crew considered too old or unfit, or who were unwilling to go on the voyage. In their place he was able to recruit volunteers from other vessels at the Nore, among them HMS *Zealand*; on this occasion

a strong instance was given of the spirit of enterprise prevalent amongst British seamen. About three hundred disposable men were called up, and placed on one side of the deck; and after the nature of the voyage, with the number of men wanted, had been explained to them, those who volunteered were desired to go over to the opposite side. The candidates were not less than two hundred and fifty, most of whom sought with eagerness to be received; and the eleven who were chosen, proved, with one single exception, to be worthy of the preference they obtained.[30]

The *Zealand* was a depot ship, holding a large number of landsmen of the type Flinders did not want. The volunteers, Ingleton suggests, were perhaps 'promised recreational visits to South Sea islands and the excitement of exploring an unknown coast' in place of risking death on a ship of the line. Instead they experienced the endless drudgery of survey work, disease, deprivation and despair. There were nineteen deaths, and the voyage ended in shipwreck. For the officers, opportunities for promotion were restricted while their commander was held captive on Isle de France.

Flinders's scientific staff

While Flinders busied himself selecting the crew and refitting and equipping the ship, Banks set about organising the expedition's scientific program with the vigour and enthusiam of a man half his age (though he was nearing sixty and unwell, suffering agonies from the gout that crippled him in later years). Everything to do with this side of the voyage was left to him: 'Any proposal you may make will be approved,' wrote the Secretary of the Admiralty, Sir Evan Nepean. 'The whole is left entirely to your decision.'

Banks was asked to design the structural changes needed to accommodate the collections of shrubs, plants and other specimens on board ship. Repeatedly Flinders sought his help in intervening with the naval bureaucracy to hurry along the preparations for departure. Most importantly, it was left to Banks to recruit the 'scientific gentlemen' for the voyage. Here his immense prestige as president of the Royal Society and his wide acquaintance among the leading scientists of the time were invaluable. Apart from the astronomer, John Crosley, appointed by the Admiralty and the Board of Longitude, he chose the naturalist, the two artists, the gardener and the miner.

As naturalist Banks chose Robert Brown (twenty-seven), a young Scottish botanist who had studied medicine at Edinburgh, and was then serving with the militia in Ireland. Brown came highly recommended by a botanist friend as a man 'fitted to pursue an object with a staunch and a cold mind', and Banks wrote offering him the post. He accepted at once, but his colonel objected, and it needed Banks's intercession with the Lord Lieutenant of Ireland to secure his release.

The position of natural history artist went to Ferdinand Bauer, a talented botanical draughtsman from Austria whose brother Franz was chief artist at Kew Gardens (where Banks was honorary director). After two failed approaches to established landscape artists, Banks settled on William Westall, nineteen, a student at the Royal Academy School, who gladly accepted the post.

Peter Good, another young man, was appointed gardener. He had recently returned from a voyage to India with Christopher Smith, Bligh's botanist in the *Providence* and a friend of Flinders, and had useful experience in the care and maintenance of plants at sea. No mineralogist was appointed, but Banks chose a miner, John Allen, to fill the post.

Apart from Good and Allen, who were not classed as 'gentlemen' and messed in the gunroom with the junior officers, the members of the scientific staff shared the captain's table with his two lieutenants. Before embarkation they were required to sign an agreement (drafted by Banks) binding them

> to render voluntary obedience to the Commander of the ship in all orders he shall from time to time issue for the direction of the conduct of his crew, or any part thereof.[31]

(It was Baudin's misfortune that similar restrictions were not imposed on the scientists voyaging with him.) Among other directions, Brown and his companions were also required to deliver up 'all journals, remarks, memorandums, drawings, sketches, collections of natural history', and so on, upon the ship's return 'to such persons as their Lordships shall direct to receive them'.

Banks also had a hand in selecting the surgeon, recommending Hugh Bell, lately of HMS *Seagull*, for the appointment. He thought Bell well suited to the post, but Bell and Flinders proved incompatible, and he was the only member of the staff to incur the captain's dislike. The fault, in Ingleton's view, did not always lie with Bell. 'Bluff in character, [Bell] did not hesitate to take Flinders to task for the state of ill-health of the *Investigator*'s crew.'

Banks's correspondence includes an engaging letter from Peter Good to his patron William Aiton, the superintendent at Kew:

> ... I accept with cheerfulness your liberal offer of going as Gardener & assistant to the Naturalist in a Voyage of discovery and render every assistance in my power to the expedition. I can at present only offer my sincere and humble thanks to you & Sir Joseph for the great preference and attention which you give me, and assure you that it shall be the business of my life to merit so particular a distinction.[32]

Good was the only member of the team who did not survive the voyage, dying from dysentery at Sydney the day after the *Investigator's* return from circumnavigating Australia.

Sheerness, March 1801. Investigator*'s refit is at last complete, stores and provisions for a six-month voyage have been taken aboard, and she is ready for sea. On the 27th she slips her moorings and moves down river to the Nore, to await further orders. On the same day, at Port Louis, Isle de France, some Danish captains provide Nicolas Baudin with the necessary funds to continue his voyage to New Holland.*

Notes

1. Quoted in O'Brian, 1987, pp. 255-256.
2. Baudin, 1974, pp. 7-9.
3. M. Ly-Tio-Fane, 1982.
4. Horner, 1987, p. 33.
5. Ageorges, 1994, p. 31.
6. Horner, op. cit., p. 42.
7. Baudin, op. cit., p. 7.
8. Brosse, 1983, p. 98.
9. Horner, op. cit., p. 100.
10. Péron, 1809, p. 14.
11. Baudin, op. cit., p. 5.
12. Ibid., p. 2.
13. Ibid., p. 3.
14. Horner, op. cit., p. 141.
15. In his biography of Flinders, *My Own Destroyer* (1962, chapter 1).
16. The development of accurate chronometers enabled the longitude of a ship at sea to be readily established by comparing local time and the corresponding Greenwich Mean Time.

17. Scott, 1914, p. 35.
18. Flinders, 1814, vol. I, p. xcvii.
19. *The Australian*, 16 April 1998.
20. Flinders, op. cit., pp. xcix-xc.
21. Ibid., p. cxix.
22. Bass sailed for Port Jackson in 1801 with a full cargo of general merchandise, expecting a good profit on his investment. The venture failed, and he tried his luck in the profitable, but dangerous, contraband trade with South America. *Venus* sailed for Chile in February 1803, and disappeared without trace. A rumour persisted that Bass and his men had been seized by the Spaniards and sent to the silver mines; his friends preferred to think the vessel had foundered in mid-Pacific, with the loss of all hands.
23. Ingleton, 1986, p. 50. Captain Williamson's navigation had its shortcomings, as James Mack points out. On this course he would have sailed overland from the south coast near the present SA-Victoria border to the Gulf of Carpentaria.
24. Ibid.
25. Ibid., p. 94.
26. Ibid, p. 51.
27. Mack, 1966, p. 46.
28. Ingleton, op. cit., p. 99.
29. Ibid.
30. Flinders, op. cit., p. 4.
31. Mack, op. cit., p. 67.
32. Ingleton, op. cit., p. 101.

CHAPTER   THREE

⚓

The Voyage Out

THE FRENCH EXPEDITION, 1800 TO 1801

Prologue: Banquet at the Hotel de la Rochefoucauld

Paris, 7 Fructidor, Year 9 of the Republic [24 August 1800]. The Societé de l'Afrique Intérieure hosts a farewell dinner for Capitaine de vaisseau Nicolas Baudin, with members of the Société des Observateurs de l'Homme and the Institut National as invited guests. The guest of honour is seated next to explorer and navigator Admiral Louis-Antoine de Bougainville, who so nearly beat Captain Cook to the discovery of New Holland's east coast, and who has entrusted his young son to Baudin's care on the voyage.

Musicians of the garde des consuls *contribute to the brilliance of the evening, and Citizen Brielle sings several pieces suited to the occasion. The toasts are delivered with great feeling, and electrify and melt every heart:*

Citizen Levaillant, a director of the Société: 'To the vessels Le Naturaliste *and* Le Géographe, *under the command of Captain Baudin. May they travel without danger to the ends of the earth!'*

Citizen Baudin: 'To Bonaparte, first consul of the French Republic, protector of the arts and sciences. To the hope that I may once more, returning from my expedition, be in the same room with the same people.'

Citizen Jussieu: 'To the progress of the physical and natural sciences, to which the voyages of Baudin and Levaillant will make a valuable contribution; and above all to citizens Maugé and Riedlé, who accompanied Captain Baudin on his expedition to the West Indies, and who would follow him to the end of the world ...'

Citizen Jauffret: 'To the progress of anthropology. May the Société des Observateurs de l'Homme be one day honoured by the useful researches of its corresponding members!'

Citizen Millin: 'To the islanders who will be able to value the good works of the intrepid mariners, who are going, at the peril of their life, to bring them civilization, useful arts, and the love of humanity.'

Citizen Leblond: 'To the Institut National and to all the learned societies of Europe. May the union of their enlightenment and of their labours efface to the last traces all political dissension.'

The immortal memory of La Pérouse is honoured in silence; and the last toast of the evening, appropriately, expresses the wish that the whole company will reassemble on the expedition's return, 'inspired by the purest zeal for the progress of the sciences and of enlightenment'.

Meanwhile, Citizen Jauffret junior has sketched a portrait of Citizen Baudin, in post-captain's uniform, which all present applaud as an excellent likeness. The finished portrait, now in the Musée de la Marine in Paris, bears the inscription:

> *De Cook, de Bougainville émule généreux*
> *Sur leurs traces Baudin va marcher à la gloire,*
> *Et, dans les fastes de l'histoire,*
> *Clio marque déjà sa place a coté d'eux*

> *[Of Cook, of Bougainville the noble rival,*
> *In their steps Baudin goes marching to glory*
> *And, in the records of history,*
> *Clio already marks his place at their side]*

The author is named as Péron, one of the expedition's zoologists.[1]

Last days in France

Le Havre, Sunday, 19 October 1800.

> *Our departure from the port of Le Havre will no doubt be a memorable event in the history of our nation for the impressive display which accompanied it. May the recollection that I shall keep of it render to the people of that town all the thanks we owe them for the interest they showed in us! Their wishes and blessing for the success of our undertaking can only bring it good fortune, and I shall undoubtedly have the satisfaction of handing back to them, on my return, all those of their acquaintance who have voluntarily followed me. [Baudin, 1974, p. 12]*

While Baudin was making his round of farewells in Paris, and received his final briefings from the Ministry and the professors at the museum, at Le Havre captains Hamelin and Le Bas supervised the commissioning of the ships. On the orders of 'the august chief under whose auspices this important voyage was planned', nothing was neglected 'that might ensure the health and safety of those who were engaged, assist their labours, and everywhere ensure their independence', wrote François Péron in his official history:

> Our numerous instruments, astronomical, surgical, meteorological, geographical, &c., had been constructed by the most celebrated artists of the capital. Everything necessary for chemists, painters, draughtsmen, were carefully selected; a numerous library, composed of the best works on marine subjects, astronomy, geography, natural history, botany, and voyages, was collected for each ship. All the instructions relative to scientific researches were written and prepared by a committee of the Institute ...[2]

The junior officers, midshipmen, and *savants* gathered at the port during September, in readiness for sailing at high tide on 3 October. Delays in the delivery of stores, the late arrival of the junior zoologist Péron, persistent onshore winds and rough seas led to successive postponements, allowing the young men the chance to get to know each other before they sailed. Fears by some of the scientists that the officers might turn out to be 'nothing but bears' were soon laid to rest in social get-togethers – 'the perfect concord between [us] was cemented daily by bowls of punch', the naturalist Leschenault noted in his journal.

'All went wonderfully well, and in the most friendly free-and-easiness', until the commandant's arrival put a dampener on the merriment – as Admiral Charles Baudin recalled a half-century later in his memoirs:

> Everyone tried at first to jump on his shoulder and eat out of his hand, but Captain Nicolas Baudin did not take to such courtesies, and was determined to establish discipline and hierarchy on a proper basis. He soon became the *bête noir*, and as we [midshipmen] were the youngest, we were not the last to take a dislike to him. Soon he began to detest cordially the scientists, the officers and the midshipmen, [who] reciprocated with all their heart. It must be said also that, if we were all tolerably unreasonable, he for his part totally lacked the kindliness of character and style which is necessary in leaders in order to win respect for their authority.[3]

Probably the commandant's mood had not been improved by two late orders forwarded by Minister Forfait. The first instructed him to make 'a special [natural history] collection for Mme Bonaparte, wife of the First Consul', in addition to those he was to make for the Paris museum:

> You will make up this collection of living animals of all kinds, insects, and especially of birds with beautiful plumage. As regards animals, I don't need to tell you how to choose between those intended for the menageries and those for a collection of pure pleasure. You will appreciate that it must comprise flowers, shrubs, seeds, shells, precious stones, timber for fine works of marquetry, insects, butterflies, etc.[4]

More worrying, no doubt, than this unwelcome complication of his task was Forfait's pointed reminder that

> at all times, Citizen, it has been expressly forbidden for officers of the Navy to engage in any kind of commercial speculation or to embark ... on their own account any merchandise or other articles of exchange.

It was vitally important to ensure that the expedition's protected status was not compromised by illicit trading at a foreign port. So that the instruction should be known to all on board, Baudin was ordered to have it read out aboard both ships, and to inform Forfait that he had done so. It did not escape his officers' notice that many crates with the initials 'N.B.' were stored on the *Géographe*'s deck.

More bad weather delayed the departure, and the ships' companies grew restless; officers and men alike took full advantage of shore leave. Determined to impose his authority, Baudin drafted a set of by-laws to be observed on board, the last of which caused particular resentment among the staff:

> In pursuance of the orders of the First Consul and the Minister of Marine, any person, whether officer, naturalist or other member of the staff, who during the expedition attempts to disturb the peace or to act insubordinately will be judged in accordance with regulations and disembarked at the first port of call.[5]

Sunday, 19 October. The weather clears, and with wind and tide favourable the blue peter is hoisted at daybreak to signal the ships' departure; Baudin requests the port admiral to announce their sailing throughout the town. The drumbeats recall the officers and men from leave ashore, and they are at their posts by 8 a.m. – certain proof, thinks the commandant, that all are 'setting off happy to be engaged upon so glorious an expedition'.

Large crowds throng the quays to cheer them on their way. The ladies gaily wave their handkerchiefs; the town band strikes up the chant du départ, *anthem of the army volunteers, and other patriotic airs; the fort's cannon thunder, the ships' topsails fill, and the officers and crews of both vessels salute Maritime Prefect Bertin with three rousing cries of* 'Vive la République!' *At nine o'clock, recalls François Péron,*

> *we passed the tower of François the First; a band played on the summit, and cheered our departure; an immense crowd from all parts covered the shore, and with one voice and gesture each of the spectators addressed us with their last adieus and wishes for our safety. All seemed to express: 'Ah, may you, more fortunate than Marion, Surville, St. Allouarn, La Peyrouse [sic] and Dentrecasteaux[6] return again to your country and the gratitude of your fellow citizens!'[7]*

Clearing the breakwaters at ten o'clock, the ships heave to and take their powder on board. A general muster of the two crews finds that one midshipman and a seaman are missing from the Géographe, *and two men from the* Naturaliste – *Captain Hamelin reports that he has replaced them with six stowaways found hiding in the hold. The ships dismiss their pilots and head out to sea, taking advantage of the offshore winds.*

To the Canaries

Off Le Havre, Sunday, 19 October 1800.

> *Having had in sight since morning the English frigate stationed off Le Havre, I directed my course in order to speak her, and at two o'clock, being within call, I brought to and went on board her. The captain received me politely, and after he had read the passport from the English Admiralty we were friends. I remained aboard his vessel about half an hour and we drank to the success of the voyage. At half past two I took leave of the captain, who expressed the desire to come aboard ... I showed him the interior of the vessel and he seemed satisfied. Upon his departure, I begged him to accept a medal struck to commemorate the voyage, which he did with pleasure, and then we parted. [Baudin, 1974, p. 12]*

With the English blockade there had been no opportunity for sea trials of the two vessels. In the open waters of the Channel it soon became clear the *Naturaliste* was an inferior sailer, and the difference in the sailing speeds of the two ships greatly worried Baudin:

> The advantage in speed that we had over this vessel pointed to the loss of much time and we judged that we should often have cause to regret that she was not a better sailer.[8]

He and Hamelin strove to develop a routine that would allow the two ships to remain in touch, but with little success. All too often contact was lost during the night, or the *Naturaliste* fell so far behind that the *Géographe* was forced to shorten sail and fall off to enable Hamelin to rejoin her.

Their course was set for Tenerife in the Canary Islands. With the Channel now behind them they ran into heavy seas coming from the west. 'The sea was extremely turbulent, and the ship, buffeted by big waves … laboured greatly with the pitching she experienced.' The fo'c'sle head was constantly awash, and the landsmen, 'not yet steady on the vessel or accustomed to her movements', took refuge in their cabins. 'As it was the first time that we had struck high seas and the ship was really tossed about,' Baudin noted somewhat smugly in his journal, 'all the naturalists, the midshipmen and a few other members of the company were sea-sick.'

Not all the first-voyagers are victims of mal de mer. *Bory de St Vincent, zoologist on the* Naturaliste, *escapes it, but observes its horrific effects on his friends Bernier and Dumont – particularly the latter, who for three days lies on a mattress beneath a table, and cannot even summon the energy to cry out when a passer-by's foot accidentally makes contact with his face. Obliged by the rough seas to keep the deadlight of his cabin closed, and unable to read or write in the semidarkness, Bory finds his mind wandering towards his native country, and the friends left behind. To shake off the despondency, he takes exercise on deck; it gives him a voracious appetite, and apart from a slight headache leaves him free of all complaint.*

François Péron is another unaffected by the malady. Although appointed to the voyage with special responsibility for comparative anatomy, he has a passion for scientific investigation extending far beyond the limits of his nominal discipline. Resolved to take full advantage of the lengthy sea passage, he begins a detailed study of the oceans – including a comparison of the temperature of the atmosphere in differ-

ent latitudes; the variations of the barometer and hygrometer in similar circumstances; the temperature of the sea on its surface, compared at different times of the day and night, with that of the atmosphere; the heat of the ocean at great depths below the surface, the relative saltiness at different depths; and so on.

Whatever the weather, at midnight and noon, at six in the morning and again at evening, he goes on deck to take his measurements. Day after day, as they sail through the storms and fogs of the Bay of Biscay into warmer waters, he gathers the data for 'a physical and meteorological chart of the sea', the elementary data missing from the narratives of previous voyages. Indifferent at first, the younger officers come to interest themselves in his research, and offer advice. Even the seamen, rough and semi-barbarous as most of them are, come to feel an affinity with this young intellectual who has lost an eye in battle against the Republic's enemies.

Though they sailed with safe-conducts from the British Admiralty, Baudin and Hamelin well knew there were captains on both sides who attacked first and asked questions afterwards – these were the privateers (*corsaires* to the French), masters of privately owned armed ships licensed by letters of marque from the national governments to attack their opponents' merchantmen. They kept a sharp lookout for hostile vessels as they neared Gibraltar.

The first alarm came on 22 October, in the blurred darkness of early morning. Emerging without warning from the gloom ahead, a large ghostly shape swept past the *Géographe* on the port side, so close the French ship was forced to bear away to the north-west to avoid collision. 'What we found so extraordinary about this meeting,' wrote Baudin,

> was that the other vessel [most probably an English frigate] did not speak to us and in fact did not pay us the least attention, continuing her course without disturbing herself in the slightest.[9]

A week later, at dawn, there was another encounter, this time with several large ships to leeward. There were eight of them in formation, heading north, and they were first taken for warships, but on drawing closer they were seen to be Swedish merchantmen. The two French ships hoisted their colours and passed them without speaking.

On the 31st, a day's sailing from the Canaries, a large cutter was seen ahead. Suspecting from her manoeuvres that she was a corsair on cruise,

Baudin gave the order to heave to, port to windward, to allow the *Naturaliste* to catch up before the corsair could launch an attack. The tactic worked, the cutter changed course and also hove to. Signalling to Hamelin to stay on course without shortening sail, Baudin continued his own cat-and-mouse game with the cutter. After some twenty minutes he bore away to rejoin Hamelin; the cutter followed, holding off at double the range of the ship's cannon. Towards evening she hoisted English colours and fired a broadside shot to windward, but kept her distance.

Baudin replied with a flag of truce, and again hove to. The corsair edged to just within cannon shot, then withdrew; probably, seeing so many men aboard, her captain decided not to risk an attack. She followed the French ships during the night, and next morning resumed the chase, but without gaining ground. As on the previous day, her captain manoeuvred to separate the Frenchmen, perhaps hoping to attack the *Naturaliste* on her own, while Baudin sought to block him. Finally, persuaded that only a show of force would deter the corsair, he ordered the cannon unlashed and their muzzles placed in the ports, where they could be clearly seen by the cutter's crew.

Within half an hour the corsair called off the chase and was not seen again. The *Géographe* and the *Naturaliste* resumed their course for the Canaries, now just beyond the western horizon.

The pleasures of Tenerife

Aboard the Géographe, *Saturday, 1 November 1800.*

> *The moment we sighted land, all the scientists and even most of the officers were so overjoyed that they behaved like madmen. Each one called out to his friend or his neighbour in such a way that pandemonium reigned on board. If a stranger had witnessed what went on and had not known when we had left Europe, it would have been impossible for him not to think that we had just completed a voyage lasting at least six months and that we had been without all essential provisions. Towards evening, when the general curiosity had been satisfied somewhat, everyone went off to get his portfolio and his pencils, and, to fore and aft of the ship, there was not a soul to be seen who was not busy sketching. [Baudin, 1974, p. 20]*

The island of Grand Canary was sighted soon after midday, some 10 leagues off. Both ships crowded on sail, and by six o'clock the high peak of Tenerife was clearly visible ahead:

Every eye was fixed on this stupendous mountain, its large base ... enveloped in cloud while its top, illuminated by the last rays of the setting sun, appeared majestically above them.[10]

At first light they stood in for the land, sailing close in to the steep, blackish shore, beyond which high mountains rose one above another, and dropped anchor at 10 a.m. in the bay of Santa Cruz. Baudin and Hamelin, with their officers, went ashore to pay their respects to the Spanish governor; to the commandant's annoyance the scientists insisted on accompanying them:

> More anxious than the rest, they had pestered me from the moment we dropped anchor to allow them to go ashore, and I had been obliged to give my permission in order to get rid of them. I must say here ... that those captains who have scientists aboard their ships must, upon departure, take a good supply of patience. I admit that although I have no lack of it, the scientists have frequently driven me to the end of my tether, and forced me to retire testily to my room. However, since they are not familiar with our practices, their conduct must be excusable.[11]

The governor received them cordially, assuring Baudin that 'everything under his control on the island was at our service'. After exchanging the customary compliments, Baudin took his leave and visited the French consul, M. Broussonet, to seek help in replenishing the expedition's stocks of wine: 'he replied ... that he would not lose a moment in getting it for me'.

Baudin learns from M. Broussonet that his visit is most opportune; a Spanish ship is about to leave for Gibraltar, under flag of truce, with a hundred or so English prisoners. They are held at the town of La Laguna, inland from Santa Cruz, and Baudin offers the use of his two longboats to ferry them out to the cartel.

The operation degenerates into farce. The garrison troops guarding the captives on the way from La Laguna allow them to drink so copiously that many collapse by the roadside. Those who are not so drunk as the others are obliged to wait at the pier until sunset, while a search is made for their companions. Of necessity, the longboats' crews are forced to stay ashore the greater part of the day until the missing men arrive.

Two midshipmen have been assigned to each boat, but they cannot prevent the crews visiting the portside tavern where, following the example of the English, they too become tipsy. When, late in the evening,

the last boatload is taken aboard the cartel, about a quarter of the pris-
oners are found to be missing. She sails without them, but carries seven
men from the Naturaliste *and two from the* Géographe, *judged unfit to*
continue on the voyage. Baudin replaces one of these, a carpenter from
the Naturaliste, *with an English volunteer from among the prisoners.*

A few days later two longboats crowded with men arrive from the
Grand Canary. They belong to the cartel's crew, and report that the
English prisoners, not wishing to rejoin their fellow-countrymen at Gi-
braltar, have seized the ship and intend sailing her to New England –
'which, however, she may have difficulty in reaching, having little food
[on board] and leaking considerably'. The nine men Baudin planned
to return to France are among those in the longboats.

After a week the consul had still not procured the promised wine; it
was in short supply, and could only be purchased at an inflated price.
Baudin had no alternative but to limit himself to 'taking only three
months' drink; three quarters in wine and the rest in English beer, which
came from a prize in port'. This and other problems kept the ships at
Santa Cruz until mid-November, and he bitterly regretted his call at
the port, which had been of no benefit to the expedition.

Not so the naturalists, who welcomed the chance to explore the
countryside away from the commandant's disapproving eye. The painter
Jacques Milbert met Anselm Riedlé and Stanislas Levillain returning from
a field trip:

> We saw coming towards us through these charming wilds the good M.
> Riedlé, the head gardener of our expedition … weighed down by his
> ample harvest, which he set down nearby, displaying its glorious rich-
> ness. [Next came M. Levillain]. His hat, completely covered with insects
> pinned to it, gave him a comic aspect; his specimen case was similarly
> well furnished.[12]

Riedlé's botanising nearly ended in tragedy. Baudin and Hamelin
were dining with the consul when the gardener's assistant arrived, a
mass of cuts and his face covered in blood. He and Riedlé had been
searching for a rare plant that grew only in inaccessible places; while
climbing a high rock to reach a specimen, the rock had moved and
they had both fallen some 30 feet. The assistant's fall had been less
severe than his chief's, and he had walked a league or more to seek
help, leaving Riedlé 'shattered and unable to move'.

Baudin and his companions rushed to the scene, pausing only to collect a stretcher from the hospital as they went. To their relief they met Riedlé on the path, supported by a native on either side. Over his objections, they made him lie on the stretcher while they examined his injuries.

We saw firstly that he was badly bruised on the back of his head and neck. He was even worse bruised down his back and had several more or less serious cuts, but fortunately no fractures. His ghastly pallor told us how much he was suffering … from his fall. Nevertheless, he did not want to be carried back on the stretcher, and walked with the aid of his companions.[13]

Back on the ship, Baudin placed the gardener in the care of the surgeons, who bled him and dressed his wounds. He was distressed by the accident, which might easily have cost him not only a friend but a key member of the expedition's scientific staff. 'His loss would have been especially unfortunate,' he wrote, 'as I think I can say in advance that he will serve as an example to the others in his devotion to his work. It remains to be seen whether they will want to emulate him.'

The bustling crowds on the streets and squares of Santa Cruz are full of interest to the savants. On one of their trips ashore, Bory de St Vincent and Delisse, botanist in the Naturaliste, *'wishing to form some judgement of the manners of the Spaniards, the greatest devotees in Europe', decide to attend a church service and procession for the souls of those in purgatory.*

Though lit by wax tapers and ornamented by gilding, the interior of the church strikes them as very gloomy. During the sermon, 'delivered extempore and in an emphatical tone', the men stand or sit upon benches, while the women huddle together in the lower part of the church, squatting upon the floor – a custom which indicates to the Frenchmen 'a want of politeness'. Nor do they observe in the congregation 'that air of contemplation which is suitable to such a place'.

Afterwards they walk through the streets to the harbour mole. Here crowds of prostitutes, muffled in tawdry mantles, ply their trade, using every effort to attract the notice of visiting sailors – all the while carrying a chaplet in their hands. During their stay at Tenerife, Bory warns, ships' captains cannot take too many precautions to prevent intercourse

*between their men and these females: ' venereal disease and the itch
are prevalent maladies'.*[14]

Though fully aware of the health risks to his men, Baudin had no illusions about their behaviour ashore; he noted in his journal

> the majority of the crew will be very sick as a result of their conduct with
> the native women, most of whom are infected with scabies and other
> diseases even more disagreeable.[15]

Aside from disease, delays, and concerns over Riedlé's health, however, he had various administrative and disciplinary matters on his mind.
Officers and scientists alike had complained about their living conditions and the quality of food provided, and demanded 'more particular treatment'. The artist Lebrun and Dr Lharidon had abused each other 'in a most ungentlemanly manner', and had almost come to blows. Sublieutenant Henri de Freycinet was reprimanded for inviting a guest of 'ill-repute' to dine at the captain's table. And last, as they were about to sail, a dispute blew up with the Spanish authorities over four deserters from the garrison, claimed to be hiding on the ships; a search of both vessels could find no trace of them. Finally, on 14 November, all was ready for departure. Scarcely able to contain his impatience, Baudin ordered the scientists below, their presence on deck so obstructed the crew as they prepared to weigh. That evening, far out to sea, the *Naturaliste* signalled the four deserters had been found on board.

King Neptune's rites

At sea, November 1800.

> *Doubtless if we had only to compare the absolute distance of these two
> courses, we should not hesitate to choose the coasting voyage along the
> shores of Africa; but the well-informed navigator takes into his calcula-
> tion other circumstances than the idle consideration of relative geographi-
> cal positions; he is not ignorant that the most considerable distances in
> appearance make little against him if he is but favoured by the wind and
> tides; that the shortest passage, on the contrary, may be retarded for weeks
> and months, if the same winds and currents oppose its progress, or what
> may retard him still more, obstinate calms, which keep his vessel almost
> immoveable on the surface of the waves. [Péron, 1809, p. 27]*

Fleurieu's 'Plan of Itinerary' for the voyage calculated that Baudin should reach Isle de France by the end of January 1801. His timetable was always optimistic; now it was hopelessly compromised. Delays at Le Havre had cost more than two weeks, another eleven days had been wasted at Tenerife, and Baudin looked for a fast passage to the Indian Ocean colony to retrieve some of the lost time. Two days after doubling Cape Verde on a night 'so dark that we could barely make each other out on the quarterdeck', he signalled a course south-south-east 'in order to cross the line between 12° and 13° of West longitude [of the meridian of Paris], if the winds would permit us'.

The decision astonished the more experienced officers aboard, wrote Péron in his account of the voyage. It was common knowledge that such a course along the West African coast exposed the 'ignorant and timid commanders' who chose it to storms, contrary currents and calms. Most captains preferred the *route du large*, crossing the equator in mid-Atlantic between 25° and 35°W, following a south-westerly course, then picking up the prevailing trades for the run to the Cape of Good Hope. Baudin was not alive to contradict him; had he lived, he could have pointed to the historical precedents of Cook and d'Entrecasteaux, both of whom had chosen the coasting route. Cook on his second voyage had crossed the line at 13°W, the very point proposed by Baudin – who, in more than thirty years at sea, had rounded the Cape many times.

Whatever his reasons, Baudin's hopes for a speedy passage were soon dashed. For two weeks the ships lay almost immobile in the belt of calms along the line, caught in the westerly drift of the equatorial current and rolling helplessly in the oily swell, the pitch bubbling in the seams under the searing sun. Then the weather turned squally and wet, heavy rainstorms requiring the crew's quarters to be kept closed for days on end; in the oppressive heat and humidity the bilge water gave off vile odours. But on 10 December the weather cleared, a slight breeze veered to the west, and they could again steer southwards. The ships were aired and disinfected, clothing was hung out to dry, and 'everybody was in spirits' as they neared the equator. Ironically they crossed the line at 23° west of Paris.

At dinner on 12 December a messenger, grotesquely attired, approaches the captain's table and asks permission to hold the traditional ceremony next day. Baudin is happy to consent, 'so that our gentlemen might obtain some idea of what it is like'.

The first voyagers among the savants and midshipmen look forward to the baptism with a mixture of interest and trepidation. For days the petty officers and old hands have regaled them with tales of previous crossings, relating in lurid detail what lies in store. None of it sounds salubrious.

At 7 a.m. the novices gather on the deck, the steady beat of the drum announcing the hour of baptism has arrived. Over the bow clamber King Neptune, his wife Amphitrite, and the royal attendants; clad in barnacle-encrusted skins and shouting wildly they make their way aft over the fo'c'sle head and down to the main deck, where Neptune's chariot – rudely knocked together from old casks and cases, and mounted on skids – awaits them.

Neptune, traditionally the oldest seaman in the ship, and his wife, one of the youngest, take their seats, their courtiers form a guard of honour, and the procession moves towards the quarterdeck, drums still beating. The captain formally welcomes them aboard, and invites the royal couple to be seated on their thrones – built overnight by the carpenter and his mates.

The old hands thrust the initiates forward for baptism. Savants *and* aspirants *are spared the worst indignities, but the sailors, seated on a stool before the king and his consort, are less fortunate. Lathered from head to waist with an evil-smelling mixture of tar, bird's droppings and other noxious matter, they are then shaved with a piece of rusty iron or tin, and pushed backwards into a canvas bath, to be held under by their jeering shipmates for as long as safety allows. Later they cluster around the water-casks, trying with little success to remove the foul mess from their hair and torsos.*

Celebrations aboard the *Naturaliste* were more subdued. Hamelin had no time for the 'ridiculous ceremony', and refused his permission on the grounds that it was apt to create disputes on board. Neptune's messenger returned disconsolate to the lower deck, and to cheer up the crew Bory and his companions took up a collection for them.

> We then took from our private stores a few bottles of Bordeaux wine, and some excellent liqueur of Maria Brizard, which we emptied in honour of the friends we had left in another hemisphere; and with whom we had now nothing in common, not even the seasons.[16]

Picking up the trades just below the line, the ships made steady southing. Spirits rose after the weary weeks in the doldrums, and 'everyone considered we were half way to Isle de France'. Baudin could not share in the general euphoria: 'I am far from believing that we shall take less time to get there than we have taken to get where we are now.'

To keep the men active and occupied during their leisure hours he ordered dances and other entertainments to be held on deck in the evenings when the weather allowed. Most of the crew took part and their health improved. On Christmas Eve, however, a number of the seamen could not resist the chance to resort to more traditional celebrations. Baudin was woken at dawn with the news that most of the port watch had got drunk during the night; helping themselves to a cask of wine belonging to the artist Lebrun, some

> were in such a confused state that two had come on to the quarterdeck, and in the presence of the officers on duty ... had set to work to smash in the chest containing the midshipmen's provisions.[17]

Brought before the commandant, all members of the watch were condemned to drink nothing but water until they had foregone the quantity of wine needed to replenish Lebrun's cask. Those guilty of drunkenness were sentenced to an extra eight days of water – a very light sentence by Royal Navy standards.

Scientists at sea

Aboard the Géographe, *Wednesday, 19 November 1800.*

> *From the note that Citizen Péron gave me on his observations concerning the temperature of sea-water, it appears that he would never have suspected that it must be warmer than that of the atmosphere. Yet the many experiments that have been carried out ... should have convinced him of the validity of this fact, although the reason is perhaps unknown. Nevertheless, as he is not convinced of it yet, he is waiting until he has collected a greater number of observations before accepting a truth which to him is still most uncertain, although not to anyone else. [Baudin, 1974, p. 42]*

Calms, contrary winds and currents, and occasional storms notwithstanding, the naturalists kept busy during the long passage south. Péron, indefatigable as always, was on deck at six-hourly intervals taking his readings of the sea and the atmosphere. On fine days he relished the gentle sounds aloft as sails, spars and running gear worked to every movement of the ship, while the wash and fall of the sea murmured under the bows. In rough weather the task was less enjoyable. On one occasion Baudin was watching the young naturalist at work in the port head when, without warning,

> a wave washed right over him and carried away his observation book as well as his thermometer. This accident, caused by the very heavy seas, did him no apparent harm, but he thought he was drowned beyond hope. So when the water which had entered the head ran out again, he was quite amazed not only to find himself still alive, but still in the same place, for he had thought himself washed right out to sea.[18]

Anselm Riedlé, recovering from his fall at Tenerife, resumed the care of his on-board garden, tending plots in the holds and on deck. Daily he sprinkled his beloved plants with a measured ration of fresh water – to control their growth in the humid heat of the tropics and to conserve the water supplies. Already the fruit trees – apricots, plums, pears and apples – were covered in flowers; even olives grew at a prodigious rate. In other plots he grew salad vegetables, supplying the officers' table with fresh lettuce, radish and watercress as a precaution against the ever-present threat of scurvy.

A group of seamen was dancing on deck one evening when a porpoise was caught forward. The quadrille ceased abruptly, as all rushed to the bows to see the animal hauled aboard. The *savants* crowded round, each man wanting to make immediate use of the carcase; some called for lanterns, some for papers, and others for instruments. Baudin was forced to intervene, taking possession of the catch to keep the peace. Instructing the sailors to suspend the body from the mainstay, he requested the gentlemen to 'moderate their ardour' – they would have time enough the following day.

A squall of wind and rain sent the scientists scurrying below, leaving the seamen free to trim the sails without interruption. At sunrise there was a further outburst of argument as naturalists, anatomists and artists all claimed first use of the carcase – the sketchers wanted it laid on its belly, the anatomists demanded that it be placed on its back for dissection. The latter decisively ended the dispute; while the artists were noisily complaining to Baudin the surgeons seized the opportunity to open the creature up, and when their companions returned, the lines needed for their sketches had been lost. To calm them the commandant promised that the next porpoise caught would be theirs.

Péron is in his element. At night in the tropics he slings his hammock on deck and wills himself not to fall asleep; lulled by the gently heaving waters around him into a feeling of oneness with the universe, his thoughts crystallise into new patterns. The star-studded sky above seems but a pale imitation of the sea, where shining stars in their thousands dart from below the water; flaming red balls 20 feet in diameter, fiery parallelograms, inverted cones of light twirling on their points, shining garlands and luminous serpents eclipse all the artificial fireworks of Paris. Sometimes the ocean itself seems as though ornamented by an immense steep of moving light, its undulating action reaching to the edge of the horizon.

Neither his fascination with the phenomena of phosphorescence nor his meteorological and physical observations are enough for Péron's hyperactive mind. Aided by his colleague René Maugé and 'assistant gunners' Charles-Alexandre Lesueur and Nicolas-Martin Petit, both fast becoming accomplished artists, he commences a study of zoophytes – tiny marine creatures long neglected by naturalists, which from the strangeness of their form, the singularity of their organisation, the beauty of their colours, and the variety of their character, so well merit the attention of the natural philosopher.[19]

Less driven than Péron, Bory de St Vincent on the Naturaliste *strives to occupy his mind – observing the flying fish breaking the surface and skimming low over the waves; the dolphins sporting under the ship's bows, their fins tinted a beautiful azure; and the ever-present birds, mostly albatross and pelicans, occasionally a majestic frigatebird suspended in space high overhead. It does little to relieve the tedium of the passage. He writes:*

> *It is impossible to form any idea of the many hours that remain unoccupied, or to conceive how much pleasure and amusement is afforded …
> by the appearance of a ship, a fish, or a bird, in such a situation.*[20]

For the mineralogists, geographers, botanists, and the official artists, with still less to fill their time, the stress was worse. Mostly there was only the emptiness and immensity of the sea, and the water-washed, indifferent sky. Quarrels erupted over the slightest incident. Lebrun, disliked by all for his sour, unsociable character, argued violently with his fellow artist Milbert and threw a glass of water in his face. The officers intervened to prevent a brawl.

Milbert, by far the worst-affected, fell into a profound melancholia:

Surrounded on all sides by a fluid that seems like a vast lake of oil reflecting the colour of an overcast grey sky reaching to the horizon, we are drawn from our gloomy reveries only by the ocean surge that sluggishly raises the ship [but fails] to give it any direction.[21]

His illness was aggravated by gossip among the staff that the commandant had censured the *savants* in a report, sent from Tenerife, to the Minister. Fearing a total breakdown, Dr Lharidon urged Baudin to talk to the artist and reassure him. Milbert seemed to respond, leading Baudin to believe that the interview had helped to restore his spirits. In fact the effect was quite the opposite, as Milbert explained to Lieutenant Gicquel:

The captain appeared very sorry for the distress he has caused me. I began to believe that he might be a good man after all, when he checked that impression by giving me a glimpse of the strangeness of his character. I mentioned my wife, whom I had left for what I thought was her own good, and said how much I missed her! 'And I', he said, 'do you not think I didn't leave a mistress behind?'[22]

The affair added to a growing campaign of rumour, innuendo and slander against the commandant. Boredom, contrary weather, and crowded quarters nourished the climate of disillusion and complaint. Some junior officers and many of the *savants* began plotting to leave the expedition at Isle de France.

Mission at risk

From the Cape to Isle de France, February to March 1801.

From [our commander's] preposterous obstinacy, which was necessarily followed by a consequence plain to foresee and easy to evade, he was forced from the beginning of the voyage to disturb and discompose all the regularity of the operations which had been prescribed for him to follow. Thus, in the execution of the most important undertakings, the slightest faults produce consequences at once grievous and irreparable! [Péron, 1809, p. 29]

The Cape of Good Hope was sighted on 3 February, eighty-two days out from Tenerife. The whole Atlantic run from the Channel to the Cape (excluding the stopover at Tenerife) took ninety-four sailing days. James

Cook, also with two ships, had completed the passage in 109 days in 1772 (including a brief stop at Madeira). Nine months after Baudin, Flinders with one ship would take eighty-six days. But for Péron and his companions the time-consuming passage south confirmed their view of Baudin's incompetence – a sailor 'barely fit to command a canal barge', as Ernest Scott phrased it a century later.

With the Cape behind them, Baudin noted mockingly that 'the date of arrival at Isle de France was already fixed, and nobody doubted any longer that [we] must have favourable winds all the time'. Hopes were again shattered and calculations upset by unseasonal headwinds, frequent squalls, and rough seas. Night-lights were kept burning at the *Géographe*'s masthead, and rockets fired, to signal their position to the *Naturaliste*, only to find next morning that she still lagged far behind. Frustration at their slow progress further strained ragged nerves and tempers.

For the seamen, there was little relief from the monotonous daily routine. On deck duty, rain-laden mists saturated their clothing, and the constant humidity so penetrated every part of the ship that Baudin redoubled precautions to prevent an outbreak of disease. In the unpleasant conditions

> most men [became] so peevish and depressed that they lost their tempers with each other over the smallest thing. I had a talk with them and urged each one to be patient since there was no other remedy for the situation. Moreover, everyone was well and meals were tolerably good, although not as copious as everyone would have liked.[23]

Quarrels flared between the officers and *savants*. Seizing the opportunity to re-establish discipline and a respect for his authority, Baudin assembled his staff in the great cabin and read aloud his instructions from Minister Forfait – stressing those sections requiring him to maintain among his officers 'subordination and exactitude in their duties', and to show 'the greatest firmness on all points'. He warned that he would discharge at the next port anyone refusing to mend his ways.

The good-natured Riedlé had kept aloof from his colleagues' backbiting and intrigues, and was moved by his captain's comments:

> The commandant observed to us that only good order and understanding would enable the voyage to succeed, that all Europe had its eyes on us, and if the voyage failed it would be due only to the bad understanding

and the discord that were arising between officers and naturalists, and on our return to France we would be blamed by the whole nation, but if we succeeded we would acquire glory and esteem.[24]

To the consternation of his listeners, Baudin closed his address with words not heard on a Navy ship since the Revolution: 'In the name of the Father, the Son, and the Holy Ghost, Amen!' The invocation was followed by several moments of stunned silence, before one of the group brought up the purpose for which they had been brought together. The meeting ended with a general exchange of apologies and offers of peace.

The uneasy truce did little to ease the tensions on board. Rather, Baudin's threat to discharge 'ill-natured persons' on arrival at Isle de France forced many of the staff to face up to difficult personal decisions. Most were feeling the physical and psychological effects of a long and exhausting sea passage, in oppressive quarters, and under an exacting but inconsistent captain. Several, like Gicquel and Milbert, were already casting around for an 'honourable' way to desert at Isle de France. The commandant's words provided an incentive for others to join them.

The weather gives the doubters further cause to worry. South of Madagascar the winds veer suddenly south-west and blow a squally gale. Black clouds cover the horizon, and once the wind's first force is spent the rain is so continuous and torrential 'that I cannot recall ever having known so great a downpour'. The rain continues throughout the night, with the wind blowing in squalls and unusually strong gusts.

At daybreak the winds again strengthen, visibility is little better than on a moonlit night. On the quarterdeck of the Naturaliste, *Commander Milius peers out into the gloom, eyes smarting from the vicious spray, watching for some sign of the storm abating. Across the waste of sea he dimly makes out the* Géographe, *green water cascading across her decks, heeled over so far in the screaming wind that her keel is exposed, at times vanishing in the chaos of tumbling waves.*

The hurtling seas cover his own ship from stem to stern as she lifts and plunges, lifts and plunges. Bory de St Vincent, immune to seasickness even in this weather, joins him on the pitching, juddering deck, hanging on desperately as yet another mass of water crashes on board. The sea's appearance fascinates and terrifies:

*the waves seemed to accumulate around us, and to form either moun-
tains, which threatened to dash us to pieces, or valleys which were ready
to swallow us up. They fell upon the sides of the vessel with such force,
that each shock was as great as if we had struck upon a rock ...*[25]

The damaged rigging, the masts stripped of their sails, smashed hen-
coops and other paraphernalia present a spectacle of utter desolation.
Scarce anyone is on deck, and the vessel seems left to itself. When a ship
lies in this condition, Milius explains, few of the sailors need to keep the
deck. 'Most of them, therefore, go below to their hammocks, and wait
patiently till the sea either becomes calm, or swallows them up.'

The storm blows itself out and the sea calms. Under a serene and
cloudless sky the ships head for Isle de France, now less than a week's
sail away. For many on board the island takes on the appearance of
the promised land – a haven offering sanctuary from a further two years
or more on a floating coffin cruising unknown and savage coasts, with
death as likely as not the outcome. Most, like Bory, are so anxious to
land that they spend a sleepless night after Port Louis is sighted. 'Before
daybreak,' he later recalls, 'I went upon deck, and walked in the fresh
air. Never in my mind did the time appear to proceed so heavily ...'. In
a state of 'cruel uncertainty', he struggles between 'the desire of accompanying my friends, and the fear of falling a sacrifice ... to an expedition so ill directed'.

Isle de France – deceptive Eden

Port North-West, Sunday, 15 March 1801.

> *After so long a voyage, the sight of any portion of land is doubtless de-
> lightful to the traveller; but how much more does it appear interesting,
> when he knows that he shall find on it the men, manners and language
> of his native country. Besides, the picturesque appearance of the Isle of
> France, the singular shapes of its mountains, the verdure which clothes
> the whole surface of the island, the numerous habitations which he dis-
> covers at a distance, all contribute something to the charm of having
> reached the first goal or resting place of his voyage. [Péron, 1809, p. 42]*

'Land ho!' At the lookout's call everyone rushed to the rails, watching
as a long white cloud to the east slowly diffused into a chain of wooded
mountains sloping towards the sea. The ships made for the harbour of

Port North-West (the republican name for Port Louis), but the winds veered southerly and failed before they reached the anchorage. They remained in the roadstead throughout the night, with lights burning, in full view of the forts. When the pilot arrived next morning he explained, apologetically, that they had been taken for enemy vessels; in fact, wrote Baudin,

> our appearance had caused such alarm, that all the troops and national guards had been ordered to their posts, where they had spent the night in order to oppose the raid, which, it was supposed, we intended to make.[26]

He drily observed that ships with a hostile intent would not ordinarily lie at anchor all night, with lights at the masthead, under an enemy's guns.

Next aboard were four members of the island's Intermediary Commission – a representative body set up by the colonists – sent to enquire about the latest news from France and the reasons for their visit. The commissioners particularly wished to learn 'the Government's intentions towards the colony [and] if there were any secret agents on board entrusted with putting into execution the decree concerning the liberty of the slaves'. Blacks greatly outnumbered whites on the island, and the colonists dreaded a repeat of the recent massacres on Haiti, which had followed the freeing of the slaves.

The commissioners demanded the expedition's papers and correspondence, official and private, 'as security for the safety of the colony'. Very reluctantly Baudin agreed, not wishing to antagonise his visitors, and accompanied them ashore to meet the colonial authorities. Mobbed in the street by crowds 'drawn by curiosity or the desire to learn the news' from the homeland, he assured them that 'everything was in the best possible order, and they would soon learn that a glorious peace would ensure the stability and prosperity of the Republic'.

Received warmly by the governor, General Magallon, and by members of the commission, Baudin then called on the intendant (administrator) of the colony, M. Chauvalon, where 'he did not find ... the polite and gracious welcome that I had met with everywhere else'. Later he learned that Chauvalon 'was offended by my going to the Commission after the Governor, instead of going to pay my respects to him'.[27] The unintended slight had profound consequences, as Baudin soon found.

His immediate concern was the revictualling of the ships in readiness for the next stage of the voyage, for which the intendant's cooperation was necessary. He carried a letter from Forfait, Minister of Marine and

Colonies, directing the colonial administration to supply the expedition with all needed provisions, and also to provide a small vessel which could support the two corvettes in their surveys of New Holland and New Guinea. However, when he presented Chauvalon with a list of his requirements, he met with an outright refusal: the colony had received no advance notice of the expedition, the island was under British blockade, its resources were stretched to the limit, and fresh food and other supplies in the quantity required were not available. Baudin argued in vain, pointing to

> the inconvenience of his refusal, and the discontent it would cause amongst the crew. It was no good, and I gained nothing by declaring that there was no money in the cash-box to pay the butcher and the baker [and] that it would cost him less to feed us on fresh meat than to replace the biscuit and salt meat which he wanted to make us eat. His reply was that we could eat what we liked, but that he would replace nothing for us.[28]

Complaints to the governor had little effect, Magallon assuring him soothingly that 'he would do his utmost to prevent the unpleasantness which I was … beginning to foresee and which I did not conceal from him'. Returning to the ship Baudin came face to face with a large notice on a street corner: *Voyage de Découvertes Manqué* (*The Failed Expedition*). Suspecting that Chauvalon's influence lay behind the protest, he feared that Fleurieu's directive 'to stay no longer than a fortnight in this colony' would prove impossible to achieve.

Escaping the stifling atmosphere of their tiny cabins, the savants hurry ashore to look for lodgings. Within a few days most are comfortably boarded in and about Port North-West, a town of some 5000 white inhabitants and double that number of blacks. The pungent scents of tropical plants, mingled with dust, dung, fish, fruit, and human sweat, engulf them; the low wooden houses and the sight of blacks running naked along the unpaved streets impart an air of misery to the whole place. The forest-clad mountains enclosing the little town, the luxuriant vegetation and rich volcanic soil, the climate and geology of the island draw them towards the countryside.

    Away from the ship, Milbert's depression quickly lifts, and he fills his sketchbook with enchanting landscapes, gracious mansions, shadowed rivers winding beneath fantastic trees, their limbs draped with twining creepers, picturesque slave huts, and his companions hunting. Invited to share the table of a well-to-do planter, he repays his host with a sketch of his home. At Pamplemousses he is entranced by Céré's 'garden of faerie', with its limpid waters, shaded pathways, and plants and trees transplanted from the most remote regions of the world.

Riedlé is fortunate; soon after landing he falls in with the Abbé Hoffman, parish priest and amateur naturalist, who professes his liking for Germans (Riedlé is from Alsace), and insists that he stay with him. Together they go botanising in the foothills above the town; drenched to the skin by a sudden downpour, the new-found friends 'make up for all that over a good dinner'.

Bory and a fellow zoologist, Désiré Dumont, make an excursion to a plantation in the district of Flacq, a day's journey from the port. The planter sends horses to carry his visitors, and two blacks to carry their baggage, but their passion for natural history and shooting is such that they alight and proceed on foot. The track passes through thick forests, 'only recently penetrated for the first time by the inhabitants', and exhibiting 'the majestic rudeness of nature'. After a convivial dinner with their host and his family, they retire early to bed, tired out by the journey. Wrenched awake by the groans of his companion, Bory finds him 'seized with a violent pain in the stomach, accompanied with ardent thirst and vomiting'. He helps him as best he can in the dark, not choosing to disturb the household. Next morning they suppose that Dumont's illness was most likely due to the ragout served at dinner, which had been cooked in a copper vessel 'doubtless impregnated with verdigris'.

After several days exploring in the 'wild woods' they return to the port, both feeling 'much worse than when we landed'. Dr Bellefin, the surgeon of the Naturaliste, *admits them to the hospital, where several of their friends are already in residence.*

For the crew there was little respite. Sails and rigging badly needed repair, the seams recaulking, and the woodwork damaged in the recent gales had to be repaired or replaced; the holds and living quarters required a thorough fumigation. In greater need of 'rest and recreation' than the staff, a dozen seamen a day were rostered on shore leave. A few days after landing, the petty officers came to Baudin with complaints about the men's rations in port – after five months at sea it was intolerable that they should still be served salt meat and worm-eaten biscuit. The same day eight sailors came down with severe colic and fever and were taken to hospital. Worried that 'the situation could not fail to serve as a pretext for desertion', Baudin appealed again to M. Chauvalon, but to no avail: 'Nothing could shake him in his resolution to give me nothing.'

In desperation he visited the government butcher, and quickly came to an agreement to purchase fresh meat for cash from his own pocket; at the bazaar he bought fish and vegetables.

> Nevertheless, I had the vexation of seeing many of my men go into hospital, where it is costing the government three times more to treat them as patients than it would to have kept them in the good state of health in which they came to the colony.[29]

Baudin had no doubt that his difficulties were a deliberate tactic by the authorities 'to force my crew to give up the expedition' – a suspicion confirmed by a member of the commission, an old acquaintance, who came to tell him that 'the constant fear of an attack by the English made it desirable that I should remain longer so that my men could be conscripted'.

Foreign aid and private enterprise

At Port North-West, Friday, 27 March 1801.

> *On this day four men were missing and the third petty officer was one of them. At eleven o'clock I went ashore and visited the hospital to see the officers and naturalists who, according to the certificates given me, were to be found there. The trip was quite useless, for I found nobody. The Sisters of Charity who run the hospital assured me that these gentlemen sometimes returned in the evening and even rather late. They usually spent the day in town, where, no doubt, their amusements were more varied than they would have been had they remained in such a residence. [Baudin, 1974, p. 127]*

Later that day Baudin dined with the Danish consul, the Chevalier Pelgrom, an old friend and a business associate of his own brother Augustin, captain of a Danish merchantman. Pelgrom had invited a dozen Danish and three Swedish merchant skippers to join them.

> We sat down at two o'clock and although the dinner was most magnificent, I only wanted the moment for dessert to come. It came at last. The first toast was to the complete success of the voyage that I was about to

undertake and in which the European nations had taken so keen an interest. This toast was proposed by the consul. The second was to the
health of all those who contributed in giving us all the help we might
need and without whom the expedition could not be carried out.[30]

The colonial authorities were unwilling to support the voyage, the
consul told his guests, their excuse being 'that the colony was menaced by an imminent English attack and the government needed all its
resources. It had, moreover, no funds at its disposal'. He then proposed
to the captains present that those among them with remittances to send
to Europe should offer Baudin 'the sum of ten thousand [Spanish]
piastres at the ordinary rate of exchange in the colony', suggesting that
'the remittances would be more agreeable and more favourable to their
principals in this way than in any other'.

This seemed so reasonable that all the captains 'wanted to make [the
loan] up in equal portions', but after much discussion agreed to accord
it to the one with the greatest need to get money to Europe. The loan
was secured by bills of exchange drawn on the French government, at
$33\frac{1}{3}$ per cent interest – the colony's usual rate, even in peacetime. Surprised by the proposal's 'complete and rapid success', Baudin made a
point of thanking each captain individually for their support – though
it pained him 'that the French, and especially the merchants with capital, would be shamed in the eyes of all Europe for having refused me
so moderate a sum'.

Within a few days stocks of meat, fish, fresh vegetables, beer, wine,
and arrack (the local spirit) were being loaded aboard both ships. To
these were added thousands of bushels of grain, which the intendant
could no longer withhold now that the expedition could pay for them
– though he still refused permission for it to be ground in the colony's
mill, which he claimed was not working. 'Not for us,' Baudin observed
sarcastically, after he had found the mill busily at work for the administration. Fortunately this was not a major problem, as the ships had
their own on-board facilities for grinding flour.

Apparently there were other matters requiring Baudin's personal
attention. The name of Louis Petitain appeared on the *Géographe*'s
complement as 'the commander's secretary', but he does not seem to
have once put pen to paper on the voyage out. On arrival at Port North-
West the 'secretary' declared himself sick and disembarked, according
to Bory de St Vincent; he and Lieutenant Gicquel linked Petitain to

'eighty large bales and boxes' marked B (or NB) that were landed at the port, along with other items. Gicquel noted that the contents were 'being retailed at the store of Citizen Maurice [who] has an assortment of all kinds and particularly of fashion goods'. Bory went further, writing

> I saw all these articles exposed for sale in an auction room, called the *Lighter Magazine* ... It contained European merchandise to the value of 300,000 francs, which, it was said, had been landed from our vessels, and on which a profit of from two to three hundred per cent was gained. It was remarkable that about this time all the coquettes of the country began to imitate the dress which was worn by the fashionable females of Paris at the period of our departure.[31]

Nor were fashion items and other retail articles the only goods landed. Bory claimed an entire printing-press was taken ashore:

> Before we arrived there was only one fount of types on the island, and they were completely worn out. The presses we brought out were soon the only ones employed, and in a short time produced considerable profits to their proprietors.[32]

He did not name 'the author of this speculation', but allowed his readers to draw their own conclusion:

> Several people did not doubt that [Petitain] remained in Isle de France to supervise the sale of the merchandise ... and of the products of the printery. Be that as it may, he could only have been the supervisor, or at least an associate; for if he had been the owner, how could the commandant of an expedition of discovery have permitted him to embark on his sole account, on the ships of the State, eighty large cases which encumbered the ship.[33]

Lieutenant Gicquel had decided to leave the expedition well before their arrival at Port North-West. These claims of double-dealing would provide a useful defence against a possible charge of desertion, and he wrote to friends in the Ministry at Paris outlining his allegations against the commandant, as well as more general complaints about the expedition. He attacked Baudin – '[he] has no qualities, either moral or social' – and Hamelin – 'no more a mariner than [Baudin]; he is a coward!' With such leaders the expedition must necessarily fail, and Baudin would then blame his officers.

Lacking Gicquel's connections, Bory de St Vincent vacillated to the end. It was only when he learned that the ships would sail the following day that, still irresolute, he 'proceeded to the port to bid farewell to my companions, who were collected there, and departed for the country without informing them that I would see them no more'.

As it happened, neither Gicquel nor Bory need have worried about their careers. No sooner had Baudin sailed than Governor Magallon appointed both of them to his staff on 'special duties'. By the end of the year Gicquel was on his way back to France, bearing dispatches from Magallon for the Ministry of Marine and the first consul. Bory left on a mission to the neighbouring island of Réunion, and visited St Helena on his return voyage, before returning to France in 1803. He too carried dispatches, which he presented to the first consul in person; he then busied himself writing an account of his adventures. Published in 1804, it repeated for public consumption the allegations against Baudin that Bory and Gicquel had already spread within the Ministry.

Isle de France – exodus

Port North-West, Tuesday, 7 April 1801.

> *At midday I went ashore to attend a large dinner being given for me by my old friends in the colony. We drank deep to the success of the voyage and everyone congratulated me on having obtained the replacements that I needed through Pelgrom, the consul. But they warned me to be on my guard with respect to the crews, and assured me that several people had already taken steps to persuade them to desert and that I might, perhaps, lose them when I least expected it. [Baudin, 1974, p. 131]*

Unaware that Bory and Gicquel were methodically recording his entrepreneurial activities (there seems little question of his involvement in the venture), Baudin put his worries aside and pressed on with preparations for sailing: 'I looked upon my business as already completed,' he wrote, 'and the continuation of my voyage as assured.'

Teams of caulkers provided by the port authorities recaulked the seams of the ships' timbers, black work-teams scraped the interiors clean, the crews repaired the masts and rerigged both ships, water and firewood were replenished, and the holds readied for stowage. On shore, meanwhile, great preparations were being made to meet the English, who were reported to be fitting out a sizeable fleet in India, the destination of which was unknown.

Confident that he could sail within two weeks, Baudin was in a buoyant mood at the farewell dinner with his friends on the 7th, dismissing their warnings of his men defecting as 'ill-founded', since 'both crews had in general strayed little from their work'. Two days later, however, he heard from another source that the Intermediary Commission planned 'to retain the ships if the English should happen to come, and to distribute the crew members to various posts'. Though Governor Magallon doubted that 'the commission would go to that extreme', the commandant hastened his preparations for departure. On the morning of 14 April, to his great satisfaction, 'both corvettes were swung, and in the afternoon anchored in the roads outside the flags'.

> I was safe then and had nothing more to fear from the intentions of the Intermediary Assembly, should the English chance to appear. It remained only for me to recall the men in hospital and the deserters, whom I thought it would be easy to find. I drew up a list of descriptions and sent it to the enrolment office and the police.[34]

The shore authorities still had some cards to play. Despite his precautions, more desertions were reported, six from the *Géographe* and three from the *Naturaliste*. On the 15th, Baudin visited the hospital to inspect the sick, and as usual found none of the naturalists present, 'for they went there only when they were not on some outing, or dining elsewhere'. More worryingly, he was told that eight of his best seamen had absconded over the wall during the night. From another informant he learnt that 'far from climbing over the walls, they had gone out through the main gate, on a permit from the commissioner in charge of that department'. Clearly the administration still intended to put off his departure, which doubtless 'was presumed impossible through lack of men'.

Well-versed in the ways of the colonial bureaucracy, Baudin now played his trump card. In a letter to Magallon, the intendant, and the commission, he made it clear that the natives still working on board the ships – numbering thirty or so – would not be allowed ashore until his deserters – by now at least forty – had either been returned or replaced with a similar number of men. Assembling the officers of both ships in council, he obtained their support to send off the letter, 'convinced it could only have a good effect'.

The response came that evening. A port officer boarded the *Géographe*, bringing a dozen men from the prison hulk, six for each ship;

he handed Baudin a letter from the intendant, asserting that as he 'no longer had any reason for retaining the native workers, [he] felt sure that [the commandant] would make no difficulty about returning them'. Baudin stood his ground – he held on to the twelve men, but refused to release the port workers, convinced that 'if our deserters had not been given refuge, it would have been impossible not to find them very quickly'.

Two days later, in a heated interview with the commissioners, he argued it was their responsibility to return his missing men: 'I had it on good authority that when they wanted to deport someone, they had found ways and means of catching him, even in the depths of the forest'; surely it was far easier 'to find forty or more missing men than a single individual'. The same afternoon the local police seized twenty-two of his deserters, found among the crowd watching the public execution of two runaway slaves. However,

> the guard which was to take them to prison was made up of publicans and canteen-keepers who allowed fourteen to escape en route, and brought a mere eight to the port. It appears that those whom they allowed to escape were their debtors, for amongst those whom they escorted back, not one owed them anything.[35]

The governor invited Baudin to call, and again raised the question of the thirty natives still detained on the corvettes – he had heard from the intendant that 'I meant to carry them off'. Well aware that 'this would greatly have annoyed some individuals in his suite, or perhaps him even, for they know very well how to make a considerable income from them', Baudin assured Magallon that though he had no such idea, he intended keeping the natives aboard until he had a sufficient crew to sail. It was up to the intendant

> to obtain men for me … He could give me from the hulk those who were being kept there for the manning of a corsair in which he was not uninterested, and which was only waiting for my departure before going privateering with a party of my deserters.[36]

Again his stand produced results. Another dozen men, all sailors, were sent out from the hulk. One of them, a Spaniard, offered to reveal the whereabouts of sixteen of the deserters for a payment of 20 piastres; Baudin promised 30 if he would lead the master gunner with a police guard to their hiding place. They returned with twelve men, all but three from the *Naturaliste*. On being questioned about their motives for deserting, 'they declared that the labour contractors for the corsair ... had offered them 150 piastres in advance, and they had not been able to resist such a tempting proposition'.

With the crews now close to full strength it was time to recall the officers and *savants* from shore leave. If Bory is to be believed, Baudin had 'no expectation' of the mass defections that followed. The veteran botanist Michaux (aged fifty-four) – 'our senior', wrote Bory, '[whom] we all respected as the chief of the expedition' – was the first to leave, 'convinced that he could be of no service to the expedition, managed as it was'. He considered himself as only a passenger on board the ship, he informed Baudin:

> his zeal for natural history was sufficiently known to protect him against the suspicion of having quitted the expedition from any other view than that of promoting ... the science to which he was attached.[37]

He intended to remain at Isle de France and write the natural history of Madagascar.

The botanist Delisse, the zoologist Dumont, two junior gardeners and the artist Garnier followed Michaux and disembarked from the *Naturaliste*. Bory de St Vincent 'left for the country', telling no-one of his intentions. From the *Géographe* there were three defectors – the astronomer Bissy and the artists Milbert and Lebrun. Apart from Michaux and the two young gardeners, they all claimed to be in 'ill health'. To their surprise, Péron refused to join the 'rebels', insisting that he would continue on the voyage – not even his growing dislike of Baudin could interfere with the discoveries in the natural sciences that lay ahead.

The artists and scientists were joined by four officers and six midshipmen. The former included lieutenants Gicquel and François Baudin – both of whom had been planning to defect for some time – and Lieutenant Bonnié, a capable officer who was genuinely ill. Each ship lost three midshipmen, swayed perhaps by their friends among the dissidents. In his memoirs, Charles Baudin (one of the 'loyalist' midshipmen) recalled that the commandant

> was well known at Isle de France, and had many enemies there; half through animosity against him, half through interest towards us, a kind of general conspiracy arose right from the outset to keep on the island all the members of the expedition – scientists, officers, and even midshipmen.[38]

Baudin professed unconcern at losing so many of his staff. Writing to Fleurieu he claimed, somewhat hypocritically, that

> the greater part of the scientists and even of the naval officers have been frightened by my resolution [to continue], and have disembarked. The former will be no loss for the success of the expedition, and the working of the ships will suffer little from the absence of the others. All are in hospital for appearances sake ...[39]

Michaux was the only one whose loss he regretted, he informed Jussieu: 'I discussed with him in the most friendly fashion the inconvenience which could result [from his departure], but to no avail.' It seemed the botanist was unwilling to surrender to the government the collections he would have made on the voyage: 'That is the only reason he abandoned us.'

As for the seamen, Baudin consoled himself that although most of the deserters had been replaced by 'admittedly inferior men', he nevertheless had enough to work the ships. To these were added at the last moment six Malays from the hulk, three for each ship, handed over by the administration for return to the Dutch at Batavia. 'In this way I had one hundred men all told, and the *Naturaliste* eighty-five. We were thus in a position to put to sea without worry.'

Baudin's last duty in the colony was to attend the governor's farewell dinner for the expedition. On 24 April he and Hamelin, both in full dress, sat down to a splendid meal at Government House, hosted by Magallon and attended by the intendant and other leading members of the administration. Forgotten were the recriminations of the past weeks:

> The whole occasion passed off in the best possible fashion, and there was not a single reference to the little difficulties that I had experienced during my stay in the colony. I was simply congratulated on having had the good fortune to find the resources to enable me to leave. Throughout it all, the Intendant protested his regret at having been prevented from helping us by his lack of means.[40]

Baudin handed Magallon two copies of his dispatches for Paris, and later gave his friend Pelgrom two further copies. The hazards of the long voyage back to France in wartime required that copies of important papers must be sent by several different ships to ensure that one at least reached its destination.

He returned on board 'firmly resolved not to set foot ashore again until my return'; but for the calm weather he would have sailed at once. At seven next morning

we felt a breeze and, although it was still weak, weighed anchor immediately and to my great satisfaction, for until then I had always been anxious, not only as to the continuation of the voyage, but also as to how I
would leave a colony where I had known so many different vexations.[41]

His friend Riedlé noted in his journal: 'But for the commandant's actions the ships would have remained in port at Isle de France'.

On the *Naturaliste*, Milius disapproved of the desertion 'of several members of the expedition' and complained of the poor quality of the provisions taken on board. 'If the staff left Isle de France with many regrets,' he wrote,

this was not true for the commandant. He departed full of disgust and
sorely embittered both by the obstacles placed in his way and the total
lack of consideration given him [by the administration]. All these circumstances combined to provide us with nothing but disagreements and many
privations to be endured.[42]

THE BRITISH EXPEDITION

Prologue: The Portsmouth prophet[43]

Portsmouth, July 1801. John Thistle celebrates his warrant as master of
HM Sloop Investigator *in a dockside tavern. Recently returned from the*
colony of New South Wales as master's mate in HMS Buffalo, *he toasts*
his good fortune at finding Investigator *in port and in want of a mas-*
ter. He knows her captain well; he has sailed as an able seaman with
Flinders on the Norfolk, and before that was with George Bass on his
whaleboat voyage to Bass Strait. Besides, he has his own reasons for
returning to Port Jackson.

  Downing his ale in the dark, smoke-filled snug, Thistle is joined by
an old man named Pine, a former sailor by the look of him, who offers
to tell his fortune for a drink or two. The cunning man begins his tale,
predicting that he will soon leave on a long voyage, and that his
ship will meet another vessel at her destination. This much, Thistle
knows, is common knowledge in the port, and he begins to regret the
waste of a pot of toddy. But, superstitious as any sailor, he is shaken by

the soothsayer's next words: he, John Thistle, will be lost with several others before the voyage is completed. As to the manner of his death, Pine will not, or cannot, say.

'Forget the sea, sailor,' keens the drab at the tavern door. 'She'll have you soon enough.'

Thistle pushes past, hurrying back to the ship. Ill at ease but resolved not to show it, he jokes at the mess table about the old man's prophecy, the warrant officers' servants taking in every word. Rumour runs like quicksilver through the crowded lower deck. The captain's boat's crew, on their next trip ashore, hasten to consult the prophet, plying him with grog to get the full story. He gives them the same account of a long voyage, but adds they will all be shipwrecked, though not in their present ship. Asked if they will survive the wreck and return to England, Pine again replies that he cannot say.

Superstition grips the crew, several of the men complaining that the captain was wrong to have brought his wife on board; all know this is an unlucky omen for the voyage. Some purchase the tail feathers of a wren as a safeguard against shipwreck, a few advertise ashore for the caul of a newborn child, a sure protection against drowning; a few attempt to desert.[44] *The officers do their best to reassure the men, Thistle asserting that Pine is no more than a waterfront charlatan after their pay. Few if any believe him – they doubt he believes it himself.*

The voyage or the wife

The Nore, March to May 1801.

> On the 14th [of March], the guns, twelve six-pounders, with their ammunition and a chest of fireworks were received; and the provisions and stores being all on board on the 27th, and the ship ready for sea, we dropped out to the Nore. I was anxious to arrive upon the coasts of Terra Australis in time to have the whole of the southern summer before me; but various circumstances retarded our departure, and amongst others, a passport from the French government, to prevent molestation to the voyage, had not arrived. [Flinders, 1814, vol. I, p. 5]

The *Investigator* remained at the Nore for two months, as the Royal Navy again reshaped the balance of power in Europe. Pitt's ministry had fallen, and Admiral Sir John Jervis (now Earl St Vincent after his great victory in 1797) had taken office at the Admiralty as First Lord.

He lost no time in dispatching the North Sea battle fleet to Copenhagen, with the object of forcing the neutral Danes to allow British ships free access to the Baltic. The resulting battle on 2 April was the fiercest Nelson ever fought, but after four hours the Danes agreed to a cease-fire and eventually admitted defeat.

Meanwhile Flinders impatiently awaited his sailing orders. At the beginning of April a letter came from Ann, raising once more the question of marriage. Matthew replied on 6 April, addressing the letter via her stepfather:

> My dearest friend – Thou hast asked me if there is a *possibility* of our living together. I think I see a *probability* of living with a moderate share of comfort. Till now I was not certain of being able to fit myself out clear of the world. I have now done it; and have accommodation on board the *Investigator*, in which as my wife, a woman may, with love to assist her, make herself happy. This prospect has recalled all the tenderness which I have so sedulously endeavoured to banish. I am sent for to London, where I shall be from the 9th to the 19th or perhaps longer. If thou wilt meet me there, this hand shall be thine forever.

From her parents he required 'nothing more than a sufficient stock of clothes, and a small sum to answer the increased expenses that will necessarily and immediately come upon me'. He specified the dowry at £200 (or £150 'if great inconvenience will result from advancing it'), and added:

> it will, I trust, be sufficient for me to say that I see a fortune growing under me, to meet increasing expenses. I only want to have a fair start, and my life for it we will do well, and be happy.[45]

As a postscript, he warned her 'to keep this matter entirely secret. There are many reasons for it yet, and I have also a powerful one. I do not exactly know how my great friends might like it'.

The letter does not sit comfortably with Flinders's general character. It is secretive, if not devious, and suggests a man fearful of losing the woman he loves, and in consequence losing his judgement.[46] To invite Ann to share his cabin on such a long and potentially dangerous voyage of discovery, involving a running survey of more than 1500 miles of unknown coast, was unusual enough; to do so without telling her it was against regulations was at best misleading; and then to imagine that their marriage and her presence on board could be concealed from his 'great friends' smacks of the irrational. Not only is he deceiving Ann, for there could be little doubt how Banks and the Admiralty would view the situation, he is deceiving himself into believing the impetuous scheme could succeed. In the event, Ann was badly hurt, and the attempted deception of his superiors almost cost him his command.

He spent the following week in London on official business. On the 15th, a Wednesday, he sought a week's leave to remain in the capital, claiming his presence there would be more useful than at Sheerness. That day he left by coach for Boston, and thence to Partney, and on Friday morning was married to Ann in St Nicholas Church by her step-father, the Reverend William Tyler. After the ceremony, bride and groom walked from the old stone church to the reception in the rectory. 'Never man more happy than poor Matthew,' wrote Ann's half-sister Isabella many years later, '& he determined to be so, in spite of the Lords of the Admiralty & Sir Joseph Banks – Yes, of all the merry group none more merry than he.'

On the afternoon of her wedding Mrs Ann Flinders writes to her friend Elizabeth Franklin, a distant cousin of Matthew. The wedding has been arranged in such haste that Elizabeth has not been informed of their plans:

> *My beloved Betsy – Thou wilt be much surprised to hear of this sudden affair; indeed I scarce believe it myself, tho' I have this very morning given my hand at the altar to him I have ever highly esteemed, and it affords me no small pleasure that I am now a part, tho' a distant one, of thy family, my Betsy. It grieves me much thou art so distant from me. Thy society would have greatly cheered me. Thou wilt today pardon me if I say but little. I am scarce able to coin one sentence or to write intelligibly. It pains me to agony when I indulge the thought for a moment that I must leave all I value on earth, save one, alas, perhaps for ever. Ah, my Betsy, but I dare not, must not, think [that]. Therefore, farewell, farewell. May the great God of Heaven preserve thee and those thou lovest, oh, everlastingly. Adieu, dear darling girl; love as ever, though absent and far removed from your poor – Annette.[47]*

Next day Matthew and Ann travelled to Donington to visit his family. His relations with his father were strained – Dr Flinders, beset by family and financial worries, had refused his eldest son a loan of £200 (sought after Flinders learned of his coming appointment as *Investigator*'s commander), and Matthew had replied to the effect that he should now consider he had only four children. Knowing nothing of their marriage plans, Dr Flinders could not conceal his dismay:

> Matthews Marriage – with concern I note that my son Mattw. came upon us suddenly & unexpectedly with a Wife on Sat. April 18 & left us next day – it is a Miss Chapple of Partney. We had known of the acquaintance, but had no Idea of Marriage taking place until the completion of his ensuing Voyage. I wish he may not repent of his hasty step.[48]

By Monday the newlyweds were back in London. The following day, Matthew wrote to a cousin, he presented himself to Sir Joseph Banks 'with a grave face as if nothing had happened', and continued with business as usual. He and Ann stayed in town until the weekend, and then went aboard the *Investigator*.

The futility of trying to conceal the marriage quickly became clear. Banks read of it in a Lincoln paper, and also learnt from the Admiralty that 'Mrs. Flinders is on board, [and] you have some thought of Carrying her to sea with you'. If that proved to be the case, he wrote to Matthew,

> I beg to give you my advice by no means to adventure to measures so Contrary to the regulations and the discipline of the Navy … their Lordships will, if they hear of her being in New South Wales, immediately order you to be superseded, whatever may be the Consequences, and in all Liklyhood [sic] order Mr Grant to Finish the survey.[49]

The blunt warning came as a rude shock to Ann, though Matthew cannot have been too surprised. He replied at once, putting the best gloss he could on his case, and pleading for Sir Joseph's help in changing their Lordships' opinion. If they persisted, his mind was made up; whatever his disappointment,

> I shall give up the wife for the voyage of discovery; and I would beg of you, Sir Joseph, to be assured, that even this circumstance will not damp the ardor I feel to accomplish the important purpose of the present voyage; and in a way that I shall preclude the necessity of anyone following after me to explore.[50]

Before dawn on 26 May, with Ann still on board and perhaps nursing a faint hope that the Admiralty might yet relent, the *Investigator* weighed anchor, under orders to proceed down the Channel to Spithead. By the time she arrived, three separate events had put an end to any possibility of Ann remaining.

Just before sailing three seamen deserted, it then being too late to replace them. The next day, at anchor in the Downs, another sailor, sent on board to take passage to Spithead, escaped ashore and vanished. Most serious of all, in fine weather and with a smooth sea, the ship gently grounded on a sandbank, the Roar, a mile and a half from the Kent coast. She floated off within two hours, with no apparent damage, but, as Flinders realised, there was more than a little to his reputation. On docking at Spithead he wrote to Banks: 'I am, therefore, afraid to risk their Lordships ill opinion, and Mrs. F will return immediately that our sailing orders arrive.' Sir Joseph accepted Flinders's explanation that the first two events were beyond his control, but had no doubt where the blame lay for the ship's grounding:

> I heard with pain many severe remarks on these matters, & in Defence I could only say that as Capt. Flinders is a sensible man & a good Seaman, such matters could only be attributed to the laxity of discipline which always takes place when the Captain's wife is on board, & that such lax discipline could never again take place because you had wisely resolved to leave Mrs Flinders with her Relations ...[51]

Banks's intervention was probably crucial. 'I have had a long conversation with [Secretary] Nepean,' he wrote, 'on the subject of the charges brought against you, and have pleaded your cause I hope effectually.' Flinders retained his command, but for Ann it was a different matter. Faced with the certainty of a long and lonely separation from her new husband she fell ill. Matthew accompanied her to London, where she remained until well enough to return to Partney with her stepfather.

Sailing orders

Spithead, Wednesday, 1 July 1801. Captain Flinders writes in his journal:

> *During the months that the ship has been waiting at the Nore and at this place for sailing orders, I constantly ... endeavoured to bring the ship's company under good order and government, to which indeed the majority of them were well inclined, but some of the heedless occasionally fell under the lash, though not a man in the ship showed any signs of ill disposition*

to a proper subjection to their officers, or to living sociably with each other.
Some desertions prevent me from permitting any to go ashore upon leave,
since payment of the ship. [Ingleton, 1986, p. 112]

On 27 June 1801 the *Times* carried an item of shipping news from
Portsmouth:

The *Investigator* (late *Xenophon*), ordered to Botany Bay on a voyage of
discovery, is expected to sail from Portsmouth with the first fair wind.
She is admirably fitted out for the intended service, and is manned by
picked men, who are distinguished by a glazed hat decorated with a globe,
and the name of the ship in letters of gold.[52]

The notice proved premature. The ship's sailing orders were delayed,
and the crew grew restless. Flinders was called to London for talks at
the Admiralty, and Ann, deeply distressed, travelled with him; she was
to stay with friends until Mr Tyler could take her home to Partney.
Matthew returned to Spithead, heartsick at their separation: 'Can I live
without thee and be happy,' he wrote. 'Indeed I think at this moment
it cannot be.' And later:

Thou wilt write me volumes, my dearest love, wilt thou not? No pleasure
is at all equal to that I receive from thy letters. The idea of how happy
we *might* be will sometimes intrude itself and take away the little spirits
that thy melancholy situation leaves me ...[53]

On his return to the ship, Flinders senses a subtle change in morale
among the crew. They had been paid in late June, and with cash in their
pockets several seamen tried to desert. As a deterrent to further attempts
he orders the men flogged, and cancels shore leave. On 14 July a boat's
crew is sent to HMS Puissant *to attend the hanging at the yardarm of*
Hadrian Poulson, a Dane, one of the leaders of the Hermione *mutiny.*[54]

Later the Investigators are mustered on deck to witness another in-
structive example of the maintenance of good order and government
on HM *ships, designed by the authorities 'pour encourager les autres' –*
the flogging round the fleet of four seamen convicted by court martial
of serious crimes at sea.[55] *Each man under sentence is placed in a boat*
in which a grating has been lashed upright across the thwarts, and is
then rowed alongside each ship lying in harbour; as the boat approaches
the ship's drums beat the 'Rogue's March'. At each ship in turn the pris-
oner receives twelve strokes with a cat-o'-nine-tails, laid on by a bosun's

mate of the ship in question – a healthy competition is encouraged between the mates. After each infliction of a dozen strokes a blanket is thrown across the prisoner's back, and a naval surgeon in the boat ensures he is fit enough to receive the next instalment of his punishment. At large bases such as Portsmouth the accumulated floggings may exceed a thousand lashes per man.

Investigator's sailing orders, so long delayed, arrived on 17 July, directing her captain 'to put to sea the first favourable opportunity of wind and weather'. They were accompanied by a letter from His Grace the Duke of Portland instructing Governor King to place the brig *Lady Nelson* under Flinders's command upon his arrival at Port Jackson, together with the safe-conduct from the French government. Flinders glanced at the passport, noted that he should have it translated, and stowed it away with other papers. He read quickly through his instructions, then penned a letter to Sir Joseph Banks, to be sent at once:

> I am happy, Sir Joseph, in being able to acquaint you that I have received my sailing orders today, with my instructions, passport &c.; although the Admiralty have not thought good to permit me to circumnavigate New Holland in the way that appears to me best suited to expedition and safety. I propose to put to sea tomorrow morning for Madeira, where I am directed to touch, though not at Rio de Janeiro.
>
> I am happy at being able to proceed upon the voyage even at this late period, and to say that I am much pleased with my messmates, who, as far as I can at present judge, are very orderly, well inclined men, and fitted for the situation which they fill. Most earnestly praying that you may see the examination of New Holland performed in the way that will be most gratifying to you, I remain, Sir Joseph, with the highest respect and esteem, your most obliged and obedient servant – Mattw. Flinders[56]

Flinders's instructions from 'The Commissioners for executing the office of Lord High Admiral of the United Kingdom of Great Britain and Ireland' were precise and detailed – and in many respects reflected his own proposals to Banks the previous year. He was to 'proceed to the coast of New Holland for the purpose of making a complete examination and survey of the said coast', calling at Madeira and the Cape of Good Hope on the way to take on water and other needed provisions.

On reaching New Holland, he was to make a running survey of the coast from 130° of east longitude (1° east of the present Western

Australia-South Australia border) to Bass Strait – 'putting if you shall find it necessary into King George III's Harbour for refreshments and water' – and use his 'best endeavours to discover such harbours as may be in those parts'. The commissioners, however, made it clear that this initial survey was merely a preliminary to a more thorough examination of the coast, which was to be carried out by the *Investigator* and *Lady Nelson* working in consort. At Sydney, Flinders was to consult with Governor King 'upon the best means of carrying on the survey', and to take the brig under his command for inshore work. They listed the order in which he was to undertake the survey of the New Holland coasts, after refitting at Port Jackson:

First, he was to 'diligently examine the coast from Bass's Strait to King George III's Harbour' – proceeding either from east to west or the reverse, whichever seemed more expedient.

Next, he was to explore the north-west coast visited by Dampier, then 'carefully examine the Gulf of Carpentaria, and parts to the westward thereof' – taking care to do so when the seasons and winds were favourable.

He was then to 'proceed to a careful investigation and accurate survey of Torres Strait', before examining 'the whole of the remainder of the north, the west, and the north-west coasts of New Holland' – especially 'those parts ... most likely to be fallen in with by East-India ships'.

Last, his instructions ordered him to

> examine very carefully the east coast of New Holland, seen by Captain Cook, from Cape Flattery to the Bay of Inlets; and in order to refresh your people, and give the advantage of variety to the painters, you are at liberty to touch at the Feejees, or some other of the islands in the South Seas.[57]

Banks's influence is apparent in this and other sections of the instructions. The use of the *Lady Nelson* for inshore surveying would free the *Investigator* to sail directly 'from one harbour to another as they shall be discovered' – thus giving the naturalists 'time to range about and collect the produce of the earth', and allowing the artists 'to finish as many of their works as they possibly can on the spot where they may have been begun'. In the off-seasons, when weather prevented the survey continuing, the expedition was to return to Sydney to rest and refit.

Though he has reservations about the itinerary prescribed by the Admiralty, Flinders is generally satisfied with his instructions. It is a demand-

ing program, but achievable, given a sound ship, capable officers, a hand-picked crew, and a scientific staff chosen by Sir Joseph himself. Perhaps, even at this first reading, there is some unease at the back of his mind about the logic of omitting Leeuwin's and Nuyts' Lands (the coast from King George Sound to 130°E) from the preliminary running survey. He puts the thought aside, and ponders Captain Baudin's voyage.

The Frenchman is now nine months at sea – ample time to have completed the discovery of the south coast, if that is indeed his intention. His passport is for a scientific voyage around the world, his present whereabouts unknown. Both Sir Joseph and Their Lordships fear that his real purpose is espionage – to ascertain the state of the colony of New South Wales, and perhaps to raise the French flag over some convenient port in New Holland. Though his orders are silent on the matter, Flinders knows that the Investigator *alone has the capability to track Baudin's movements in these distant waters.*

Unaware of the British expedition, the French ships are making their way northwards along the continent's west coast. Separated in gale-force winds, the Géographe *is off North-West Cape, the* Naturaliste *at anchor in Shark Bay.*

From the Channel to the Cape

At sea, the Western Approaches, Tuesday, 21 July 1801.

> *On July 18 we sailed from Spithead; and in the afternoon of the 20th, having a light breeze from the eastward, with fine weather, our departure was taken from the Start, bearing N. 18°W. five or six leagues. On the following day we fell in with Vice-Admiral Sir Andrew Mitchell, with a detachment of four three-decked ships from the grand fleet cruizing before Brest. [Flinders, 1814, vol. I, p. 17]*

On the fourth day out they sighted a squadron of four ships of the line – *Windsor Castle, Temeraire, Formidable* and *Atlas*, all ninety-eight guns – patrolling the approaches to Brest. Signalled by the flagship *Windsor Castle* to come aboard, Flinders ordered a boat lowered and was rowed across in the choppy seas. Vice-Admiral Mitchell received him affably in the great cabin – a room dwarfing the *Investigator*'s cramped quarters – and listened with interest, interjecting frequently, as the young commander summarised his plans for the expedition. The admiral wished Flinders success, observing as he left that after seventeen weeks at sea in all weathers, there was not one scorbutic man on board the flagship. Flinders took it as a good omen for his voyage.

Next day two luggers flying English colours approached the ship; the first fired a shot across the *Investigator*'s bows, but both broke off and disappeared when Flinders hoisted his ensign and cleared for action. There was another alarm a few days later, when a strange sail ignored his signal. Again the sloop's guns were run out, but this time Flinders fired first. The stranger hove to and identified herself as a Swedish brig bound for Stockholm; after an exchange of courtesies, both vessels resumed their course.

More worrying than these chance encounters was the discovery, as they cleared the Channel, that the ship was leaking badly through the side planking – as much as 3 inches of water an hour when heeled to a beam wind. Once the weather eased, Flinders lowered a boat and went around the ship with the carpenter to examine the seams close-up. Those under the counter and at the butt ends appeared sufficiently bad to explain the leaks, and he hoped the cause did not rest lower, under the copper sheathing.

Other defects showed themselves. Several yards and spars, and parts of the rigging, were found faulty and needed to be replaced. One beam was so rotted that it carried away when a seaman leant on it. Flinders decided on a thorough overhaul at their first port of call, Funchal on the island of Madeira, which they reached on 3 August.

Funchal Bay was full of shipping, but it was HMS *Argo* that caught the commander's eye as they entered the harbour – nine days before she had convoyed a British expeditionary force to the island to forestall a possible French occupation. Flinders called first on the commanding officer, Captain Bowen, then presented his respects to the Portuguese governor, who readily gave permission for the officers and scientists to take lodgings ashore, and make excursions into the interior.

While the carpenter and his mates set to work recaulking the seams above the copper sheathing with oakum, others of the crew checked the masts and rigging and replaced defective parts. Many of the faults found were due to dockyard negligence or worse, and should perhaps have been picked up during the refit at Sheerness. Supplies of water, wine, and fresh meat were taken aboard and stowed in the hold. The wine 'was charged at the enormous price of 5s.8d. per gallon, and the beef at 10d. per pound', and Flinders as ship's purser 'therefore took only small quantities of each'. Fresh fruit and onions, however, were in abundance, 'and probably were not of less advantage to the health of the people than the more expensive items'.

Flinders is the guest of Mr Pringle, the British consul, and thus avoids 'experiencing the accommodation afforded to strangers at a house in the town, dignified with the name of hotel'. Robert Brown and his colleagues complain 'of its being miserable enough, even without the swarms of fleas and other vermin' that molest them. This is not their only misfortune. Free to botanise again after more than a fortnight at sea, they make an excursion to Pico Ruivo, the highest point of the island's great central ridge, some 15 to 20 miles inland. They find lodging overnight in an empty room of a country priest's house, but there are no refreshments, and the only bedding available is two vermin-infested mattresses laid on the bare floor.

Rising at first light the naturalists set out for the peak, but the path over steep hills and down deep valleys is difficult and tiring, and the weather uncomfortably hot. Unable to reach the summit they turn back, finding some relief at the church, where two ladies serve them a meal of boiled potatoes, butter and excellent water, on which they feast heartily. At sunset, writes Peter Good,

> *[we] arrived at Funchal much fatigued we had some dinner and went to the Shore and hired a Boat to take us to the Ship for which they made us pay two Dollars before they would put off – as soon as we got in and the Boat shoved off the violence of the surf dashed us again on the Beach & beat full over us so that we all got compleatly ducked, however we all reached the shore in safety but most of us lost some articles of apparel ... we returned to the inn and went to bed but were much disturbed by Bugs and fleas.*[58]

Next morning, Flinders sends a boat to embark them, but the surf is still running high and they are soaked a second time; 'as we had not been Dry since last night it was not regarded'. They get aboard about 8 a.m., and all is bustle till noon, when the ship gets under way.

They sailed from Funchal on 7 August and headed south. Past the Canary Islands, the prevailing south-westerlies drove Flinders eastwards towards the African coast, and brought with them a heavy swell from the south, which caused the ship to plunge considerably. The oakum in the seams worked loose, and the old leakiness increased, amounting to 5 inches of water an hour by early September. In an effort to make the vessel less top-heavy, two carronades, the spare rudder, and other stores were removed from the deck and stored below. Afterwards, Flinders noted,

> the tremulous motion caused by every blow of the sea exciting a sensation as if the timbers of the ship were elastic, was [much] diminished; and the quantity of water admitted by the leaks was also somewhat reduced.[59]

This fresh evidence of weakness in a ship 'destined to encounter every hazard' further confirmed 'the unfavourable opinion of her strength' given him by the dockyard officers at Sheerness. He had made no formal complaint at the time, he wrote in the *Voyage*, because it had been made clear she was the only vessel available, and 'his anxiety to complete the investigation of the coasts of Terra Australis did not admit of refusing the one offered'.

Investigator crossed the line (the equator) on 7 September. Flinders readily agreed to the ceremony of shaving and ducking 'to be performed in its full latitude'. Afterwards the men were given as much grog as they could drink, 'the ship having been put under snug sail' – on a small ship with a picked crew there was no need for the strict discipline of a three-decker.

Few had crossed the line before, and most were baptised according to the time-honoured ritual – probably it followed similar lines to that on Cook's *Endeavour* in 1768, described by Joseph Banks. The victims were held fast in a stout frame and hoisted in turn on a long rope drawn through a block at the end of the main yards; from this height they were ducked three times in rapid succession into the sea. Banks and Solander opted out, 'giving the duckers a certain quantity of brandy for which they willingly excused us the ceremony'. The duckings lasted until nightfall:

> and sufficiently diverting it certainly was to see the different faces that were made on this occasion, some grinning and exalted in their hardiness, whilst others were almost suffocated and came up ready enough to have compounded after the first or second duck, had such a proceeding been allowable.[60]

After a three-day diversion in search of St Paul's Rocks, thought to lie near the equator about 18°W, the passage to the Cape resumed. On fine evenings 'the drum and fife announced the forecastle to be the scene of dancing', and on occasion the seamen were allowed 'other playful amusements which might be more to their taste'. These pastimes were all part of 'the beneficial plan first practised by the great Captain Cook' to maintain good health at sea, followed by Flinders throughout the voyage.

Prior to Cook, scurvy had taken an appalling toll on long ocean voyages, and Flinders was determined to keep the crew free of the disease. In the tropics lime juice and sugar were added to the daily grog ration as an antiscorbutic; in the higher latitudes 'sour krout' and vinegar were substituted. The mess decks were cleaned and aired daily. The men were forbidden to sleep on deck or to lie down in wet clothes; their hammocks, chests and seabags were aired regularly. Twice a week they were mustered on deck, every man appearing clean-shaven and in clean linen. In consequence, he reported with pride at the Cape, 'we had not a single person on the sick list, both officers and men being fully in as good health as when we sailed from Spithead'.

Cape Town

Off Cape Town, Friday, 16 October 1801.

> *At one o'clock we hauled round the rocks which lie off Cape Point, and steered into False Bay. Near these rocks were two whales; and one or more of what seamen call thrashers [swordfish] were engaged in a furious combat with them, at a less distance than half a mile from the ship. The sinewy strength of the thrasher must be very great; for besides raising his tail high out of the water to beat the adversary, he occasionally threw the whole of his vast body several feet above the surface, apparently to fall upon him with greater force. Their struggles covered the sea with foam for many fathoms round. [Flinders, 1814, vol. I, p. 38]*

Investigator dropped anchor in Simon's Bay alongside ships of the British squadron lying in the port, and Flinders went ashore to present his orders to the commander-in-chief, Vice-Admiral Sir Roger Curtis. British forces had seized control of the Cape colony from the Dutch in 1795, and in the ship's weakened condition he welcomed Sir Roger's assurance that the expedition would be given all possible assistance. The most essential work – a thorough recaulking inside and out – began the next morning, with a gang of caulkers provided by the squadron.

The ship was repainted from stem to stern, spars and rigging received a thorough overhaul, and the sails were taken to the flagship, HMS *Lancaster*, for repair. Shortages in supplies and provisions were made good, and though fresh fruit and vegetables were not to be purchased, the naval hospital sent across what could be spared (including oranges and lemons) from its stores. When the time came to sail, wrote Flinders, 'the state of the ship and of our provisions and stores were as complete as when leaving Spithead'.

The admiral also allowed Flinders to exchange four seamen, whose fitness and character 'were not proper for so long a voyage', with four volunteers from the flagship. One of the 'young gentlemen of the quarterdeck' who wished to return to England was also discharged, and his place taken by Midshipman Denis Lacy from the *Lancaster* – 'a red hot volunteer for the service', Matthew wrote approvingly to Ann.

The astronomer Crosley set up his observatory on the south side of the bay, guarded by a party of marines. Flinders put his brother Samuel in charge of the camp, appointing him Crosley's assistant in making and calculating the observations. The tents were sheltered from the prevailing south-easterlies, but even so 'the quantity of sand put in motion by every breeze was a great molestation, and proved injurious to the instruments'. The location posed a further annoyance that neither Flinders nor Crosley had foreseen. It was on the road leading from Simon's Town to a place called the Company's Garden, and quickly became a local tourist attraction:

> ... this was the sole ride or walk in the neighbourhood, which the inhabitants and the gentlemen belonging to the ships in the bay could enjoy. From those of the first rank, who took their morning's ride, to the sailor who staggered past on a Sunday, and even the slave with his bundle of fire wood, all stopped at the observatory to see what was going on ... Some wanted information, some amusement, and all would have liked to see how the sun appeared through the telescope.[61]

Crosley had frequently been unwell on the passage from Madeira, and after a few days ashore came shamefaced to inform Flinders that he felt unable to continue on the voyage. His loss was a severe blow; Flinders had relied on the scientist's more accurate observations for his surveys, and no replacement was available at the Cape. From now on he and Samuel would have to share Crosley's responsibilities, even though 'the duties of commander joined to the occupation of surveyor left little time for other employment'. He wrote to Sir Joseph informing him of Crosley's departure:

> I am sorry that this takes place, but I shall endeavour that the injury which the service must sustain by this misfortune may be as much reduced as our remaining joint abilities in this department can make it.[62]

The timekeepers and other instruments supplied to the astronomer by the Board of Longitude were transferred to Flinders's care. He assured Banks that, despite his limited experience with 'the astronomical clocks and the Universal theodolite', he was 'by no means without hope of fulfilling nearly the instructions from the Board to Mr. Crosley'.

After their wasted excursion on Madeira, the botanists spent every possible moment exploring on Table Mountain and the adjoining ranges. 'They were almost constantly upon the search,' noted Flinders, 'and their collections, intended for examination on the next passage, were tolerably ample.' The young gardener Peter Good was out every day, rain or shine:

> October 27. Breakfasted at 6 in morning ... Mr. Brown and I ascended to the Top of the Table Mountain in two hours – but very unfortunately as we came to the Top a thick fogg and rain came on which continued all the time we were on the top – and we had not descended far when the day cleared up and became fine – we continued several hours ranging the skirts of the Mountain and found some fine heaths – ... but very few things we had not before found – we then descended a very steep bank [and] near the bottom we found ourselves in a wood of some extent of natural Timber of a good size ... which detained us 3 to 4 hours ... – at length we fell in with a foot path and set out a Brisk walk for Tokay but were benighted some miles distant, however we found our way though not the most direct arrived time enough to meet a kind reception and partake of an elegant supper ...[63]

Called back on board after no more than a fortnight's stay, Brown and his companions grumbled among themselves. Flinders wrote:

> In taking so early a departure, I had to engage with the counter wishes of my scientific associates; so much were they delighted to find the richest treasures of the English green house, profusely scattered over the sides and summits of these barren hills.[64]

Time was running out if he was to catch the summer on New Holland's southern coast.

The commander's days were fully occupied: regular calls on Sir Roger Curtis; taking observations with Crosley and Samuel at the observatory; checking and rechecking the rates of the timekeepers (critical for accurate surveying); discovering a new source of fresh water in the sandy hills north of the bay – the quality was superior and it would keep better at sea; consultations with Surgeon Bell to discuss diets for the crew on the coming voyage; daily meetings with Lieutenant Fowler and Thistle to keep track of progress with the repairs and reprovisioning. By month's end the ship was ready for sea.

The nights are different. The hours of darkness drag interminably – is it only four months since Ann shared his sleeping cabin? He is filled with longing, and with guilt. How long before he sees her again? He recalls their wedding day, the picnic on the Essex side of the river, that awful parting in London, her uncontrollable weeping. Tonight he cannot sleep. He rises, lights a lantern, and takes up his pen. The men on watch nudge each other as they see the light flickering on the cabin window.

Write to me constantly, write me pages and volumes. Tell me the dress thou wearest, tell me thy dreams, anything, so do but talk to me and of thyself. When thou art sitting at thy needle and alone, then think of me, my love, and write me the uppermost of thy thoughts. Fill me half a dozen sheets, and send them when thou canst. Think only, my dearest girl, upon the gratification which the perusal and reperusal fifty times repeated will afford me, and thou wilt write me something or other every day. Adieu, my dearest, best love. Heaven bless thee with health and comfort, and preserve thy full affection towards thy very own, Matthew Flinders.[65]

It will be another eight months before he hears from her, at Port Jackson.

Fresh gales blowing from the south-east kept the *Investigator* in port until 4 November. She sailed at daybreak, but was soon becalmed under the high land, and had not yet cleared the bay at noon. The sea for a great distance round the ship appeared tinged in alternate stripes of red and its natural colour, each stripe from 10 to 50 or perhaps 100 yards wide and stretching to the horizon. The lead was hove but found no bottom. Some water was taken up in a bucket for examination; to the naked eye it appeared perfectly clear, but under Robert Brown's microscope was seen to be full of minuscule red insects. After some

Detailed sketch of HM *sloop* Investigator, *by R.T. Sexton.*

hours the colour gradually disappeared, and a light breeze enabled them to make the open sea. Next day

> the report of the guns fired by the squadron in Simon's Bay, to commemo-
> rate the escape from gunpowder treason, was distinctly heard at one
> o'clock, when we were occupied in making sail to a fine breeze which
> had sprung up from the south-westward. At six in the evening it blew fresh
> with cloudy weather ... and we took our departure for New Holland.[66]

Notes

1.   Degérando, 1969, pp. 22-24.
2.   Péron, 1809, p. 14.
3.   Horner, 1987, p. 4.
4.   Ibid., p. 82.
5.   Ibid., p. 85.

6. Commanders of five previous French expeditions of discovery, all of whom died on their voyage.

7. Péron, op. cit., p. 16.

8. Baudin, 1974, p. 13.

9. Ibid., p. 14.

10. Péron, op. cit., p. 17.

11. Baudin, op. cit., pp. 21-22.

12. Brosse, 1983, p. 98.

13. Baudin, op. cit., pp. 29-30.

14. Bory de St Vincent, 1805, p. 16.

15. Baudin, op. cit., p. 35.

16. Bory de St Vincent, op. cit., p. 47.

17. Baudin, op. cit., p. 64.

18. Ibid., p. 37.

19. Péron, op. cit., pp. 35-37.

20. Bory de St Vincent, op. cit., p. 37.

21. Brosse, op. cit., p. 98.

22. Horner, op. cit., p. 104.

23. Baudin, op. cit., p. 106.

24. Horner, op. cit., p. 111.

25. Bory de St Vincent, op. cit., p. 54.

26. Baudin, op. cit., p. 121.

27. Administrative authority was split between the governor – responsible for defence and the military – and the intendant, controlling the civil administration, justice and finance.

28. Ibid., p. 123.

29. Ibid., p. 125.

30. Ibid., p. 127.

31. Bory de St Vincent, op. cit., p. 68.

32. Ibid., p. 69.

33. Horner, op. cit., p. 129.

34. Ibid., p. 132.

35. Baudin, op. cit., p. 134.

36. Ibid.

37. Bory de St Vincent, op. cit., p. 70.

38. Horner, op. cit., p. 124.

39. Ibid., p. 131.

40. Baudin, op. cit., p. 136.

41. Ibid., p. 137

42. Milius, 1987, p. 5. (Passage translated by J. Treloar.)

43. Flinders, 1814, vol. I, p. 136.

44. Kemp, 1976, pp. 847-848 ('Superstitions of Sailors').

45. Retter and Sinclair, 1999, pp. 17-18.

46. Ann's letter has not survived, but it appears from Matthew's reply that she sought a firm commitment from him before he sailed, or she would call off the engagement.

47. Scott, 1914, p. 109.

48. Ingleton, 1986, p. 105.

49. Ibid., p. 106.

50. Ibid.

51. Ibid., p. 110.

52. Ingleton, op. cit., p. 111.

53. Scott, op. cit., p. 193.

54. The *Hermione* mutiny in 1797 was the bloodiest in Royal Navy history. Captain Hugh Pigot and nine officers were killed, and the ship handed over to the Spaniards.

55. Kemp, op. cit., p. 317 ('Flogging round the Fleet').

56. Ingleton, op. cit., pp. 112-113.

57. Flinders, op. cit., pp. 8-12.

58. Good, 1981, p. 37.

59. Flinders, op. cit., p. 32.

60. Lyle, 1980, p. 50.

61. Flinders, op. cit., p. 39.

62. Ingleton, op. cit., p. 123.

63. Good, op. cit., p. 44.

64. Flinders, op. cit., p. 43.

65. Scott, op. cit., p. 194.

66. Flinders, op. cit., p. 44.

CHAPTER FOUR

⚓

New Holland and Timor,
April to November 1801

Leeuwin's Land

On board the Géographe, *Saturday, 25 April 1801.*

> *We were scarcely under sail, when we were informed by our commander, that from that time we should have but half a pound of new bread once in ten days; that instead of the allowance of wine, we should have three-sixteenths of a bottle of bad rum of the Isle of France, bought at a low price in that colony; and that the biscuit and salt provisions should be our general food. Thus, from the first day of a voyage which must necessarily be both long and difficult, we were abridged all at once of bread, wine, and fresh meat – a sad prelude, and chief cause of all the miseries we in the end experienced. [Péron, 1809, p. 54]*

On their departure from Isle de France, both vessels were undermanned, although Baudin had recruited enough 'admittedly unwilling men' to replace most of the seamen who had deserted. The loss of the officers and scientists was of less concern – many, he believed, were malcontents responsible for much of the disharmony on board, and he professed to be 'very glad' at leaving them behind.

The passage to New Holland passed without incident. They sighted the coast at daybreak on 27 May – 'a blackish stripe from the north to the south was the humble profile of the continent', wrote Péron. At first identified from St Allouarn's charts as Cape Leeuwin, the landfall was later found to be slightly to the north, near the present Cape Hamelin. Winds and currents defeated all attempts to close the land that day. The following two days were spent coasting northwards along the low, sandy and sterile-looking shore, darkish in colour, and dotted with occasional whitish specks; low hills, also dark and barren, and covered with a type of heath, ran down to the sea.

Before leaving Isle de France, Baudin had discussed with Hamelin his plans for the next stage of the voyage. The season, he felt, was now too advanced 'to allow us to follow the Government's intentions to the letter'. He proposed instead to sail north, examining the continent's west and north-west coasts before calling at Timor for refreshment. Rottnest Island and Shark Bay were to serve as the first and second rendezvous in case of separation.

The decision dismayed Péron, who blamed Baudin for the delays on the outward voyage and at Isle de France that 'had lost us part of the favourable season for our expedition'. It appeared that his captain's timidity would now lose them more precious time:

> Our commander feared to be driven towards Diemen's land, and therefore resolved to begin his exploration by reconnoitring the N.W. of New Holland, reserving for the ensuing spring the voyage to the South. This important determination gave us much concern, because it was not absolutely necessary from our actual situation. The season, though advanced, was not so much so as to prevent us from doubling the South Cape; and
> ... it appeared to us more prudent to pay respect to the instructions we had received from government[1] ... We shall see in the end the consequences produced by this first deviation from our orders.[2]

After some thirty years at sea, Baudin found it extraordinary that not a single landbird visited the ships, while two Cape pigeons were the only seabirds seen. He explained the oddity by 'the aridity of the land and the few resources that the sea can offer'. He dismissed as a mirage the lookouts' claim to have seen several large quadrupeds moving in and out of a small cluster of trees beyond the shoreline: 'They said they were cattle, or at least appeared to be.'

On 30 May the ships doubled a hilly cape ahead 'from which projected a reef where the sea broke with violence', and found themselves in a great bay stretching deep inland and extending far to the east. It was the first significant discovery of the voyage, the old maps showing only an unbroken line running more or less due north from Cape Leeuwin. With a steady southerly blowing, the vessels steered warily into the bay, coming to anchor at dusk several miles offshore. To mark their visit the bay was later named after the *Géographe*, and the cape and another dangerous reef in the middle of the bay after the *Naturaliste*.

Boats were hoisted out to sound around the ships, and reported an excellent sandy bottom. The weather remained mostly fine overnight, but shortly before midnight the *Géographe* was caught by a sudden squall; she dragged her anchor, almost colliding with her consort. They did not touch, but it was a very close thing.

Two boats are readied to reconnoitre onshore in the morning. Sublieutenant Picquet leaves first for Cape Naturaliste, where the astronomer Bernier and Boullanger the geographer are to take bearings to determine its correct geographical position. Before his armed longboat reaches the rock-bound coast, the weather changes, whipping up heavy breakers that crash violently upon the rocks. He can find no way through; the shore remains inaccessible. For the rest of the day and throughout the night the men in the boat battle the waves without being able to regain the ship, 'from which the wind had incessantly driven us, carrying us out to sea'. It is midmorning before they return, 'overcome with fatigue and drenched with sea-water'.

In the second boat Sublieutenant Henri de Freycinet has better fortune. His passengers, Depuch the mineralogist and the gardener Riedlé, land south of the anchorage in a sheltered cove, but even here they are forced by the breakers to wade ashore up to their waists in water. Their findings are disappointing: there is no fresh water, a nearby creek bed is dry, pebbles and small boulders glint white in the sun. The only traces of living creatures are of animals – some burrows freshly dug in the parched earth, the print of a 'cloven hoof' on the beach, and animal droppings like horse dung lying everywhere. Their one worthwhile prize is a small lizard caught in a small crevice in the rocks, and killed.

The ships crisscrossed the bay for three days, sounding and surveying the coast. At night they anchored out to sea in deep water, Baudin having no wish to be caught on a lee shore if the winds turned northerly. The scientists' requests to go ashore were ignored; watching the glint of fires in the dusky woods behind the shore, their frustration and discontent festered anew.

Baudin relented on the fifth day. Boats from both ships landed, north and south of the present Wonnerup Inlet (near the town of Busselton). Baudin went ashore with Bernier, Boullanger, Depuch, and his friends Maugé and Riedlé, to examine the countryside and collect some specimens. As they beached, they saw their first Aborigine, up to his waist in water and busy spearing fish. At first the native 'appeared to pay little attention to us and concerned himself solely with his fishing', but as they approached closer 'he began to shout violently, signalling to us to go back'. Finally he picked up three spears and, 'still yelling, fled precipitately, without having made any offensive movement towards us'.

Péron, meanwhile, landed from another boat further along the coast. Once ashore he lost little time attempting to make contact with the people whose fires he had watched at dusk from the ship:

> I ran towards the interior in search of the natives, with whom I had a strong desire to be acquainted. In vain I explored the forests, following the print of their footsteps, [but] all my endeavours were in vain ...[3]

Tired and footsore, he arrived back at the beach three hours later, to find his companions about to mount a search for him. Wind and tide were against them, and it was already night when they regained the ship.

Hamelin had also sent several boats ashore to reconnoitre the coast. The commander of one of these, Sublieutenant Heirisson, reported on his return that he had found a large body of water, a river or perhaps a lake, which teemed with birdlife; he had followed it some distance inland without coming to the end. The water seemed mostly brackish, but at one point he had come across a well on the bank, dug by the natives, where it was clear and fresh.

The news determined Baudin to remain another day at the anchorage to investigate the discovery. Three hours before dawn the *Géographe*'s longboat, well-manned and commanded by his second captain Le Bas, and two boats from the *Naturaliste*, with Hamelin in command, set out for the shore. Running before a fresh north-easterly, they made the coast close to the mouth of the inlet. The longboat's draft was too deep to cross the bar, and it was moored offshore in the charge of some sailors. The *Naturaliste*'s boats entered the river and separated, Hamelin taking his to explore upstream as far as his men could row.

Péron and his fellow naturalists follow the banks of the stream, collecting as they go. Soon, however, drawn by 'the desire of meeting the inhabitants of these regions', he sallies off on his own. Coming to a swamp 'on whose salt waters several companies of black swans [were] sailing with great elegance', he strips and wades across, ignoring the pleas of his friends to return. Footprints in the mud show that several natives have recently passed this way. Alone and unarmed, he plunges into the woods on the far side. It is about eleven o'clock, the sky is serene, and the air pleasant:

with these circumstances in my favour, and full of the hope of soon meeting with the natives of these shores, I endeavoured to follow their steps, when a singular discovery stopped my course.[4]

On the banks of a small stream he comes upon a half-circle of melaleuca trees, gleaming white in the sunlight; all have been stripped of their bark, leaving the wood beneath a shining white. Within this half-circle lie three more semicircles, all with their 'concavities turned towards the banks of the stream'. The first is a green bank of fine herbage, about 2 feet in breadth and raised 6 inches or so above ground level; the second is a semicircular clear space about $2\frac{1}{2}$ feet in breadth and covered with black sand. The third and innermost half-circle is covered with smooth white sand, in which are planted numerous rushes, 'so distributed as to form a succession of figures'; these, though coarsely executed, have 'much of design and originality' about them, and appear to represent 'triangles, lozenges, irregular polygons, some parallelograms, very few regular squares, and not any circles'. Moved by the beauty and serenity of the setting, Péron reflects on its likely spiritual significance:

'This charming place', I repeated to myself, 'is probably dedicated to some public or private mystery. The worship of the gods may be the particular object. It is from this river and the marshes adjoining, that the inhabitants of these shores in a great measure derive the food for their subsistence. – A new race of Egyptians, who like the ancient inhabitants of the Nile, have consecrated by their gratitude the stream which supplies their wants. Perhaps on particular solemn occasions, they assemble on its shore to pay the debt of gratitude, and offer up their thanksgivings!'[5]

The dream of a new race dissolved with the declining sun. Péron made his way back to the beach, but looked in vain for the longboat. He was relieved to meet up with Lesueur and Ronsard, fresh from an encounter with two natives, a man and a woman. As soon as the pair observed them, Lesueur told him, the man ran off into the swamp; deserted by her companion, the woman, scraggy, naked, and far gone in pregnancy, sank down on her heels in the sand and covered her face: 'she remained as one stupefied and overcome with fear and astonishment, perfectly without motion, and seemingly insensible to all that passed around her'. Unable to communicate with her, they had left her still crouched in the sand, and hiding her face.

115

Further along the beach, they came up with the rest of their party, together with Hamelin and his boat's crew. From them they learnt that the longboat had been driven ashore by the rising seas, and could not be refloated; they were trapped ashore for the night.

The disappearance of Helmsman Vasse

On board the Géographe, *Saturday, 6 June 1801.*

> *At half past nine at night … Mr. Hamelin came aboard. He had been about twenty-two [hours] at sea in his small boat with four men. During the strongest phase of the wind, which lasted from eight in the morning until four in the afternoon, he had spent the whole time at anchor on his grapnel, at the risk of being engulfed a thousand times over. The sea broke over them so heavily that he did nothing but continually bail out the water that came in. He told me that my longboat had been stranded on the beach, but did not know precisely how this accident had occurred. This news was a terrible blow to me, on account of those who had gone in it. [Baudin, 1974, p. 179]*

Baudin had ordered his boats to return by nightfall. Shortly before sunset he had seen the two smaller boats beating off the wind along the coast, but of the longboat with its distinctive sails there was no sign. Lights were kept at the masthead all night to guide them back, but without result. Next morning the strong winds and rising sea made it too dangerous for any boat to risk leaving the shore.

The weather improved during the day, and Baudin raised the signal ordering all shore parties to return It was followed by a cannon-shot fired every two hours, but still the boats did not appear. At four a small boat crossed from the *Naturaliste*, carrying Bailly the mineralogist with the news that the supposed river discovered by Heirisson was no more than a lagoon. He had returned the previous evening, and was unaware that the longboat had been driven ashore. His account convinced Baudin that the officers had been persuaded by the 'fine talk' of the naturalists to allow them extra time for their studies. He vowed this would not happen a second time:

> they shall not go ashore again, except when the ship is no longer in danger of being driven out to sea, and they do not run the risk of being dismissed on a lonely, unknown shore.[6]

It was only with Hamelin's unexpected arrival after nightfall that the plight of the men on shore became clear. Although armed, they were still, Baudin feared, 'at the mercy of the natives, who must ... become venturesome upon noticing their distress'.

The stranded men pass a hungry, sleepless night in the open, huddled together for warmth around the embers of a dying fire. Captain Le Bas, as senior officer, posts sentinels to guard against a surprise attack by the 'savages', whose howlings can be heard deep in the forest. Their only provisions are a few biscuits soaked in sea water, some rice, three bottles of arrack, and about 15 pints of water. These do not go far among twenty-five men.[7]

At dawn, Le Bas's first thought is to salvage the longboat if it is possible to do so, but they find her submerged, pounded by the surf, and half-filled with sand. The day is tempestuous and the rising wind blows directly from the north. All look in vain for a sight of the ships, or for boats coming to their rescue. Péron, prone to imagine the worst, pictures to himself 'our unfortunate ships compelled to sail, and abandon us on this inhospitable shore'.

Le Bas sends parties out to forage for food, but they return empty-handed, with just one gull and some sticks of wild celery between them. The only water found is brackish and barely drinkable. Ravenous, they boil the gull and celery with the few grains of rice in a large pot saved from the wreck, and serve it as soup. Within the hour most of the men are seized with a violent colic, retching, and agonising stomach pains. Exhausted, sick, and freezing, they endure another long and tedious night; the thunderous surf alone is enough to deprive them of sleep.

Baudin and Hamelin worked into the early morning preparing the rescue mission. Their plan was for each ship to dispatch a boat at 4 a.m. to make contact with the men on shore and if possible to bring them off, but the rising northerlies and rough seas made this too risky. At daylight Baudin signalled Hamelin to take the *Naturaliste* as close in to shore as possible without hazarding the ship, and attempt a rescue from his new anchorage; when she was at her station he would follow in the *Géographe*.

In the storm the *Naturaliste* fouled her anchors, and could not sail for some hours. Baudin weighed, and with topsails reefed the *Géographe* steered shoreward, the leadsman heaving the lead continuously in the shoaling waters; at midday she came to anchor in 7 fathoms about a mile offshore, off Wonnerup Inlet. Midshipman Bonnefoi was sent to reconnoitre the coast, with orders not to land unless he could do so in safety – unlikely in these conditions. The ship herself was in danger, rolling in a heavy swell and with a strong northerly blowing on a lee shore.

Baudin followed Bonnefoi's progress through his glass, but to his 'bitter disappointment' saw the boat drop its grapnel without anyone coming in sight. Convinced that natives had already overcome the shore party, he withdrew to his cabin to consider his next move. Even when an officer came to report that the boat's crew were apparently talking to a group of people who had emerged from the dunes, he refused at first to believe these were his men: 'I persuaded myself that they were natives trying to prevent our men from landing, and ... were no doubt chopping our longboat into pieces.'

It was past nine before Bonnefoi returned, guided by the light burning at the masthead. Though unable to land he had seen the longboat lying deeply embedded in sand, and considered it was probably beyond repair. Péron was with him, more dead than alive after being dragged out to the boat on a lifeline through the heavy surf. So wasted from hunger, colic and fatigue that his friends hardly knew him, it was some time before the naturalist could give a coherent account of the plight of the men ashore. They were starving, he reported, many were ill with colic, but none had died and there were no serious injuries. Baudin was visibly relieved; the news freed him 'from the painful existence I had led for three whole days without being able to find, even in sleep, the rest which was so necessary to me'.

Long before dawn, both ships sent off boats to rescue Le Bas and his men. The wind had eased, though the sea was still heavy and swelling. Baudin watched through his telescope, fearing another accident might yet prevent their return:

> I could see how strong the undertow was on shore. The men whom I awaited could not reach the boats without going up to their necks in water. Finally, towards midday, I saw the *Naturaliste*'s set sail and come back. It reached the ship at about two o'clock with all the naturalists ... Everyone of them was wet to the skin and perished with cold.[8]

He ordered the boat to return to help his own cutter evacuate the rest of the stranded men. The weather was fast deteriorating, and he feared damage to the masts if he remained much longer at anchor. To hasten the rescue, a cannon was fired every hour as a reminder to the boats not to delay.

The *Géographe*'s cutter returned at four, with all his men aboard. Black storm-clouds now covered the horizon to the north and west, the wind was gusting strongly, and the ship tossed alarmingly at her moorings. Still he delayed, for the *Naturaliste*'s boats had yet to return. As well as the large boat he had sent back to the rescue, a small dinghy had landed earlier with an officer aboard – doubtless for amusement, Baudin thought sourly, since it had beached at some distance from the wrecked longboat. Neither had returned by nightfall, nor had they responded to the cannon shots fired to recall them. Both ships lit port-fires and set off rockets to guide them back.

Lieutenant Commander Milius sets off in the dinghy intending to assist with the longboat's salvage and the rescue of her crew. Mist and rain obscure the beach from view, and he fails to notice the cutter take off the last of the castaways. He makes the coast safely, but as the dinghy heads in to land it is struck by a sudden squall and capsizes. Pitched overboard into the mountainous surf, drenched to the skin, bruised, and shivering in the icy wind, Milius and his men struggle ashore. By good fortune a following wave dumps the dinghy nearby.

Making their way to where the longboat lies waterlogged and half-filled with sand, the new castaways find the beach littered with abandoned tackle, utensils, cast-off clothing, even some muskets and ammunition – evidence of a confused retreat. Of the crew there is no sign. Milius thinks they have taken refuge in the woods, the seamen suspect they have been captured and eaten by the savages. To reassure them he ventures unarmed into the trees, shouting to those in hiding to come out. There is no response, only the roar of the surf and the whistling of the wind off the sea.

Twilight closes in, the wind blows stronger. The men, shivering from fear and cold, are trying to get a fire going when voices are heard hailing them from beyond the breakers – seeing their plight the Naturaliste*'s longboat has come to their rescue. Several expert swimmers on board throw themselves fearlessly into the raging seas to bring lifelines ashore. The stranded dinghy is righted and hauled out through the surf, then the castaways follow one by one. Milius is the last to leave. The scene is frightful – in the darkness the waves tower 8 feet tall, and crash with* *dreadful fury on the beach. Each man is tied to a line and dragged through the breakers by their rescuers in the boat. The operation is going well until the helmsman Timothée Vasse, a strong swimmer, loses his grip and is swept away; his shipmates search for him in vain. Later their recollections differ, and they give conflicting accounts of his disappearance.*

The survivors reached the *Géographe* towards nine that night. Milius reported that Vasse had been carried off by a wave and drowned; the fear of endangering his other men prevented a full-scale search for him. It was the expedition's first death, and Baudin held Le Bas and Milius responsible:

> Without this tragedy, I should at least have had, as consolation in my misfortune, the fact that no one had fallen victim to the rashness of two officers. But this happiness was not for me; and by some singular mischance, it must happen that the one who least deserved to feel its effects should become its victim.[9]

Postscript: The several deaths of Timothée Vasse

On Wonnerup Beach, Monday, 8 June 1801.

> *We attached ourselves one after the other to a line, and were somehow hauled out between the waves by the men in the boat. One of our seamen, Timothée Vasse, an excellent swimmer, had the misfortune to lose his life, he was dumped by a wave and buried in the sand. I exposed myself once more to danger, in a futile attempt to save him. The men in the boat … were so tipsy that I was unable to turn this to account. I was therefore obliged to abandon this coast, leaving it with regret that I had not been able to save an excellent seaman. I also lost a very good hunting dog.* [Milius, 1987, p. 9]

By his death, Timothée Vasse, helmsman second class, a native of Dieppe, won himself a sort of immortality. Louis de Freycinet named the Wonnerup Inlet for him in his atlas published in 1807; the river flowing into the inlet now bears his name. Some among his shipmates, however, did not accept that Vasse died in the surf that night. Neither the men on the beach nor those in the *Naturaliste*'s boat actually saw him die; nor in the darkness and confusion was his body recovered. After the ship's return to France in 1803, rumours circulated that he had swum ashore and, helped perhaps by natives, remained there still, nursing hopes of rescue by some passing vessel. Finally an account of his 'survival' appeared in the French newspapers, where it was read by François Péron while he was writing his *Voyage de Découvertes aux Terres Australes* (1807). Since 'every circumstance of his disappearance united to make his death inevitable', wrote Péron,

> not one person of the expedition retained the least doubt on the subject, till the time when a paragraph was published repeatedly in all the newspapers, that interested the public in the fate of the unfortunate Vasse, and awakened some hope in the breasts of his companions. It was asserted ... that having escaped as if by miracle from the fury of the waves, Vasse, after the departure of the two ships, joined the savages of that part of Leeuwin's Land, adopted their manners, learnt their language, and thus passed two or three years with them.
>
> This paragraph then made him meet with an American vessel, three or four hundred leagues south of ... where he had been wrecked; that he had been received on board this ship, which some time after fell in with an English cruiser [*sic*]; and it was even added, that he had arrived safe in England, where, contrary to the law of nations, he was detained.[10]

At Péron and Freycinet's request the Ministry of Marine investigated the rumour, and was satisfied that 'the whole of the account concerning our unfortunate companion was entirely fabulous'. They named the place where he disappeared 'to preserve the memory of his misfortune and our sorrow'.

This was not the end of Vasse's story, however. For several years past, Dr T.B. Cullity of Perth has pursued a personal interest in the helmsman's fate, and has uncovered the following versions of his 'life after death'. They are reproduced here with his kind permission.

Perth, 5 May 1838. Mr George Fletcher Moore, advocate-general of Western Australia, writes to the Perth Gazette. *He has recently returned from a visit to his friends the Bussells, settled near Wonnerup Inlet, which has enabled him to glean new facts about Vasse's fate:*

> *Poor Vasse did escape from the waves, but enfeebled as he was with the sickness and exhaustion by his struggles, exposed to the fury of the storm, unsheltered and apparently abandoned among the savages, perhaps he would have thought death a preferable lot. But the savages appear to have*

commiserated his misfortunes; they treated him kindly and relieved his wants to the extent of their power by giving him fish and other food. Thus he continued to live for some time, but for what length of time I have not yet been able to ascertain.

He seems to have remained most constantly on the beach looking out for the return of his own ship, or the chance arrival of some other. He pined away gradually in anxiety, becoming daily, as the natives express it, 'weril weril' (thin thin). At last they were absent for some time on a hunting expedition and on their return they found him lying dead on the beach, within a stone's throw of the water's edge. They describe the body as being then swollen and bloated, either from incipient decomposition, or dropsical disease. His remains were not disturbed even for the purpose of burial, and the bones are yet to be seen. The natives offered to conduct us to the spot but time pressed, we were then upon the point of embarkation and the distance was six or seven miles. The spot indicated is near Toby's Inlet at the south-eastern extremity of Géographe Bay.

Busselton, 1841. Mrs Georgiana Molloy writes from her home at Vasse River to a family friend, Captain Mangles, concerning 'Mons. Vasse':

Dr. Carr, who has lately come out with the Australindians, and is 'il Medico pro tem', has undertaken to reclaim the Bones of Mons. Vasse, the Gent^m. from whom this river takes its Name. Some society in Paris has offered a reward or present for them. These Natives know where they are, in the vicinity of Cape Naturaliste, and are now employed getting them, or for what I know, have got them. This event happened about thirty years since; this unfortunate Gent^m. came in shore to explore, was seized, strangled and the spear went in at the right side of the heart. So runs the sequel. However, until enquiry was made by Dr. Carr, he was never heard of. They represent him as being tall and thin according to the French Author's description, and when they bring the bones, he will easily be identified as their head and teeth are quite different to ours.

Dr Cullity has been unable to find any record of 'Mons. Vasse's bones', nor, despite extensive enquiries, has he been able to trace the elusive Dr Carr, 'with or without human bones'. The various versions of Vasse's fate recounted in his monograph, *Vasse: An Account of the Disappearance of Thomas Timothée Vasse* (1992), are all circumstantial, and any one of them (or none) may be true. Whatever the real facts, it is as he says a pitiful tale and it is hard not to be moved by it. The reason why it engages our sympathy is probably because it is like an existential nightmare, meaningless, hopeless, and alienated. There could be few more cruel fates than to find oneself alone on Wonnerup beach in 1801 on a freezing stormy night, with nothing, chilled to the bone, stiff, half-drunk, possibly partially scorbutic, in cold saturated calico trousers impregnated with rough sand. God knows what tooth-ache, salt-water boils and other miseries a sailor of that epoch would have had to complete the picture. His friends had departed, the savages were watching him, he had no food, no shelter, no weapons, no chance of influencing his fate, and no certainty about what games the natives were going to play with him. As a European, he seemed to be as far away from anywhere, or anybody, as it was possible to be. On the other hand, perhaps this is fanciful, and he died a decent death in the sea.

Rat Nest Island and the River of Swans

On board the Naturaliste, *Monday night, 8 to 9 June 1801.*

> *The rain was so heavy, and the mist so thick, that we could make out neither the* Géographe *nor the land, from which however we were no more than one and a half leagues distant, or two leagues at the most. One can readily judge the sort of night we spent, having to tack every two hours and the wind still aft. Although I had come on board with a very high fever, I remained constantly on the bridge and directed all the manoeuvres until the evening of the next day, by which time I could no longer speak or stand upright. The fever, anxiety and pain exhausted me and I fell into an alarming state of depression. [Milius, 1987, p. 9]*

Boarding the *Géographe*, Milius found Baudin 'plunged in the cruellest anxiety' over the position of the ships – dangerously close to the coast, the barometer dropping fast, and with a northerly gale blowing directly onshore. After a short rest to recover his strength, Milius returned to the *Naturaliste*, carrying the commandant's orders for Hamelin to sail immediately.

No sooner had he put off than a sudden gust of wind, screaming and full of vicious spray, hit the boat and almost capsized it in the monstrous seas. Luckily for Milius and the boat's crew the same squall struck the *Naturaliste*, causing her to drag her anchor and carrying her in their direction – 'I only needed to leave my craft to the wind and waves,' he wrote thankfully.

Back on board there was no rest for Milius or any of the crew. To save precious minutes, Hamelin ordered the cable cut and abandoned the anchor. With all sails set, despite the risk to sails and masts in the high winds, he fought desperately to avoid being driven ashore. For the next forty-eight hours the entire ship's company, officers and men, remained on call night and day. Kept awake by copious draughts of coffee, they tried to make headway into the north-easterlies, tacking every couple of hours in their attempts to clear the bay. Once during the night they caught a brief glimpse of the *Géographe* bearing away from them in the wind, and fired rockets to show their position. Pre-occupied with her own safety, *Géographe* did not reply to their signals.

On the third day the weather moderated, the sun broke through, and the winds turned north-westerly; of the *Géographe* there was no sign. Hamelin set his course north-north-east for Rottnest Island, nomi-nated as the rendezvous in case of separation, and arrived at the an-chorage on 14 June. Surprisingly the flagship was not there, although she was the faster ship and the better sailer. Hamelin dropped anchor in the channel between the island and the mainland, and, deciding to put the waiting time to good use, dispatched three boat parties – one to explore Rottnest (named by the Dutch for the rat-like scrub walla-bies that infested it), another the nearby islands, and the third the 'river of swans' discovered by the Dutchman de Vlamingh in January 1697.

Hamelin's instructions to Heirisson, commanding the last party, were explicit – there was to be no repeat of the mistakes made in Géographe Bay. He was to follow the river as far as he could go in the longboat, and report on the area's suitability as a possible port-of-call for Euro-pean vessels; Bailly the mineralogist would accompany him to assess its natural resources. The greatest care must be taken in any contact with the natives – on no account were firearms to be used, except in the most dire emergency. 'The French government's intention, and a desire close to my own heart,' Hamelin warned him, 'is not to shed the blood of the natives at any of the places we visit.'

Wednesday, 17 June. Heirisson and his party leave the ship before dawn. The river's mouth is obstructed by a rocky bar, on which the boat grounds three times before they can manoeuvre it across. Prodigious numbers of pelicans cruise on the lagoons beyond, large flocks of elegant, brightly

Top: *A view of the Dutch Fort Concordia at Kupang, Timor, by C.-A. Lesueur. [Kerry Stokes Collection]*

Above left: *Malay slave girl, Kupang, by Nicolas-Martin Petit. [Collection Lesueur, Muséum d'Histoire Naturelle, Le Havre]*

Above: Géographe *and* Naturaliste *weather the storm in Géographe Bay, south-western Australia. Painting by Neville Weston. [Author's collection]*

Above: 'King George's Sound', by William Westall, circa 1801. [By permission of the National Library of Australia]

Left: Vancouver Peninsula, King George Sound. [Photo by David Moore]

Below: 'Port Lincoln ... from south', by Westall, 1835. [Rex Nan Kivell Collection; by permission of the National Library of Australia]

coloured parrots dart through the trees. Cliffs on both banks of the river form a great circular wall covered with vegetation. Doubling a low point jutting far out into the lake, they camp for the night at the base of a precipitous hill on the right bank (Mount Eliza), protected from attack by the cliffs behind. From the summit, Bailly enjoys a beautiful spectacle:

> on one side one sees the upper course of the river, which ascends towards a plateau of distant mountains, and on the other one can follow its lower course as far as the ocean beach. Both banks are almost everywhere covered by beautiful forests, which stretch far back into the interior of the country.[11]

Next morning they set out early, only to run aground on a bank of soft and sticky mud. With straining muscles the sailors haul the boat over the obstacle before relaunching it in deeper water. On the far side they see their first black swans – majestic birds that make an appetising dinner; Bailly notes that after death their beaks lose their lustrous red colour, turning a dull black. Following a small tributary, Heirisson and Midshipman Moreau discover a human footprint of extraordinary size – the only sign that the country is inhabited. Camping overnight on the riverbank, they continue their voyage upriver, hopeful of reaching its source. After rowing all day they realise that the mountains are still a long way off; the riverbed is narrowing, and their provisions are running short. Heirisson decides to abandon the survey, and next morning begins the return journey.

Sunday morning, 21 June. Reaching the shallows that had hindered their ascent, Heirisson decides to avoid the obstacle by hugging the right bank of the stream. Almost at once they find themselves aground once more, and all their efforts to move the boat fail. Forced to build a makeshift raft, they load it with the boat's anchor, water casks, and other heavy objects, and try again. Everyone gets into the water, and pushing with all their strength succeed at last in refloating the boat.

Their joy is short-lived. The boat grounds again on another bank just 6 inches below the surface. Slipping and falling in the muddy water they toil for hours trying to clear it; in the end they are rescued by a sudden wind gust springing up at the critical moment. After thirteen hours all are worn out, exhausted by exertion and fatigue. Scarcely enough rations remain for a single meal, and the ship is at least twenty-four hours away. Soaked, hungry, and tired almost to death, they prepare for a few hours' uncomfortable rest:

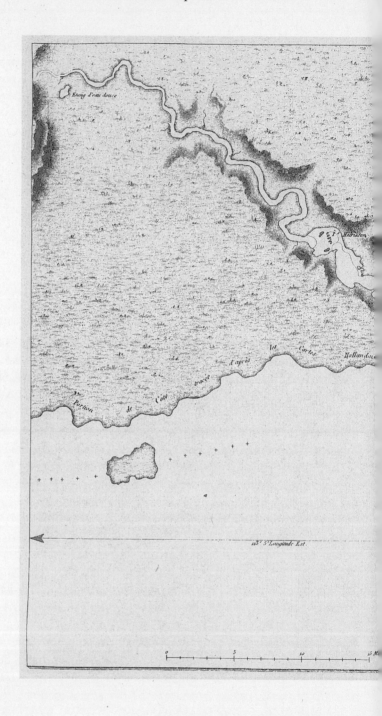

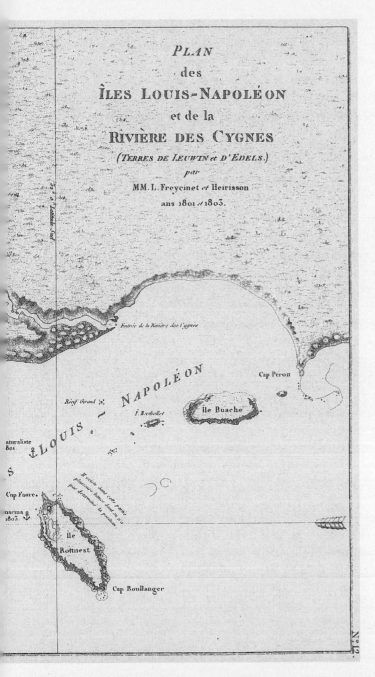

The Swan River, 1801 and 1803. Chart by Freycinet and Heirisson.
[By courtesy of Scotch College, Hawthorn]

We were about to repair to the shore to dry ourselves, and to recover our strength, when all at once a fearful noise[12] froze us with terror; it was like the roaring of a bull, but much louder, and appeared to come from the neighbouring reefs. At this sound we lost all our desire to land, and although shivering with cold, we preferred to pass the night on the water without supper, and powerless to close our eyes on account of the rain and the cold.[13]

The following day they reach the river's mouth, and go ashore to warm themselves and eat the last scraps of food before returning to the ship.

Hamelin's other boat parties fared even worse. Milius, with the zoologist Levillain and nine men, left in the longboat to survey the islands to the south of the anchorage. After two days at Rottnest he sailed for two islands lying to the south-east, and set up his instruments on the smaller of these (Garden Island). Before he could complete the survey, however, squally northerlies decided him to return to the ship, but he could make little headway into the teeth of the gale. Towards midnight the small craft dropped anchor off the mouth of the Swan River; the rising seas broke over the boat, the rain never ceased, and all aboard bailed furiously, using hats, mess tins, anything that came to hand. The hours dragged on, the boat rocking 'like a crazed pendulum', the crew miserably awaiting the dawn

When it came, there was no relief. The wind blew as strongly as before, and as they raised the sail a gust snapped the longboat's mast and they were swept landwards. Caught in the surf the boat was swamped and its occupants cast onto the beach – fortunately onto soft sand. Though all managed to scramble to safety without serious injury, they were without food and lacked the means to light a fire – and, more worryingly, any tools or materials to repair the boat. For the second time in ten days, Milius found himself marooned on an isolated beach, cold, soaked through, and hungry, and in imminent danger (as he thought) of an attack by savages.

The third party, under Louis de Freycinet, also came to grief. Sent with the geographer Faure to survey Rottnest Island, his dinghy capsized in the surf as he was about to land, and sank with the loss of almost all their provisions. For three days the little party explored the island on foot, discovering the salt lakes in the interior and also the remains of a wrecked ship, perhaps a sealer, on the rocks of the north-east coast.

With only the dinghy's sails for shelter from the driving rain they camped in the low hills behind the shore, living off the flesh of seals too slow and clumsy to escape the sailors' clubs.

Hamelin, meanwhile, waited for his boat parties to return, his anxiety growing as the days passed. All three were overdue. During the night he ordered rockets fired at regular intervals, and lights to be kept burning at the masthead – partly as a signal for the *Géographe*, if she should arrive in the darkness, but mostly out of concern for his men. 'At this moment,' he wrote in his journal, 'I would give half – what am I saying? – all that I possess on board to see these brave men return.'[14]

Separation – the *Géographe* in Shark Bay

On board the Géographe, *Thursday, 11 June 1801.*

> *The weather was even worse than the day before. We spent the day tacking to keep our course against the sea, the wind, and the currents, but despite our efforts were rapidly carried South-East... During the morning the winds changed to North-West and varied to West-North-West. They were accompanied by heavy rain and such strong squalls, that I expected every moment to lose some sails ... The sea was so rough, that when we rolled, it washed over the leeward gangways and we had all the trouble in the world to preserve ourselves from the falls which the ship's rapid movements caused us. I, for my part, had a fall, the effects of which I am afraid I shall feel for a long time. [Baudin, 1974, p. 186]*

Baudin made out to sea for deep water and relative safety. Driven south before the wind, he found himself a second time off Cape Leeuwin, and it was three days before he again entered Géographe Bay. There was no sign of the *Naturaliste*, and he concluded that Hamelin must have sailed for Rottnest Island, the agreed rendezvous.

The storm had cost them most of their live cattle – a critical loss, for the meagre rations of salt meat taken aboard at Isle de France had already rotted. A short-lived calm gave Péron the chance to throw the drag overboard; this netted several new treasures for the naturalists, among them a bright-purple sponge, which, at the slightest pressure, yielded a liquid of the same colour, resistant to the action of air when spread on various substances. On the afternoon of 18 June they arrived off Rottnest, but the weather was too uncertain to approach closer than 6 or 7 leagues:

I therefore went on the other tack [noted Baudin] in order to be able, before nightfall, to round with ease the point of a rather deep bay on our beam, the northern tip of which no doubt ends near Rottnest Island and perhaps forms the entrance to the Swan River ... In this region the sea was extremely high and swelling, which, for me, was an extra reason for not going any further North in such threatening weather.[15]

A falling barometer and gusting winds, with more heavy rain, again drove the *Géographe* out to sea. When the weather eased, Baudin set course for Shark Bay, making no attempt to find out whether Hamelin was at the anchorage. The decision, says Péron, shocked his officers:

As this was the first rendezvous ... we had always intended anchoring, either to meet or wait there for news of our consort, concerning which we were very anxious, as her being so bad a sailor subjected her the more to the dangers we had experienced in Geography [*sic*] Bay. How great then was our consternation and surprize, when at the very time that we first discovered this island, we heard our commander give orders to sail to the bay of Seadogs [Shark Bay] ... From this time we despaired of seeing any more of the Naturalist during the rest of the voyage ...[16]

Again Péron was writing with hindsight; at the time his companions saw the matter differently. Engineer Ronsard recorded in his journal: 'The weather was not right for exploring the coasts, and we could not wait for fine weather ... the season was too advanced. We had to sail outside Rottnest Island.'

Baudin also gave weather conditions as his sole reason for abandoning the rendezvous: 'Seeing no change in the weather, I decided to head North-West by North to reach a more suitable latitude for navigation and our work.' The decision was his alone, made without consultation. Possibly it stemmed from a cumulative 'campaign fatigue' – a loss of balance and judgement after a lifetime at sea, to which not even such a superlative seaman as James Cook remained immune.[17]

Keeping far out to sea the *Géographe* sailed northwards, not sighting land until 22 June, north of the Murchison River. The coast here appeared forbidding, towering cliffs banded horizontally in rust, brown and black, and Baudin continued north past Dirk Hartog Island. He anchored at last off the northern end of a long narrow island (Bernier Island), at the northern entrance to Shark Bay, on 27 June. The Dutch chart proved to be very accurate: 'Of all our maps,' he wrote 'it is the

most exact and the best drawn up.' Next morning two boats went
ashore, with Baudin and all the naturalists aboard.

The boats put off at 8 a.m., but take two hours to reach land because
none of the oarsmen chosen can row. Baudin and Maugé give chase to
the island's birds, while Péron, Riedlé and the others head inland to
examine the plant life, with orders to return by five o'clock. The sailors
cast the dragnet for the evening meal.

The savants *turn hunters. Finding that the small and beautiful band-*
ed kangaroos inhabiting the island have built narrow covered ways
through the dense undergrowth, they beat the thickets with long sticks
while others post themselves at the exits to these pathways: 'thus the
animals flying through the usual places of retreat became the victims
of enemies inevitable'. Their flesh is fine-flavoured, as tasty as the wild
rabbit but more aromatic.

Leaving the others to their diversions, Péron treks alone into the wil-
derness. He is heedless of time; the sun is setting when he turns back
towards the landing-place. Night comes quickly, the moon has not yet
risen, and he misses his way among the downs and brambles. Loaded
with specimens, he walks at a great pace until about eight, only to find
himself on the western shore, at the far side from his starting point.
Exhausted, he falls to the earth 'overpowered by weariness and empti-
ness, not having either eaten or drunk since the morning, and having
walked the whole of the day'.

Drawing on his last reserves, the naturalist drags himself to his
feet, and in the darkness retraces his steps to the east. He stumbles on
through the thickets until, towards eleven o'clock, 'entirely overcome
by fatigue and perspiring at every pore', he again sinks down, 'resolved
to pass the rest of the night on this spot, even though I might perish in
this frightful desert'.[18]

Péron awoke before dawn, half-frozen and barely able to move his numbed
limbs. As the sky lightened, the report of a musket renewed his spirits,
and soon he was back among friends. Ignoring Baudin's orders to return
to the ship at moonrise (about 10 p.m.), Sublieutenant Picquet and his
boat's crew had spent the night ashore and set off at dawn, 'determined
not to quit the island till they had lost all hopes of ever seeing me again'.

With a stiff breeze blowing, it was past midday when Picquet's boat
reached the ship. Baudin had spent a restless night, guessing from a

great fire blazing on shore that the naturalist had become lost, and called Péron into his cabin:

> he presented himself to me in the most pitiable state. He had not eaten for 24 hours and was quite worn out, having wandered about all night until, dropping exhausted, he fell into a sound sleep at the foot of a shrub … Since he is to make me a report, I shall refrain here from saying what I could say. However, I firmly promised him that when he went ashore again, I would send someone with him who would keep him constantly in sight and be responsible to me for getting him to the place of departure [on] time.[19]

Next day the *Géographe*, all sails set and sounding continuously, sailed further into the bay. Whales swarmed about the ship, providing entertainment for all on board; some were so friendly that they played alongside, 'and it would have been easy to fire on them, if a shot could have killed them'. Baudin anchored briefly off Péron Peninsula in the centre of the bay, before returning to the east coast of Bernier Island on 5 July. Here he set up his tents, and while the astronomers busied themselves with their observations, the officers surveyed the island and Dorre Island to the south.

Though critical of the latter's efforts – 'the pleasure seekers, and I have plenty of them, set off kangaroo hunting and killed about twenty' – he was full of praise for the scientists. His friend Riedlé, 'to whom every moment was precious', daily returned with specimens of new plants. Maugé the zoologist likewise made the most of his time ashore: 'he was supplied with beautiful birds; and insects and shell-fish occupied him when he saw nothing more to work on'. Even Péron came in for a backhanded compliment:

> Citizen Péron, whose extreme enthusiasm leads him to undertake everything without thought for the dangers to which he is exposing himself, went on a visit to the western part of the island … He found nothing of interest on his excursion, so, in order not to have undertaken it in vain, he went down to the beach to pick up shells there. At first he amused himself at the expense of the crabs, and taking off his shirt, tied up about fifty of them in it. This willingness of his deserved a better reward than awaited it. For, wanting to climb some rocks, where the sea appeared not to break too roughly, he was knocked head over heels by a wave which carried off most of the beautiful shells that he claims to have found.[20]

They remained in Shark Bay for more than a fortnight. Baudin prepared to sail on 11 July, but with the weather worsening once more it was midday on the 13th before they gained the open sea. The passage through the North Channel was daunting, the rain so heavy that it was 'as if we had been in total darkness'. By midday the land was lost to view, and the captain

> went to have some rest, for I had been on deck for 26 hours without a break. I noticed again at that stage that not one of the officers appeared on deck, except for the person on duty, and that our frequent tacks did not prevent their sleeping more soundly than [if] we had been in the most comfortable position.[21]

The weather cleared next day, and Baudin set his course north along the coast for North-West Cape and the Dampier Archipelago. It was 14 July, anniversary of the storming of the Bastille fortress in Paris.

Forty-nine days – the *Naturaliste* in Shark Bay

Rottnest Island, Sunday, 28 June 1801. Before sailing, Captain Hamelin leaves a note in a bottle for Commandant Baudin:

> *It was with regret that on the night of 8th June I lost sight of your light in Géographe Bay. I feared you had lost your corvette, especially the following day. I had trouble with my anchor, etc. I therefore decided to make for our first rendezvous (Rottnest Island). I arrived safely and Heirisson went 18 leagues up the Swan River. At 2 p.m. on 18th June we saw the* Géographe. *We waited ten days for you. We must be in Timor by 2nd October 1801. I have a fear we'll not meet you. We suffered a lot of wind damage. [Marchant, 1982, p. 172-173]*

The northerly gales caused grave problems for the *Naturaliste*'s crew as well as for the shore parties. Although three anchors were down, the ship was driven several cable's lengths towards the coast, and Hamelin sought a more protected anchorage closer to the mainland. That night, seeing the glow of a fire on the beach, he worried whether it had been lit by Heirisson's party or by natives – in fact it was Milius and his men. 'May the sympathetic reader judge of my concern over our absent people,' he wrote in his journal:

> I was certain that [Freycinet's] small boat was wrecked and its crew in the rain for 36 hours; it was almost certain that the large boat [Heirisson's]

was in the same position ... and the cruel doubt about whether a single man survives from the longboat [Milius], on which I had not the least information for 50 hours.[22]

His worries were soon relieved. The carpenter, sent to Rottnest with supplies on the ship's remaining small boat, was able to repair the wrecked dinghy, and Freycinet, Faure, and the five men with them returned on 20 June in good spirits, having lived 'for three days with the seals, the spray and the rain'. The following day another rescue party was sent to investigate the fire seen on the beach – 'to communicate with the savages', Milius commented, 'because they thought us already lost'.

Milius, meanwhile, had organised several excursions to search for food, one of which had found a species of almond about the size of a walnut. When roasted in the ashes they tasted like chestnuts, and the hungry men tucked in; almost at once those who ate them were attacked 'by alarming and painful vomiting', and thought themselves mortally ill. Luckily the symptoms slowly eased, and the victims, though weak and wan, were able to swallow the food brought by their rescuers.

Milius, one of the worst affected, sent the relief boat's crew back to the ship to inform Hamelin of their dire situation. It returned the next day, bringing food, a change of clothing for Milius, and most importantly some carpenters to repair the longboat. In twenty-four hours it was seaworthy, and after five days of hunger, thirst, and cold, Milius and his men boarded the *Naturaliste*; they found Heirisson and his crew also safely back on board, though 'cruelly harassed with fatigue, and almost famished'. Despite the dangers all had encountered, not a single man had been lost from any of the boat parties.

Hamelin's anxieties now focussed on the safety of the *Géographe*, and grew as each day passed. 'What could be the reason that prevents her from coming to the rendezvous?' he wrote. 'Would Commandant Baudin have gone to the second rendezvous at Shark Bay? – and why?' He waited another six days, then headed north along the coast; expecting to rejoin his consort at Shark Bay, he limited himself to 'making such surveys of the coast as were necessary to correct the Dutch manuscript chart ... which in many respects we discovered to be very erroneous'.

The Abrolhos group of islands were sighted on 8 July. Hamelin planned to sail between them and the mainland, but contrary winds drove him seaward. Arriving off Dirk Hartog's Island a week later, he coasted it at a distance of 2 miles before anchoring, in the strait (Naturaliste Channel) separating it from Dorre Island to the north, on 17 July – just three days after Baudin had left Shark Bay via the Géographe Channel. Parties sent ashore to find whether the *Géographe* had left any sign of her passage returned empty-handed. (Baudin had indeed left a message, but on a remote island at the northern entrance to the bay, and it was not found).

Convinced that the flagship had not visited the bay, Hamelin called the officers together to discuss their next move. Most felt they should wait for eight to ten days, then continue the voyage if the *Géographe* had still not arrived. Hamelin disagreed, as his instructions from Baudin 'left no room for deliberation' – they must 'wait in Shark Bay, till he should come there and join us'. The decision, in Freycinet's view, doomed them 'to waste our time on these desolate shores …'.

Three men are landed at the northern end of Dirk Hartog's Island to set up signals for the Géographe. *Scrambling through sand and scrub they make their way to the summit of a rocky cape at its north-west tip. Here they find a rotted oaken post; nearby, half-buried in the sand, lies a flattened pewter plate, much corroded, bearing a roughly carved inscription in some foreign language. The boat's coxswain carries it back to the ship.*

Hamelin recognises the writing as Dutch, and with the aid of a Dutch-speaking crewman deciphers the worn inscription – in fact there are two, one dated 1616 and the second 1697. The first reads:

> On the 25th of October arrived here the ship Eendraght, *of Amsterdam:*
> *first supercargo Gilles Miebais Van Luck; captain Dirck-Hartighs, of Amsterdam;*
> *She again set sail on the 27th of the same month. Bantum was second supercargo;*
> *Janstins first pilot; Pieter Van Bu …; in the year 1616.*

The second records the arrival on 4 February 1697 of the ship Geelvinck *of Amsterdam, Commander Willem de Vlamingh, and two support vessels. It ends: 'Departed from hence with our ships and sailed again from the southern shores, being bound to Batavia.' Vlamingh has had the plate engraved and left as a record of his visit.*[23]

More at home on the quarterdeck of a warship than a vessel of discovery, Hamelin is nonetheless affected by the find: it would be 'sacrilege to carry away this plate, which had been respected for near two centuries by time, and by all the navigators who might have visited these shores'. Overruling the protests of Freycinet, who argues so valuable a trophy should be returned to Europe, he carries the plate ashore and erects it at the exact spot where it was found, on Cape Inscription. Another post, part of a studding-sail boom, is driven into the sand nearby with a second plate attached, recording the name of Hamelin's ship and the date of their arrival:

REPUBL^que FRA^nce

Expedition de decouvertes

sous les ordres de Capt^ne de V^au

BAUDIN

La Corvette Le Naturaliste

Capt^ne HAMELIN

27 messidor an 9e

[16 July 1801]

Hamelin and his men remained in Shark Bay for forty-nine days, from 17 July to 4 September, surveying, mapping, and collecting specimens of natural history. By good fortune their work neatly complemented Baudin's investigations, which had concentrated on Bernier and Dorre Islands and the bay's northern reaches. The *Naturaliste* found a safe anchorage near the northern tip of Péron Peninsula, in the centre of the bay; tents were set up just south of the present Cape Péron, one serving as a hospital and others housing the observatory.

The shortage of drinking water was now acute, for none had been found on the islands or the arid peninsula. The alembic still (salt-water purifier) sent from France was taken ashore, and was soon producing forty to fifty pots a day, more than enough to meet the needs of the thirty or so men at the encampment. The longboat damaged in the surf near the Swan River was beached and the broken timbers replaced. For those on board, the schools of whales sporting on either side provided a fascinating spectacle; swimming past in pairs, the creatures would make sudden amazing leaps and synchronised turns. Far less welcome were the swarming sharks, which never left the ship:

> We could not make a trip ashore without being escorted by a hundred or more of these ravenous beasts. It would have been very dangerous to fall into the water or to bathe, since they all seemed very hungry.[24]

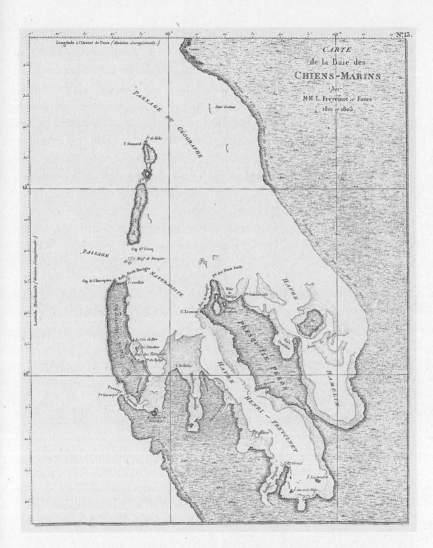

Shark Bay, 1801 and 1803. Chart by Freycinet and Faure.
[Courtesy of Scotch College, Hawthorn]

Hamelin dispatched boat parties to explore the waters lying east and west of the peninsula. In fifteen days Louis de Freycinet charted the whole of the western shore, from the tip of Dirk Hartog's Island round the gulf to Cape Péron, and found that the so-called Middle Island was in fact a peninsula; he later named the gulf for his brother Henri. The parts of Shark Bay to the east of the peninsula were surveyed in the same thorough fashion by Midshipman Moreau and the geographer Faure, and given the name Hamelin Pool. The large island at the entrance to the pool – named after Faure – yielded a rich harvest of turtles, happily taken on board to supplement the rations.

At length, Hamelin could no longer withstand the urging of his officers. The shore parties were recalled, the boats embarked, and he sailed directly for Timor to keep the next rendezvous, set for mid-October.

'So dangerous a navigation'

Off de Witt's Land, July to August 1801.

> *All that part of New Holland, which from the cape N.W. extends as far as the cape N. of this vast continent ... was first discovered by William de Witt, a Dutch navigator, who gave it his name; but the precise time is not generally ascertained: some make it as far back as the year 1616; others bring it to the year 1623, or even to 1628 ... In 1699 Dampier appeared on these shores; but repelled by the same obstacles that so soon multiplied around us, he was compelled to quit them ... From this epoch, a century had passed away since any European vessel had appeared in these seas, and we shall soon be able to judge, that it was not without reason, that voyagers had so long abandoned so dangerous a navigation. [Péron, 1809, p. 102]*

The *Géographe* rounded North-West Cape on 22 July, and began a survey of the continent's dangerous and inhospitable north-west coast. The work took nearly a month, as Baudin cautiously negotiated the multitude of uncharted offshore islands, coral reefs, banks and shoals barring his passage. At 16 feet, the ship's draught was too deep for the shallow coastal shelf, while the fluctuating 30-foot tides and swirling currents made navigation still more dangerous. Mostly the ship kept at such a distance from shore that it proved impossible to distinguish island from mainland. Shipwreck on this coast risked a lingering and unpleasant death for all on board.

Nonetheless, numerous islands and coastal features were discovered and mapped, adding to and correcting the rough outlines of the old Dutch charts. Many of the expedition's names still remain in use – among them the Forestier Archipelago, capes Borda, Depuy and Leveque, the Lacepede Islands and the Bonaparte Archipelago, named 'in honour of the august protector of our expedition'.

Confined to the ship, the naturalists kept busy observing the rich marine life of the region. The net dragged up enormous jellyfish, many of them $2\frac{1}{2}$ feet in diameter and weighing 50 pounds or more. Sharks, whales and tortoises swarmed around the vessel, and two new species of sea snake were observed, one of them from 8 to 10 feet long.

One landing was made, on Depuch Island in the Forestier Archipelago, but, says Péron, 'though permission was earnestly entreated, none of the naturalists were permitted to go on shore'. After the last episode in Shark Bay, Baudin's impatience with Péron's continuing indiscretions must have reached its limit. Depuch, the mineralogist, was the only scientist allowed to accompany Ronsard and the shore party.

The island, unlike others they had seen, appeared 'entirely volcanic; prisms of basalt, generally with five sides, and heaped one on another … constituted the entire mass of the soil'. At one point they were arranged in a 'basaltic pavement, similar to those of the famous Giant's Causeway'. The black colour of the rocks added to 'the melancholy aspect and monotony of this little island', and made walking uncomfortable. According to Péron,

> M. Ronsard supposed, from the general conformation and colour of part of the neighbouring continent, that it was of a similar nature, and also volcanic. This would have been so much the more important to ascertain, as till that time we had seen nothing of the kind on the shores of New Holland … but our commander paying little attention to a phenomenon which belongs essentially to the geography of New Holland, gave orders for us to continue our course.[25]

Baudin continued the survey for a further three weeks, but the ship's supplies of firewood and water were fast becoming exhausted; no-one on board had eaten fresh food for more than a month, and several of the crew had come down with scurvy. The naturalists, for their part,

> who had not had fresh food for more than a month, were sighing for the call at Timor, and although they said nothing to me about it, they murmured amongst themselves in such a way that their conversations reached me.[26]

Furthermore, without a longboat he lacked the means to reconnoitre close in along the coast. On 19 August, 'seeing that I could do nothing, I decided to head North to the anchorage that everybody hankered after'. The news so stirred the sick men 'that several of them found the strength to come up on deck in order to make sure that it was really true'.

Two days later, the high green mountains of Timor came into view.

Kupang

On board the Géographe, *Kupang Bay, Friday, 21 August 1801.*

Probably it would be difficult to find a more beautiful and picturesque situation than that which we then enjoyed; surrounded on all sides by the land, we seemed as if in the middle of a beautiful lake, on every side clothed with the richest colours ... [Wherever] we turned our eyes, the picture of the most amazing fecundity seemed to be renewed with additional charm and interest. How great was the contrast between the beauties of such a situation, and the sterile and monotonous neighbouring shores of north-west New Holland. [Péron, 1809, p. 113]

The *Géographe* cast anchor below Fort Concordia, which guarded the entrance to Kupang Bay, on 22 August. The little town contained no more than eighty houses, mostly inhabited by Malays and Chinese; the finest of them belonged to Governor Lofstett, Dutch East India Company officials, and Malay dignitaries. The Dutch, political allies of the French Republic, promptly provided two large houses for use by the expedition. Baudin, with Bernier, Faure and the two artists, occupied one – 'a superb dwelling which, if transported to a European colony, would look graceful there' – and the naturalists the other. A large airy building owned by the company was made ready as a hospital to receive the scurvy cases.

Town and fort both showed the scars of war. British troops had taken Kupang a year or two before, easily overcoming the handful of Dutch defenders, but had failed to pacify the tribes in the interior. The port pilot, a Frenchman in the company's service, related the story to Péron:

Fort Concord, into which [the invaders] had retreated, having been taken by assault, 70 or 80 Englishmen had been cut in pieces and eaten by the Malays; and from that moment the most implacable hatred had subsisted

Right: *Flinders Monument, Stamford Hill, Port Lincoln, South Australia. First erected by Sir John and Lady Franklin in 1844, it was rebuilt in 1867. [Photo by Colin Gill]*

Below: *'View on the north side of Kangaroo Island', by William Westall, engraved by William Woolnoth, 1814. [In* Views of Australian Scenery, Rex Nan Kivell Collection; *by permission of the National Library of Australia]*

Left: 'Kangaroo Island, sailors & servants', by William Westall, 1802. [By permission of the National Library of Australia]

Below: Mount Brown, in the Flinders Ranges, South Australia. Flinders named the mountain after the naturalist Robert Brown. [Photo by R. Marshall]

among the whole Malay nation towards the English, and towards all that could remind them of these conquerors.[27]

The British survivors had retreated to their ships under cover of darkness and sailed away. Later the Dutch had returned, restored the fort, and made their peace with the local rajahs, whose rights they respected. The latter, in return, 'maintained a submissive attitude [among] their compatriots, who appeared to have a natural inclination to be restless as well as a likelihood to revolt'.

Baudin's residence belongs to the widow of a former governor, Madame van Esten, and is exquisitely appointed. She is a Malay by birth, a rajah's daughter, and sole heir to her late husband's wealth. She refuses to accept any rent for the period of his stay – she is 'always delighted to be of help to foreigners, and to Frenchmen above all'.

The commandant, accompanied by Governor Lofstett, an interpreter, and several of his own staff, hastens to call on Mme van Esten to thank her for her kindness. Her country estate is situated by the seashore, amid 'a most delightful country watered by running streams on every side'. On arrival, Baudin is astonished to find more than 100 neatly dressed slaves, male and female, drawn up on the galleried veranda awaiting them. The lady of the house, elegantly clothed in a rich and beautiful Malay dress, stands at the head of the steps to receive them – the governor has sent notice of their coming and she wishes to be seen at her best. Despite her noble bearing, she is by no means reserved.

Within the mansion, tea and coffee are served, together with fruits, pastries, preserves and all varieties of sweetmeats, prepared under her roof and offered on finely wrought silver plates by about twenty young girls; all are clad in cotton sarongs and white bodices, with their long black hair platted and folded round the head, and most are very pretty. Their way of serving this collation, their graceful motions, the regular ceremonials they perform in quick succession, and their profound silence enchant the visitors. The gathering is a 'lovely sight', and Baudin regrets neither artist is present, 'for undoubtedly he would have made a fine picture of it'.

It is past nine when they prepare to leave, but the Commandant and his companions can see as clearly as by day. In the garden, red-cloaked slaves are drawn up in readiness for their departure – each carries a wax or candlewood torch, coated with resin to produce a wide arc of light; some walk ahead and others behind the visitors to light their way on the path back to town. The dazzling procession, thinks Baudin, is not unlike a scene at the Paris Opera – the visitors descending with Orpheus to the underworld, and the slaves resembling the devils of the opera. The governor escorts Baudin home, and they bid each other a friendly goodnight.

The opportunities for rest and relaxation ashore did nothing to calm the festering tensions between Baudin and his subordinates. Any hope of settling the problems ended with further confrontations at Kupang, which saw two of his officers discharged and left behind in Dutch hands – one under arrest and the other in hospital, wounded in a duel.

At Géographe Bay, at Shark Bay, and at sea the commandant had repeatedly reprimanded Sublieutenant Picquet for slackness, disobedience, and 'insouciance' on duty. Shrugging off the rebukes, Picquet 'continued to behave badly on board', and Baudin relieved him of duty. On arrival at Kupang, he wrote to the young officer, ordering him to find lodgings ashore since 'you are no longer a member of my staff'; he was to await the *Naturaliste*'s arrival and place himself under Captain Hamelin's orders, 'if it still suits [him] to take you on board'.[28] Moments later Picquet, 'frothing with rage', burst into the cabin, crying out '*Foutre*, sir, it's absurd that you should write to me like this!' Baudin ordered him out and shut the door, determining to ask the governor to hold him under arrest in the fort.

The scene was repeated two days later, this time in Baudin's house ashore. Picquet arrived in uniform, his sword at his side, and asked to speak with the commandant in private; once inside, he again demanded an explanation for his treatment. Baudin gazed at him without speaking, further infuriating the young man, then asked him to repeat what he had said.

'*Foutre*, if you haven't understood, follow me,' Picquet shouted, drawing his sword. 'If you dare come near me, you'll soon see what I use this for!' Running onto the veranda, Baudin ordered the first person he saw to bring some soldiers from the fort to arrest the sublieutenant. Still trembling with rage, Picquet stood cursing and swearing for two or three minutes, then 'vanished in a flash'.

Baudin could no longer avoid decisive action, and ordered his second captain on the *Géographe*, Sainte-Croix Le Bas, to place Picquet under arrest:

When the Government learns of this, it will no doubt blame me for not having had him shot then and there ... But I am determined to endure this reproach, not on his account (for he deserves no consideration), but on his family's, for I know it to be respectable.[29]

Péron, the expedition's 'anthropologist', has looked forward to meeting the Malays, descendants of 'the ancient conquerors of the grand archipelago of Asia, [who] still preserve their original character of independence, boldness, and ferocity ...'. He and his friends, on an excursion into the countryside, are invited to enter a Malay dwelling for refreshments. The head of the household, a man of substance, welcomes them warmly: 'Doudou, doudou, baé oran di France' ('Sit down, sit down, good men of France').

Thirsty after their walk, the visitors ask for fresh coconut milk (safer to drink than the water from the stream flowing past the house). In a trice a young man climbs a lofty tree overhanging the courtyard, gathers four nuts, and gripping two between his teeth and with the others in one hand descends as agilely as he had ascended. Meanwhile the men of the household observe their visitors keenly: 'our physiognomy seemed to please them, and our youth to interest them much in our favour'. 'Baé oran mouda' ('good young men') passes in a kind of whisper from one to another.

Inquisitive as always, Péron takes up a sagaie, *or spear, to examine it, then asks one of the Malays to demonstrate its use. The man, clearly reminded of the bloody events following the invasion, brandishes the spear, crying out in a furious passion: 'Oran ingress, oran bounou!' ('Englishmen assassins!'), and 'Oran djahat!' ('wicked men!'). Snatching a coconut, he places it on the point of the spear, his 'unequivocal gestures' leaving the visitors in no doubt that having cut off the heads of the English, they had carried them about on the points of their spears; the war dance had then been danced round them, and afterwards having cut in pieces the bodies of these unfortunate Europeans, they had then devoured them.*

The grisly demonstration gives way to a scene of charming domesticity, as their host, 'more and more pleased with his new acquaintances', invites the young women of the house to join them. At the arrival of the Frenchmen they have retreated to an inner room where, 'more curious than timid', they peep at the foreigners through gaps in the bamboo screen. The visitors have 'naturally directed our eyes often towards the harem'. There are five women, the eldest no more than twenty-five; all

are well proportioned, easy in their bearing, with fine features 'express-
ing that affectionate softness ... and beauty belonging to the women of
these shores'.

The sight of so many young strangers makes a lively impression on
the women, but, says Péron, 'they soon overcame their natural timidity
to receive the different presents we offered them'. No longer fearing 'to
lift up their large black eyes to regard us with kindness', each in turn
offers a small present as their visitors prepare to depart. The French-
men regretfully take their leave and return to the town.

In the hospital, the change in diet – fresh fruit, meat and vegetables in
place of the 'detestable' victuals brought from Isle de France – quickly
restored the scurvy patients to health, but on board ship 'all the crew
were incommoded by the water'. A week after their arrival in the port,
two more men were admitted to the infirmary, one with dysentery and
the second with a high fever. Others soon followed, and within three
weeks, according to Péron, 'eighteen men were confined to their beds,
all severely and dangerously ill with a most cruel dysentery'. Depuch,
Maugé, and the 'good and active' Riedlé were among them. Baudin
came down with a severe intermittent fever (probably malaria), and
was treated by Dr Lharidon at his house; for a while he seemed unlikely to recover.

The artist Lesueur narrowly escaped an agonising death from snake-
bite. Bitten in the heel while chasing a troop of monkeys on the river
bank, he first felt a numbness in the whole of his leg; very quickly it
became stiff and swollen, and he could scarcely walk:

> To retard the action of the venom, he bound his thigh tight round above
> the knee, but this ligature had little effect; the thigh itself swelled to such
> a degree, that it was as much as my poor friend could do to reach the
> house ... there he laid himself down on his bed, overcome by fatigue
> and pain, and already experiencing all the symptoms of a violent fever.[30]

Fortunately Lharidon was close by. He hurried to his friend's bed-
side and cauterised the bite very deeply; next, 'applying a compress,
wetted with ammoniac, he gave a strong dose of the same drug to the
sick man, recommending him to keep perfectly still and quiet'. The
treatment brought on a profuse sweat, and the pain soon abated; 'and
in a few days M. Lesueur felt no more of the wound, except a stiffness
and difficulty of bending his knee, which remained a long time ...'.

'These destructive shores'

Kupang, Monday, 21 September 1801.

> *Our anxiety for the fate of the* Naturaliste *increased every day ... This cruel uncertainty grieved us all; we began to lose all hope, and to despair of ever again seeing our friends, when on the 21st of September, in the morning, a signal was made that the* Naturaliste *was entering the bay. The joy was general, and we were soon among our companions, who not having found us at the two rendezvous, were not themselves without great anxiety on our account. [Péron, 1809, p. 131]*

At midday the *Géographe*'s boat pulled alongside the newcomer, bringing orders from Baudin and the unwelcome news that many of their friends ashore were dangerously ill; the commandant was so weakened by fever that he could not leave his bed, and Captain Le Bas was in temporary command of the ship. The *Naturaliste*'s crew was in far better shape; only two men aboard were sick, both with early symptoms of scurvy – the result, wrote Péron, 'of their long stay on shore at different places, and the particular care [Captain Hamelin and surgeon Bellefin] had taken of the men's health'.

Hamelin hastened ashore to report to the commandant. Shivering with fever, and worried for the expedition's future, Baudin told his second-in-command that, in the event of his death, Hamelin should move to the *Géographe* and assume the command. He added, mysteriously, that

> he would find in the place I designated a particular instruction to which he would have to conform strictly, or he would be personally responsible to the Government and to posterity for departing from it.[31]

The *Naturaliste*'s officers and scientists moved ashore while the ship underwent a much-needed refit. Milius took on the task of replenishing her supplies, aided by those crew members who spoke enough Malay to be understood. He soon found that the laws of supply and demand operated as effectively in this remote outpost as in Europe:

> On our arrival in Timor we were paying no more than 12 pins each for chickens; a few days later, you had to give a six farthing knife, and by the time we left, the Malays were demanding double this amount, so that poultry could cost two or three sous apiece. Turkeys, ducks and geese

were sold in proportion to this price. We bought many pigs that we paid for with iron pots, or with saws, hatchets, adzes, etc.[32]

One day, while bartering with the local merchants, the French sailors began talking of their nation's leader, bragging about the first consul's victories 'and the superiority of his genius over all others'. Their stories

> seemed to make a very strong impression, and as a way of expressing their admiration, they asked us if General Bonaparte was the son of a crocodile [a beast they held in great veneration]. This question pleased us enormously, as it showed that they had a very high opinion of our First Consul, for according to them the soul of such a great man could belong to no other than a crocodile.[33]

Milius finds another type of commerce is also as widespread here as in more civilised nations:

> *The women are beautiful and walk with a voluptuous gait. Some go naked, except for a loincloth; others are swathed from neck to lower calf in a piece of blue-checked cotton. The chiefs' wives take a little more care with their dressing, wearing quite good quality jewellery which is sold to them by the Dutch. Mostly, the women have big, expressive eyes; their appearance is quite seductive overall. They don't take much care of their bosom, and their breasts are flat and not a very attractive shape. But they take the greatest care of their hair which is very long and a beautiful black; they rub it with coconut oil … which gives it a most unpleasant odour …*
>
> *Although the men are generally very jealous of their wives, they are not above prostituting them to us provided we give them some knives or similar bagatelles for the price of winning the favours of these new Cyrenes. One can see among the Timorese some men who, just as in the civilised nations of Europe, have no compunction about carrying on a trade which is both shameful and degrading to the human race. Every night these wretched men come to trade with us for their wives' favours – as well as their daughters'. I don't need to make the observation that we complacently avail ourselves of this type of commerce, and that as a result, several of us have had cause to repent of these services which were certainly not very expensive at the time, but which cost us, by their consequences, some burning regrets.[34]*

Baudin's fever worsened by the day. He ceased keeping his daily journal on 5 September, not resuming it until the ships sailed from Kupang on 13 November. From 25 September to 1 October, wrote Péron, the commandant,

> who had been ill for some time with a dangerous ataxic fever, experienced successively three such violent attacks, that for some hours he was thought to be dead. There was not a moment to lose in giving him the bark, in large doses; but as that belonging to the ships was of a very inferior quality, I shared with him the small quantity I had brought from Europe for my own use.[35]

The effect of the bark (no doubt quinine) was miraculous: 'it stopped this terrible fever, and in all appearance saved the life of our commander'. Dr Lharidon nursed Baudin devotedly, added Péron, 'but to say what was the reward for his care and attention, would shock every reasonable mind'.

Every day more men took sick, as dysentery and fever ravaged the crews of both ships. Frantz, one of the *Géographe*'s gunners, was the first to die, on 12 October; another man died on the 18th. One by one the scientists and their servants came down with the 'dreadful malady', excepting Péron:

> In the midst of such sorrow, and among so many disasters, I was in perfect health, and I was the only one up among those who lived in the same house. This precious advantage was certainly not produced by repose, for no one ... exerted himself with more zeal, or suffered more fatigue, than I did ... and with still stronger reasons I could not ascribe my health to the strength of my natural constitution, for it was weak and delicate.[36]

Riedlé, the gardener, 'although much broken down by the distemper [dysentery]', ignored the pleas of Péron and Lharidon to rest from his labours. 'Every morning at daybreak he set off to make new collections, without seeming to care at all about his disorder, entirely absorbed by his desire to justify the First Consul's confidence'. On 11 October he became very ill, 'the inflammation having spread from the rectum to the rest of the intestines, and the pains he suffered were horrible'. Lharidon placed the dying man in his own room, 'that he might be always at hand to afford him every care'. Riedlé succumbed on the 21st, after ten agonising days.

Baudin was heartbroken by the news: 'Nobody knows how much I love him, how attached I am to him,' he wrote. 'I feel in advance how much the expedition will lose ... No one aboard can even partly replace him'. Too ill to leave the house, he gave orders that his old friend should be buried with the same honours due to himself. Learning from Péron that David Nelson, Bligh's gardener on the *Bounty*, was buried in the cemetery, he arranged for Riedlé to be buried next to the Englishman, and had a monument erected over both. The ships lowered their colours, the guns fired every quarter hour, several volleys were fired over the grave, and the soldiers of the fort stood to arms.

Sublieutenant Picquet is held a prisoner in the fort until a vessel sails for Batavia. An officer from each ship visits daily to keep him company and share a meal with him – they consider it 'a sacred duty', Baudin notes sarcastically in his journal. Fearing the consequences for their friend if he returns to France in disgrace, the officers of both ships meet at their lodgings ashore. All are 'eager to give him letters and proper attestations to refute the calumnies that might be repeated to his prejudice', says Péron. The senior officer present, Captain Le Bas, promises that, should the letters fail in their purpose, he will himself denounce the commandant when they return to Europe. The comment is brought to Baudin's notice.

After Picquet's departure on 7 October – 'a day of sorrow and affliction for both ships', according to Péron – Baudin moves to deal with Le Bas's disloyalty without, he hopes, bringing about further confrontations with his staff. Privately he informs Le Bas that, on surgeon Lharidon's advice, he is to remain at Kupang when the expedition sails – Le Bas's health will be endangered if he continues on the voyage. Should he refuse, there will be no alternative but to report his conduct to Paris. With bad grace, Le Bas eventually agrees.

Baudin's satisfaction at so neatly settling the problem proves premature. Lharidon arrives in haste with news that Le Bas and Engineer Ronsard have fought a duel. Le Bas has been wounded and is in hospital – it emerges that he challenged the engineer for reporting his remarks at the officers' meeting to the commandant. The injury is not serious, a flesh wound in the arm, but it requires treatment. Baudin's anger soon subsides – Le Bas's precipitate action has provided the ideal solution. Surgeons Lharidon and Bellefin both sign the health certificate: apart from his wound, Captain Le Bas is suffering 'syphilitic symptoms' and 'arthritic lumbago', and is unfit to contine on the voyage.

The preparations for departure were interrupted on 23 October – an English frigate was reported in the Semau Strait, heading for the anchorage. Governor Lofstett immediately ordered his troops to defend the fort and the roadstead in the event of an attack. The Malays also mobilised, determined to resist any landing by the same enemy who so recently 'had captured the colony and treated the people like a conquered nation'.

Happily the precautions proved unnecessary. Baudin dispatched Midshipman Bonnefoi de Montbazin to present the expedition's passports to the frigate's captain for inspection. Once the enemy commander learnt the nature of their mission, wrote Péron,

> he behaved with the greatest politeness. Having heard that our commander was sick, he offered M. Montbazin some excellent bottles of wine for him, which he did not think himself authorised to accept.[37]

According to Bonnefoi, the captain had heard rumours of the French visit to Kupang, and supposing the ships to be merchantmen, 'had resolved to make prizes of them, in spite of the Dutch cannon'. The town itself, he said, was not worth the trouble of an attack. To Governor Lofstett's relief the frigate departed without firing a shot. 'In thus abstaining from all hostility,' Péron concluded, 'the English captain gave us a particular mark of his esteem and consideration for the object of our voyage.'

The following day Milius picked up a deserter from the frigate. The man 'had cast himself into the water … and had swum more than one and a half leagues [$4\frac{1}{2}$ miles] to reach the shore'. From him they learnt that her crew, too, was greatly weakened by dysentery. The English sailor was promptly recruited as a replacement.

On 12 November the governor farewelled Baudin and his staff at Fort Concordia, presenting the commandant with a superb Malay costume. Meanwhile, the last of the livestock – pigs, goats, sheep and fowl – were embarked. Riedlé's collection of living plants (which, Baudin reflected sadly, might well have cost the gardener his life) were also taken aboard.

The ships sailed before dawn. Baudin, shivering with fever, 'was too impatient to leave the country where disease had dealt us such a blow' to delay his departure any longer. Apart from Riedlé, he had lost six men from two ships, all from dysentery. Many more remained stricken with the deadly sickness – against which, wrote Milius, it seemed nothing could prevail. A dozen sailors died during the next ten days at sea.

Notes

1. Péron is writing here with the benefit of hindsight. In *The French Reconnaissance*, Horner (1987) points out that at the time none of the officers mentioned any such concern in their journals.

2. Péron, 1809, p. 55.
3. Ibid., p. 59.
4. Ibid., p. 62.
5. Ibid., p. 64.
6. Baudin, 1974, p. 178.
7. Péron's figure. Other sources put the number of men on shore at nineteen or twenty.
8. Ibid., p. 182.
9. Ibid., p. 183.
10. Péron, op. cit., pp. 80-81.
11. *Western Mail*, Perth, 12 December 1913.
12. The 'fearful noise' has been variously ascribed to bullfrogs or the cry of the brown bittern, more like a boom than a cry. It gave rise to the myth of the bunyip.
13. Excerpts of Bailly's account published in the *Western Mail*, Perth, 12 December 1913.
14. Horner, 1987, p. 170.
15. Baudin, op. cit., p. 194.
16. Péron, op. cit., p. 83.
17. Especially on Cook's third and final voyage, which was characterised by a series of misjudgements, and culminated in his murder at Kealakekua Bay, Hawai'i, in February 1779.
18. Ibid., p. 98.
19. Baudin, op. cit., pp. 208-209.
20. Ibid., p. 215.
21. Ibid., p. 219.
22. Horner, op. cit., p. 170.
23. Péron, op. cit., p. 153.
24. Milius, 1987, p. 15. (Passage translated by J. Treloar.)
25. Péron, op. cit., pp. 104-105.
26. Baudin, op. cit., p. 252.
27. Péron, op. cit., p. 113.
28. Picquet had been transferred to the *Géographe* at Isle de France, at Hamelin's request.
29. Baudin, op. cit., p. 259.
30. Péron, op. cit., p. 130.
31. Horner, op. cit., p. 181. The content of these secret instructions has not come to light; they may have concerned finding a site for French settlement in the region.

32. Milius, op. cit., p. 28.
33. Ibid.
34. Ibid., pp. 26-27.
35. Péron, op. cit., p. 132.
36. Ibid., p. 136.
37. Ibid., p. 135.

CHAPTER FIVE

⚓

Flinders and the Unknown Coast,
December 1801 to May 1802

Landfall

On board Investigator, *at sea, Sunday, 6 December 1801.*

> *The examination of Nuyts' and Leeuwin's Lands was not prescribed in my instructions to be made at this time; but the difference of sailing along the coast at a distance, or in keeping near it and making a running survey, was likely to be so little, that I judged it advisable to do all that circumstances would allow whilst the opportunity offered; and I had the pleasure to find this slight deviation approved at the Admiralty. [Flinders, 1814, vol. I, p. 48]*

Flinders had made the passage from Cape Town to New Holland three times, and set his course due east along the 37th parallel to avoid the roaring forties, with their heavy gales and long ugly seas. His luck held, and for nineteen days, from the Cape to Amsterdam and St Paul's islands on the mid-ocean ridge, 'the winds were never so strong as to reduce the *Investigator* to close-reefed sails; and on the other hand the calms amounted to no more than seven hours'. Fortunately there was little call upon the pumps, 'the greatest quantity of water admitted ... being less than two inches an hour'.

The boats were readied to land on the islands, but in hazy weather and drifting fog they could not be found, and the eastward course was resumed. For the next twelve days, from the longitude of Amsterdam Island to Cape Leeuwin, 'the same winds attended us; and a hundred and fifty-eight miles per day was the average distance, without lee or calm' – even in these conditions an exceptional time for a deeply laden former collier.

In passages like this, writes Flinders, 'it is seldom that any circumstance occurs, of sufficient interest to be related':

Our employments were to clean, dry, and air the ship below; and the seamen's clothes and bedding, with the sails, upon deck. These, with the exercise of the great guns and small arms, were our principal employments in fine weather; and when otherwise, we were wet and uncomfortable, and could do little ... The antiseptics issued were sour krout and vinegar, to the extent of the applications for them; and at half an hour before noon every day, a pint of strong wort, made by pouring boiling water upon the essence of malt, was given to each man ... The allowance of grog was never issued until half an hour after the dinner time.[1]

Land was seen from the masthead shortly before dusk on 6 December, directly ahead and about 10 leagues distant; with following winds the passage from the Cape had taken just thirty-two days. Flinders stood in for the coast at dawn, and judged it to be Cape Leeuwin, 'the south-western and most projecting part of Leeuwin's Land'. They coasted eastwards, Westall sketching the capes and promontories while Peter Good took in every detail of the new land:

Several reefs of Breakers along the shore and some abrupt rocks – ... in some places a low flat shore where we could see some distance inland where there appeared the greatest fertility – Towards evening the Coast became more rocky and the land terminated in Cliffs nearly perpendicular the Summit of which was covered with a fine low green verdure resembling at a distance Sheep pastures of England ... at night stood to out to Sea.[2]

Though it was contrary to his instructions, Flinders decided to commence a running survey of the south coast at this point – it was still early summer, and he had made surprisingly good time from Cape Town. The coast from Cape Leeuwin to King George Sound had been surveyed by Captain Vancouver of the Royal Navy in 1791, and the following year Admiral d'Entrecasteaux's expedition in search of La Pérouse had charted it as far east as 130° (just beyond the head of the bight). What the Admiralty now required of Flinders was a running survey of the Unknown Coast lying between 130°E and Bass Strait (at 145°E). By departing from his orders, his arrival on this coast was delayed by several weeks.

The Admiralty later approved this 'slight deviation', but the approval was of course retrospective. As Geoffrey Ingleton has pointed out, Their Lordships might well have been less forgiving had the gamble failed.

In the event, it was Baudin's delays in Tasmanian waters that allowed Flinders to claim prior discovery of the Unknown Coast for Britain.

The *Investigator* carried charts of Vancouver's and d'Entrecasteaux's voyages, and in his *Voyage* Flinders – perhaps to justify his decision as much to himself as to his superiors – took the opportunity to compare his examination with these earlier surveys:

> It will thence appear, that the employment of fifteen days in running along the coast, more than would probably have been required had I kept at a distance, was not without some advantage to geography and navigation ...[3]

Princess Royal Harbour

At sea, Tuesday, 8 December 1801.

> *The wind blew fresh at this time, and a current of more than one mile an hour ran with us, so that, by carrying all sail, I hoped to get sight of King George's Sound before dark. At seven, we passed close on the south side of the Eclipse Isles ... and at eight o'clock we hauled up round [Bald] Head, with the wind at west, and made a stretch into the sound. It was then dark; but the night being fine, I did not hesitate to work up by the guidance of Captain Vancouver's chart; and having reached nearly into a line between Seal Island and the first beach round Bald Head, we anchored at eleven o'clock, in 8 fathoms, sandy bottom. [Flinders, 1814, vol. I, p. 53]*

Flinders was authorised to call at 'King George the Third's harbour' for refreshments and water. He remained there for a month, preparing ship and crew 'for the examination of the south coast of Terra Australis'. Princess Royal Harbour within the sound offered a secure anchorage, 'where the masts could be stripped, the rigging and sails put into order, and communication had with the shore'. Leaving the ship's overhaul to Lieutenant Fowler and the warrant officers, he camped ashore with the master and the naturalists.

While Brown and his companions botanised near Bald Head, Flinders and Thistle explored the sound, finding fresh water at Oyster and Princess Royal harbours. They searched in vain for the bottle and parchment left by Vancouver in 1791, and for signs of the vegetable garden he had planted. Some visitors had called, however, for 'several trees had been felled with axe and saw'. They also came across some bark sheds, similar to the native huts in the forests behind Port Jackson, but they seemed long deserted. The mystery of the felled trees and the disappearance of Vancouver's bottle was solved by the discovery of a plot of ground 6 or 8 feet square, 'dug up and trimmed like a garden'; on it lay a copper sheet with the inscription: 'August 27, 1800. Chr. Dixson – ship *Elligood*'.[4] No trace was found of a visit by Baudin's ships.

With *Investigator* secured, Flinders fixed upon a place for the tents. These were set up under a marine guard, and the observatory and instruments sent ashore in Samuel's care. Signs of the country being inhabited were apparent everywhere, but, he wrote,

> there was nothing to indicate the presence of the natives in our neighbourhood. [It being Sunday] I therefore allowed a part of the ship's company to divert themselves on shore this afternoon; and the same was done every Sunday during our stay in this harbour.[5]

Next morning the refitting of the ship resumed. The naturalists 'ranged the country in all directions, being landed at such places as they desired', while Flinders divided his time 'betwixt the observatory and the survey of the sound'.

Peter Good has spent the Sunday ashore collecting plant specimens and seeds in the vicinity of the tents. Next day 'smokes' are seen at the head of the harbour, and he accompanies Brown, Westall and their servants to investigate the source. The fire is out before they reach the place, but on their way they see a native walking on the beach at a little distance. At first he seems unconcerned by their presence, but as they approach within 50 yards he begins to call out with great force and brandishes his spear, making signs to them to turn back. 'Seeing us persist in advancing', he retreats into the scrub behind the beach, setting fire to the bushes to bar their way. Through the flames they see a group of women and children fleeing inland.

Next morning two natives approach the tents, crying out and gesturing in a threatening manner. Tall and slender, they are naked but for a kangaroo skin loosely wrapped round their bodies, and fastened at the shoulder with a wooden skewer. They advance cautiously, the leader keeping his spear poised and ready to throw. The assistant surgeon, Mr Purdie, walks towards them unarmed, and soon wins their confidence. Brown and his group offer some red nightcaps and other trifles in exchange for a spear and a stone hatchet. After this meeting the natives and their companions return almost every day, often spending a whole morning at the tents. They will not allow anyone to go with them when they leave; the women and children stay hidden in the bush.

The 'Indians' admire their visitors' white skin: 'They rubb'd their skin against ours, expecting some mark of White wou'd appear upon theirs,' says Seaman Samuel Smith, 'but finding their mistake they appear'd surprised.' They laugh delightedly when a few beardless young sailors yield to their unmistakeable gestures to show their sex.

Flinders set off with a party of thirteen men, including the naturalists, their servants and several sailors, to explore two lagoons lying behind the cape west of the harbour. Fully armed and provisioned, they had not gone far when they met an old native who often came to the tents. Though he greeted them cordially enough he sought to bar their way, running from one to another and all the time making a great noise. Obligingly they detoured around the wood, 'where it seemed probable that his family and female friends were placed'. He kept them company for a while, for which they had cause to be grateful when 'with evident symptoms of horror' he knocked a snake out of the hands of a seaman who had just picked it up, indicating it was dangerous.

The going was difficult, and they found themselves wading through a succession of swamps and forcing their way through thick brush, until towards evening they came upon higher ground, with water at hand, and camped for the night. Next day the party turned back towards the ship, walking along a sandy ridge 'where the want of water was as great, as the super-abundance had been in the low land going out'. Towards sunset, with the tents still several miles distant, Bauer collapsed from fatigue – the effect of the exertion of the walk, the excessive heat, and thirst – and could not continue. Brown and a sailor remained with him while the others pushed on to the tents. The three stragglers arrived about midnight, and all 'slept sound till morning'. They returned on board at six o'clock.

It was Christmas Day, a Friday, and the ship followed the usual Sunday routine, with the crew mustered on deck and then allowed ashore for recreation. The officers and scientific gentlemen joined Flinders in his cabin for Christmas dinner while, according to Samuel Smith, 'the Sailors had holliday & were more regular & orderly than usual on such occasions however several got compleatly drunk'. Next day it was back to work as usual.

Wednesday, 30 December. Wooding and watering is completed, the rigging refitted, and the sails repaired and bent. 'Our friends the natives being at the tents this morning' – among them the old man who has been a regular visitor – Flinders orders the marines on shore, to be exercised in their presence:

> *The red coats and white crossed belts were greatly admired, having some resemblance to their own manner of ornamenting themselves; and the drum, but particularly the fife, excited their astonishment; but when they saw these beautiful red-and-white men, with their bright muskets, drawn up in a line, they absolutely screamed with delight; nor were their wild gestures and vociferation to be silenced, but by commencing the exercise, to which they paid the most earnest and silent attention.[6]*

The old native takes his place at the end of the rank. In his hands he holds a short staff, which he shoulders, presents, and grounds in unison with the marines as they go through their drill. Before the muskets are fired the 'Indians' are acquainted with what is about to happen; 'so that the vollies did not excite much terror'.

One of the native men submits good-humouredly to Mr Brown and Surgeon Bell as they measure his bodily dimensions, then names the different parts of the body in his language. Peter Good notes the names in his journal: Caat, *the head;* Waart, *the neck;* Taa, *the mouth;* Davaal, *the thigh;* Mat, *the leg;* Twang, *the ears;* Mite, *the privates;* Catta, *the hair.*

On board the ship, William Donovan, a seaman, is brought before the captain for repeated drunkenness and fighting. His punishment is thirty-six lashes.

The same day the tents were struck, the observatory closed down, and the instruments taken on board. On New Year's Day the *Investigator* weighed to leave the harbour, but unfavourable winds kept her within the sound for several more days. On 4 January Flinders, with Fowler and Brown, landed for the last time and left a bottle upon the top of Seal Island; within it was a parchment 'to inform future visitors of our arrival and intention to sail on the morrow'. Next day they sailed out of the sound 'to prosecute the further examination of the coast'.

Unknown to Flinders, Baudin's ships were almost due south, close to 200 miles out to sea, and heading south-east for Van Diemen's Land, away from the Unknown Coast.

Archipelago of the Recherche

At anchor in Lucky Bay, Tuesday, 12 January 1802.

> *Several seals were procured on this and the preceding day, and some fish
> were caught alongside the ship; but our success was much impeded by
> three monstrous sharks, in whose presence no other fish dared to appear.
> After some attempts we succeeded in taking one of them, but to get it on
> board required as much preparation as for hoisting in the launch. The
> length of it was no more than twelve feet three inches, but the circumfer-
> ence of the body was eight feet. Amongst the vast quantity of substances
> contained in the stomach was a tolerably large seal, bitten in two, and
> swallowed with half of the spear sticking in it with which it had probably
> been killed by the natives. The stench of this ravenous monster was great
> even before it was dead; and when the stomach was opened, it became
> intolerable. [Flinders, 1814, vol. I, p. 82]*

Prior to the *Investigator*, only four ships are known to have sailed this
coast. There may have been others – a Portuguese caravel, perhaps,
in the 16th century, or a lone whaler such as the *Elligood* in the past
year or two – but if so, no record survives. Most probably the Dutch
were the first; in 1627 a Dutch East India Company vessel, the *Gulden
Zeepard*, outward bound for Batavia, was driven south by fierce gales
in the Indian Ocean and made her landfall near Cape Leeuwin. Captain Thijssen, instead of sailing north, followed the coast eastwards for
about 1000 miles to Nuyts' Archipelago before turning back.

The second was Captain George Vancouver's *Discovery* in 1791. Sent
by the Admiralty to explore the southern coast of New Holland and
the north-west coast of North America, Vancouver, a veteran of Cook's
last voyage, discovered King George Sound and explored another
300 miles of coastline before gales forced him south and east to Van
Diemen's Land. He was closely followed by Rear Admiral Bruny d'Entrecasteaux, commanding the frigates *Recherche* and *Espérance*, dispatched
by the French government in search of the missing explorer La Pérouse.
With instructions to combine the search with the exploration and charting of the south coast, d'Entrecasteaux sailed past the sound without
landing, and entered Esperance Bay (named after the first vessel into
the bay) on 9 December 1792. The large group of offshore islands was
named after the flagship, the *Recherche*.

From Esperance the French ships headed east, the hydrographer

Beautemps-Beaupré meticulously charting the coast, and entered what are now South Australian waters at year's end. D'Entrecasteaux noted in his journal:

> The whole coast from west to east looked the same: limestone rock rising precipitously, the same height all the way ... No bird emerged from this arid coast, no smoke, everything suggested that this land was uninhabited; it seemed that its aridity had banished men and birds.[7]

Quitting the sound on 5 January, Flinders followed his predecessors' course to the east. From the outset he adopted a strict and systematic routine that produced charts so accurate that some remained in use until World War II.

In his passage along the coast, Flinders aims to keep so close in to the land that the breakers on the shore are visible from the ship's deck. This greatly reduces the degree of error in judging distance from the shore, and the chance of a river mouth or opening of any kind escaping his notice. Though not always possible, especially when the coast retreats far back, it is always attempted when it appears practical and does not involve undue danger to the ship. When it cannot be done, Flinders is commonly found at the masthead with his glass.

All bearings are laid down as soon as they are taken, while the land remains in sight; and before retiring to rest, Flinders lays down on a rough chart the coast just passed, and notes his astronomical observations and comments on it in his journal. When hauling off the land at night, he takes every care to come in at the same point next morning before resumimg his route along the coast.

The officers of the watch have strict orders to pay attention to the log – to see that the deep-sea lead is cast and the soundings marked every half hour on the log board, and that the heights of the thermometer and barometer are recorded thrice daily. His brother Samuel and Midshipman John Franklin are held responsible for winding the two timekeepers at noon each day, so that the ship's longitude may readily be found by comparing local time with Greenwich Mean Time

'This plan, to see and lay down every thing myself,' he writes

required constant attention and much labour, but was absolutely necessary to obtaining that accuracy of which I was desirous ... and it was adhered to in all the succeeding part of the voyage.[8]

The westernmost island of d'Entrecasteaux's *Archipel de la Recherche* – a vast labyrinth of some 200 islands, rocks, and shoals stretching eastwards for 125 miles and from 30 to 40 miles offshore – came into view shortly before sunset on 8 January. 'The French admiral had mostly skirted round the archipelago,' wrote Flinders, 'a sufficient reason for me to attempt passing through the middle, if the weather did not make the experiment too dangerous.' The mainland coast behind the island chain was little known.

Next day was spent exploring the islands just east of Esperance Bay. By late afternoon the ship was off Mondrain Island, one of the largest in the group, 'with no probability of reaching a space of clear water in which to stand off and on during the night, and no prospect of shelter under any of the islands'. Caught between a rock and a hard place, Flinders chose to steer directly for the mainland coast, 'where the appearance of some beaches, behind other islands, gave a hope of finding anchorage'. The gamble paid off; as night fell they entered a small sandy bay, dropping anchor in 7 fathoms. Flinders named the place Lucky Bay.

At the botanists' request the *Investigator* remained at the anchorage for five days, allowing Brown and his assistants to gather 'an abundant variety of shrubs and small plants – a delightful harvest to the botanists, but to the herdsman and cultivator it promised nothing'. Leaving the bay on the 14th, the ship returned to the maze of offshore islands and anchored off Middle Island, at the centre of the archipelago, in a snug and well-sheltered cove. Here a shore party returned with 'upwards of 2 Dozen Geese of a kind peculiar to this country & little inferior to English Geese',[9] providing the ship's company with a welcome change of diet.

The officers and scientists went ashore exploring, while the Master was sent to explore the various passages through to the eastward; he returned with another twenty-seven geese, some of them still alive. The botanists meanwhile discovered a small salt-water lake, 'tinged of a beautiful red colour with a quantity of fine white salt on its shores …', and collected some twenty species, of which twelve proved to be new. A piece of planking, with nails in it, was picked up on the shore, 'but no trace of the island having been visited, either by Europeans or the natives … was any where seen'.

The Unknown Coast

At sea, Wednesday, 27 January 1802.

The same shore like a wall composed all our view of land till about noon
it changed its appearance to a Sandy Beach and low country – The Rocky
Coast I think extended about 100 miles as also the one from 18th to the
22nd and the whole shore since the 16 had run nearly from west to East
– It now changed its direction and trended to South East The soundings
along this whole extent of Coast were very regular. [Good, 1981, p. 57]

Investigator passed the easternmost islands of the archipelago on
17 January, and followed the mainland coast at a distance of 3 to 5 miles.
The shore began to curve, taking on first a north-easterly, then an easterly direction. Day after day nothing was to be seen to port but an
extraordinary bank of cliffs, brown in the upper part and nearly white
in the lower, towering from 400 to 600 feet above the sea. There was
no break for upwards of 500 miles, and the country behind lay hidden; *Investigator*'s masthead stood no more than 100 feet above the
deck. So uniform was the appearance of the cliffs that Flinders found
very few features whose bearings could be set a second time:

> Each small projection presents the appearance of a steep cape, as it opens
> out in sailing along; but before the ship arrives abreast of it, it is lost in
> the general uniformity of the coast, and the latitude, longitude, and dis-
> tance of the nearest cliffs, are all the documents that remain for the con-
> struction of a chart.[10]

Later, when he came to write the *Voyage*, it occurred to Flinders that
the cliffs perhaps concealed an interior sea, and he regretted

> not having formed an idea of this probability at the time; for notwithstand-
> ing the great difficulty and risk, I should certainly have attempted a landing
> upon some part of the coast, to ascertain a fact of so much importance.[11]

On the 27th the cliffs at last came to an end, and some 3 leagues
later the coast trended south-east by east, forming the head of what
Flinders termed at the time 'the Great Bight or Gulph of New Holland'.
Later he changed the name to Great Australian Bight – the first place-
name of a coastal feature to include the word 'Australian'. At five o'clock
they were abreast of Cape Adieu – 'the furthest part seen by the French
admiral when he quitted the examination', and an hour or so later came
upon extensive reefs lying several miles from the land. These appeared
at the extremity of the Dutch chart, but not on the French, and with
pardonable satisfaction Flinders entered in the ship's fair logbook:

'These breakers were not seen by him.' Immediately beyond lay the Unknown Coast.

Before venturing into unknown waters, Flinders, typically, paused to pay a generous tribute to d'Entrecasteaux and his cartographer:

> From the Archipelago eastward, the examination of the coast was prosecuted by [him] with much care, and with some trifling exceptions, very closely … Monsieur Beautemps Beaupré, geographical engineer on board *La Recherche*, was the constructor of the French charts, and they must be allowed to do him great credit. Perhaps no chart of a coast so little known as this was, will bear a comparison with its original better than those of M. Beaupré.[12]

Flinders came to the limit of Dutch exploration, a 'high cliffy cape' that he named Cape Nuyts to celebrate the occasion, on the 28th, and towards evening dropped anchor in a wide bay promising shelter from the southerly winds. He called it Fowler's Bay, after the first lieutenant.

From here to Streaky Bay, 125 miles along the coast, Flinders names most features after the ship's officers and scientists: Fowler is honoured by a point as well as a bay; Master's Mate Thomas Evans and midshipmen Lound, Lacy, and Franklin, with Assistant Surgeon Purdie, give their names to islands; Midshipman Sinclair, Surgeon Bell,[13] the botanist Robert Brown and the artist Westall give theirs to points; Ferdinand Bauer, like the Dutchman Nuyts, is commemorated by a cape.

Close confinement on an overcrowded ship takes a psychological toll on nineteen-year-old Samuel Flinders. The second lieutenant confronts his brother, complaining that his duties – as acting astronomer and deck officer – are too onerous, and requests they be reduced. Matthew considers he may have erred in judgement in choosing his brother for the voyage; certainly it would be unthinkable to tolerate any laxity in his case:

> *The 2nd lieutenant not having given me all the assistance in the astronomical and surveying departments that I expected, he was ordered to keep his own watch during the night.[14]*

The decision is unfair, but out of pride Samuel chooses 'to keep it in the day also, and to continue giving me the same proportion of assistance as he had done before'.

Fowler's Bay lacked fresh water, or any wood suitable for use as fuel. Signs of native occupation were seen, but no huts or anything else suggested this was recent. The seamen caught a few fish alongside the ship, and the shore party shot some ducks and gulls, but with nothing to hold them longer Flinders sailed after midday on 30 January.

For ten days the ship tacked to and fro between the mainland and an extensive group of islands lying offshore, the largest of which he identified as those the Dutch had called St Francis and St Peter – he retained their nomenclature, but named the entire group Nuyts Archipelago. The broad but shallow inlet opposite the Isle of St Peter was given the title Denial Bay – 'as well in allusion to St. Peter as to the deceptive hope we had found of penetrating by it some distance into the interior country'. On one of the islands, says Seaman Smith, 'Boats was sent on shore with several of the Gentlemen & Kill'd plenty of Mutton Birds, which makes a good fresh meal'. When placed in the shade of a rock the thermometer showed 100°F, but in the sun the mercury rose to the top of the tube at 130°.

With the heat and a heavy head-swell, the leaks that had so troubled Flinders earlier on the voyage recurred. On 1 February the ship began making more than 3 inches an hour, and next day the carpenter reported more as oakum in the seams worked loose. Fortunately the leaks seemed confined to the upperworks, and by shifting 4 tons of iron ballast forward the ship's trim was improved and the intake of water reduced.

Leaving the archipelago, the *Investigator* sailed south-east along an exposed and dangerous coast, studded with rocks, reefs, and shoals. By 14 February she was among a cluster of islands that Flinders named the Investigator's Group – the largest received the name 'Flinders, after the Second Lieutenant', perhaps to soothe his brother's ruffled ego. Not once on the voyage did he name any coastal feature for himself.

One of the cutters was sent to sound and to kill some seals for food; the other landed the captain and the botanist on Waldegrave Island (named after a prominent Lincolnshire family), Flinders to take his bearings and Brown to investigate the flora and fauna. Flat-topped and grassy, the island did not offer 'a single novelty in natural history' to Brown's observation, and he contented himself with noting the domestic customs of the local seals:

In our walk along the beach we found several families of seals, one male to 3 or 4 females. At first we were obliged to defend ourselves from the males, but afterwards, unless wantonly attacked and even killed, several of them hobbled up to us but a blow on the nose, head or indeed any part of them was sufficient to make them retreat. The female if she had young in the neighbourhood, attacked the retreating male, biting him about the mouth. This, however, seldom induced him immediately to renew his attack. If the female was killed, the male immediately made towards the water; if the male, the female remained with her young. The cub of one we had killed we gave to another who carried it to her own. A cub being killed, two females fought about the remaining one; the male settled the difference, awarding it to her, I supposed his favourite, most of us thinking that the cub did not belong to her. The other, however, seemed satisfied or at least allowed her quiet possession.[15]

The islands lacked fresh water, and with the supply on board running very short, Flinders made again for the mainland. On the 16th the ship entered a wide bay (Coffin Bay) backed by rugged cliffs with wooded hills behind. 'Many smokes were seen … and also two parties of natives', suggesting that this part of the coast was more thickly populated than any seen since leaving King George Sound. The bay was shallow, however, and the anchorage exposed, and Flinders tacked out without landing.

The running survey continues, but there is no sheltered anchorage on these dangerous coasts. The recess of the sun to the north is a constant reminder to Flinders of winter's approach.

Unexpectedly, on the afternoon of 20 February, Investigator *rounds a headland and, for the first time since arriving in New Holland, meets a tide running from the north-east, apparently on the ebb; no land is to be seen in that direction. Conjecture grips the ship's company – large rivers, deep inlets, inland seas, and passages to the Gulf of Carpentaria are the topics of conversation that evening. The prospect of making a singular discovery infuses new life and vigour into every man on board.*

Catastrophe

At sea, Sunday, 14 February, to Sunday, 21 February 1802.

> *14th we got under Weigh & Investigated the Coast with the greatest niceness, going into Harbours, & Bays, Laying them down in all respects with Acuracy; & attended with foul Winds which caused Investigating more Dangerous. These Harbours Abounds with Birds & Fish in great plenty, from the last date nothing perticulour Occurr'd untill the 21st when we*

> *came into a Harbr afterwards named by our Captn & came to an An-*
> *chor; the next morning sent A Boat on Shore with the Master, Mr Jon This-*
> *tle, Wm Taylor Midshipman & 6 AB Seamen ... [Seaman Samuel Smith's*
> *journal, MS, Mitchell Library]*

Investigator came to anchor in a passage 4 miles wide between the mainland and a large piece of land to the east, 'whether island or main we could not tell'; Flinders tentatively named it Uncertain Island. On Sunday morning, the 21st, he and Thistle landed and climbed a sandy hill behind the beach to establish whether or not they were on an island, and to take bearings. Seals lay on the beach and traces of kangaroo were everywhere.

A causeway of white sand blown almost across the island by the prevailing winds made their path easy. On their way to the summit, carrying their muskets and survey instruments, they came across a large speckled snake asleep on the sand.

> By pressing the butt end of a musket upon his neck, I kept him down
> whilst Mr. Thistle, with a sail needle and twine, sewed up his mouth; and
> he was taken on board alive, for the naturalist to examine ...[16]

They found that the naturalist's party had already killed two others of the same species, one of them 7 feet 9 inches long.

As they proceeded upwards with their prize, a large white eagle, wings outspread, came bounding towards them, taking to the air just 20 yards distant; a second eagle swooped down as they passed beneath. Flinders conjectured that the birds had taken them for kangaroos, 'having probably never before seen an upright animal on the island'. He observed that the eagles sat watching in the trees, and when a kangaroo came out to feed in the daytime, it was 'seized and torn to pieces by these voracious creatures'.

Flinders had planned to get under way at noon, but on comparing Samuel's observations of the island's longitude with his own 'a differ-ence was found which made it necessary to repeat the observations on shore' – if the island indeed lay at the entrance to a vast strait sepa-rating the eastern and western parts of the continent, its position had to be established beyond doubt. The delay made it too late to move the ship before dusk, and he sent the master over to the mainland in the red cutter 'in search of an anchoring place where water might be procured' – their supply was now dangerously low.

Coming on deck at sunset, Peter Good observes Thistle's cutter leaving the mainland shore, some 5 to 7 miles distant. She is sailing into a stiff south-easterly, and is close hauled; the current seems to take her a little to windward of the line she is making for the ship. He watches idly for ten or fifteen minutes, when she is about halfway to the ship and just to the lee of a small island. He takes a leisurely turn of the deck, and when next he looks the cutter has disappeared. The master's mate, Evans, makes the same observation and raises the alarm. Rushing from his cabin, Flinders scans the sea with his glass in the fading light, but in vain.

Lieutenant Fowler is away with the blue cutter, not returning for another half hour. Though it is now dark, the captain points out the spot where the red cutter was last seen; Fowler is to lose no time in reaching there, to rescue if he can the people wrecked, but on no account must he risk his own boat. For two hours and more he rows and sails in every direction where wind and tide may have taken the missing men, while the watchers on the ship wait with mounting anxiety. At ten Flinders orders a gun fired to recall the blue cutter.

On his return, Fowler reports that he has seen no sign of the missing cutter or her crew; his own boat was caught in a great reef of breakers and a strong rip near where they disappeared, and half filled with water before he could steer clear. He and the captain sit talking late into the night, dissecting the tragedy that has struck the voyage. Had it been daylight, they agree, some at least of the eight men might have been saved, though only two were strong swimmers. But in the night it was too dark to see anyone struggling in the current, or over the noise of the breakers to hear any answer to the hallooing, or to the firing of the muskets. With the tide running seaward at the time the boat was missed, all are feared drowned.

For the first time Fowler tells his captain of Mr Pine's prophecies. Flinders makes no comment, but in his mind resolves to 'recommend a commander, if possible, to prevent any of his crew from consulting fortune tellers' before leaving on a voyage. In the fo'c'sle the men who visited the Portsmouth prophet swear they will stay clear of the Lady Nelson *when she joins them at Port Jackson.*

In the tight-knit community of a king's ship the loss of any man to their enemy the sea casts a pall of gloom over the entire company. The disaster stunned the *Investigator*'s crew all the more because it took place in fine weather, when 'not the least danger was apprehended'.

Thistle was one of their own, having risen from seaman to master on his merits, and filled the position, says Good, 'with Credit to himself and satisfaction to all around him'. Of the other victims, Midshipman Taylor was 'a very promising young man', and the six seamen were 'all choice sailors'.

The master's death affected Flinders deeply. Thistle had sailed with him, on and off, for eight years, and on this last voyage had become a friend as well as a trusted officer. In his logbook and journal, and years later in his *Voyage*, Flinders paid him a heartfelt tribute:

> The reader will pardon me the observation, that Mr. Thistle was truly a valuable man, as a seaman, an officer, and a good member of society. I had known him, and we had mostly served together, from the year 1794. He had been with Mr. Bass in his perilous expedition in the whale boat, and with me in the voyage round Van Diemen's Land, and in the succeeding expedition to Glass-house and Hervey's Bays. From his merit and prudent conduct, he was promoted from before the mast to be a midshipman, and afterwards a Master in his Majesty's service. His zeal for discovery had induced him to join the *Investigator* when at Spithead and ready to sail, although he had returned to England only three weeks before, after an absence of six years ... His loss was severely felt by me; and he was lamented by all on board, more especially by his messmates, who knew more intimately the goodness and stability of his disposition.[17]

Flinders was no less saddened by the deaths of William Taylor and the six seamen, all 'active and useful young men' who had volunteered for the voyage. To commemorate the cutter's crew he gave Thistle's name to the 'uncertain island' on which he and the master had found the speckled snake; a smaller island nearby became Taylor's Island; and on six islets lying close to Cape Catastrophe he bestowed the names of the six lost seamen.

Memory Cove

Monday, 22 February 1802. Captain's fair logbook:

> *At 5 hrs. 30 weighed the anchor, and steered towards that part of the main land near which the lost boat was last seen ... Seeing a little beach on the main, steered for it, and finding it was in a nice cove which was open only to the north-eastwd., stood into it and anchored in $10\frac{1}{2}$ fms sandy*

bottom ... Sent the cutter away in search of the lost boat and people, and two parties went to walk along the shores ... The cutter soon returned towing the wreck of the other boat bottom upwards; she was stove all to pieces, having to appearance been dashed against the rocks ... One oar was afterwards picked up, but nothing was seen of the bodies of the unfortunate people. Sent the cutter away again upon further search and stationed a petty officer upon a head of land with a spyglass to watch every thing that might drift past with the tide.

At 4 hrs p.m. nearly calm; shortened in the cable. At 5 hrs. the cutter returned with some small remnants of the boat found to the northward ... but nothing was seen of the officers or people ... At the back of the little beach in the cove, I found many footsteps of our people, made I presume whilst they were searching here for water before the accident happened ...

Peter Good's journal:

Got underway Early and stood across the Straight to the Main and soon perceived the chain of breakers where the Cutter was supposed to have upset – Mr Fowler went to sound and found plenty of water & that it was entirely occassioned by the violence of the current – where some Islands confine it to a narrow channel and that it frequently disappears which makes it more dangerous for boats – about 7 anchored in a snug little Bay of the Main ... – Mr Fowler went with the other Cutter to search the different Islands in case any had got on them ... – Mr Fowler soon found the remains of the Cutter near the point of the Main which sheltered the Anchorage from the South East winds which seem prevalent here – The Cutter had been on shore & broke to pieces & no part of the Crew about it – and at night when all parties had returned on board nothing was found but broken fragments of the Boat on the Beach & some oars – Mr Fowler also found the Compass and binnacle floating & unhurt – thus all hopes of finding any of the Crew alive vanished.

Tuesday, 23 February 1802. Captain's fair logbook:

Light breezes and cloudy weather ... The Commander took the cutter to search for the unfortunate people lost in the boat or for pieces of the wreck ... About 4 hrs p.m. the Commander returned, having found nothing more than a small keg which belonged to Mr Thistle, and two broken remnants of the boat. Additional Remarks: In the boat excursion of this day I followed all the sinuosities of the shore for twelve miles to the northward of the ship, thinking it likely that the bodies of some of our unfortunate companions might be thrown on shore; but we had no success in the pursuit ... some of our gentlemen from the highest land near the ship, saw an inlet going in at some distance to the westward; this I learnt on my return [Port Lincoln].

Peter Good's journal:

Captain went with Cutter to search to leeward most fragments were found in that direction – Mr Brown & I went to the Top of the hills & collected a few plants – The Shore was searched in every direction & nothing found but trifling fragments & Captain found floating a small cask in which Mr Thistle used to carry his liquor.

Wednesday, 24 February. Captain's fair logbook:

Strong breezes until 3 hrs a.m., when it moderated and the weather became fine. Sent away Lieut. Fowler in the cutter to examine the outer southern islands for the remains of our unfortunate people, but he was not able to land or obtain the least information upon the object of his pursuit ... Noon moderate breezes and fine weather. In the evening sent the cutter to haul the seine upon the beach of this little cove, which she did with such success as to give all the ship's company two meals of fish and some to be cured ...

General Remarks: ... As every search had now been made for our unfortunate shipmates, that we could think had any prospect of being attended with success, I thought it would avail nothing to remain longer on their account for there was only a small chance of obtaining their bodies when they might rise to the surface, from the number of sharks that we have constantly seen about. Even this small chance of obtaining their bodies would have induced me to wait a few days longer had not the want of water been so pressing to hurry us forward. I caused a stout post to be erected in the Cove, and to it was nailed a sheet of copper upon which was engraven the following inscription:

Memory Cove
His Majesty's ship Investigator – Matt[w]. Flinders – Commander,
anchored here February 22. 1802.
M[r]. John Thistle the master –
M[r]. William Taylor – midshipman
and
six of the crew were most unfortunately drowned near this place from being upset in a boat. The wreck of the boat was found, but their bodies were not recovered.
Nautici cavete!

Thursday, 25 February. Peter Good's journal:

> *Took our departure from this Bay which Captain named* Memory Cove *as a small tribute to the memory of the brave fellows who were lost there – he caused to be fixed on shore a plate of Copper commemorating this melancholy and disastrous event – with the date of our anchorage here &c. We stood along the Coast to the North & passed several Bays and Islands and afterwards stood to West into a large Bay with many Islands and anchored to Leeward a small hill ...*

New Lincolnshire

At anchor in Port Lincoln, Friday, 26 February 1802.

> *The port which formed the most interesting part of these discoveries I named* PORT LINCOLN, *in honour of my native province; and having gained a general knowledge of it and finished the bearings, we descended the hill and got on board at ten o'clock. The boat had returned from Boston Island, unsuccessful in her search for water; and we therefore proceeded upward, steering different courses to find the greatest depth. Soon after one o'clock we anchored in 4 fathoms, soft bottom, one mile from the beach at the furthest head of the port, and something less from the southern shore. [Flinders, 1814, vol. I, p. 142]*

Investigator sailed from Memory Cove early on the 25th, and at two o'clock entered the great bay seen by the naturalists from the hills beyond the cove. A boat was sent ashore to look for drinkable water by digging at the back of the beach, but the first attempt failed: 'the water which flowed into the pit was quite salt; and notwithstanding the many natives' huts about, no fresh water could be found'.

The following day the ship moved further into the port, anchoring at the head of the South West Bay (today's Proper Bay). Here pits were dug in clay-bottomed soil, into which 'water flowed pretty freely, and though of a whitish colour, and at first somewhat thick, it was well tasted'. The slow laborious business of watering the ship began; 'the last turn of water was received on the 4th, and completed our stock up to sixty tons'.

Climbing the ridge rising nearly 500 feet behind the southern shores of the port (Stamford Hill), Flinders saw before him a splendid natural harbour 'capable of sheltering a fleet of ships'. It was the only port of importance found since leaving King George Sound, and he spent nine days surveying and sounding the various bays within it and their surrounds. The tents and observatory were set up onshore, with Samuel charged with rating the timekeepers and taking the daily observations. A solar eclipse fell due on 4 March, observed by Flinders 'with a refracting telescope of 46 inches focus, and a power of about two hundred'. Despite the cloudy weather, the eclipse – which was 'almost Total, only a small part of the lower limb was not obscured' – was quite distinct, even to the naked eye.

The scientific party meanwhile explored the nearby hills, with little success. They found few new plants, and although coming across some well-trodden tracks they neither heard nor saw any sign of the natives. On 1 March, Brown and his party found themselves 'on the Shore of the main Ocean and the Bay which we had looked into on 20 instant [Sleaford Bay] which is large open and exposed & a dreadful surf all around'. Cast up on the shore they found the mainsail of the wrecked cutter, but no other remains. On their return to the ship with this news, Flinders dispatched Fowler to make yet another search for the bodies of any of the lost crew, but without result.

On the 4th, several natives were heard calling to a boat's crew who had just landed, but they retreated deep into the woods before contact could be made. Noting the incident in the ship's log, Flinders believed

> that their appearance on the morning the tents were struck, was a prelude to their coming down; and had we remained a few days longer a friendly communication would have ensued.[18]

The near-contact took place shortly before the eclipse, and it would be interesting to know what significance, if any, the men in the woods gave to the coincidence of the two events.

Next day, Flinders landed one last time to take bearings, then returned on board. The boat was hoisted in, 'and our operations in Port Lincoln being completed, we prepared to follow the unknown coast to the northward, or as it might be found to trend'.

Port Lincoln was not named at the time, but was shown on Flinders's rough charts as Bay no. 10. Years later he created an antipodean Lincolnshire on the map of his discoveries – translating the solitary landscape of the fens to the lonely shores of the Unknown Coast. It was not until 1814 that the extensive harbour became Port Lincoln, its southern point Cape Donington after his birthplace. The port of Boston, home of his friend Bass, from which the Pilgrim Fathers had sailed for America, was commemorated by Boston Bay, Boston Island and Boston Point. Stamford Hill, Spalding Cove, Sleaford Bay and Mere recalled the county's market towns, while the Sir Joseph Banks Group of islands honoured not only Flinders's patron, but also a dozen or so Lincolnshire parishes – including Sir Joseph's manor of Reevesby, John Franklin's home village of Spilsby, and Partney, where Flinders and Ann were married. In all, he bestowed some thirty Lincolnshire placenames on this part of the coast.

Mudflats and a mountain

At sea, Monday, 8 March 1802.

> *We had then advanced more than twenty-five leagues to the north-north-east, from Cape Catastrophe; but although nothing had been seen to destroy the hopes formed from the tides and direction of the coast near that cape, they were yet considerably damped by the want of boldness in the shores, and the shallowness of the water; neither of which seemed to belong to a channel capable of leading us into the Gulph of Carpentaria, nor yet to any very great distance inland. [Flinders, 1814, vol. I, pp. 154-155]*

The first day at sea the clothes and personal effects of Thistle and the other lost men were sold at the customary auction before the mast. Evans, the master's mate, was appointed acting master and Nathaniel Wright, a seaman, was promoted midshipman 'for meritorious conduct'. *Investigator* anchored among the islands of the Sir Joseph Banks Group, the scientists landing to collect specimens and the commander to take bearings – 'but [the islands] proved so numerous that the whole could not be completed before dark'.

Resuming their course next morning, they followed the line of the coast north-north-east, but in the afternoon the water began to shoal; next day, land was seen to starboard, extending northwards 'in a chain of rugged mountains, at the further end of which was a remarkable peak':

> Our prospect of a channel or strait, cutting off some considerable portion of Terra Australis, was lost, for it now appeared that the ship was entered into a Gulph; but the width of the opening round Point Lowly [on the western shore] left us a consolatory hope that it would terminate in a river of some importance.[19]

Soon the coasts on each side closed in, finally ending in mudflats 'little better than a drain from a swamp'. Flinders looked for an anchorage, but in a fresh gale blowing up the gulf the ship almost grounded in $2\frac{1}{2}$ fathoms on a middle bank. He 'wore round and stood back into 5 fathoms, then anchored with the best bower and furled sails'; the low shores to port and starboard seemed barely 4 miles apart.

At dawn on 10 March, Robert Brown, Bauer and Westall, with Peter Good, John Allen, and two servants, set out for the chain of mountains to the east, its base seeming no more than 5 miles distant. For the first half-mile they struggle knee-deep through mud and weed, then find their path obstructed by salt rivulets and mangrove swamps that slow their progress.

The marshy ground gives way to a parched scrub land, relieved here and there by meagre trees lining dried-up creeks. Beyond is a straggling wood; a few native huts, 'rather more comfortable than those about King George's Sound', lie scattered along its borders, but their inhabitants appear to have fled. Once through the trees the party crosses a grassy plain and makes for the highest mountain in the chain, planning to reach the summit by midafternoon. The distance from the ship proves closer to 15 miles than 5, and it is already past two when they arrive at the base of the mountains.

The servants, laden with baggage, are overcome by heat and fatigue; unable to continue, they camp at the foot of the spur, with a small spring nearby. Brown and the others start the ascent, climbing and descending a series of lightly wooded foothills separated by narrow ravines, and reach the summit shortly before sunset.

They are rewarded with the most extensive views they have yet had in New Holland – in the clear evening air they can see perhaps 25 leagues into the heart of the country. To the south lies a range of steep hills rising one behind another, cut by ravines and gullies; to the east, an almost featureless plain relieved only by one small range; and to the west, between them and the gulf where the ship lies at anchor, is more level country, stretching north and south as far as the eye can see. The line of the stream draining into the gulf gradually diminishes till it is lost in a vast plain, ending in a range of low flat hills. As the sun slips beneath the horizon the first stars punctuate the twilight.

Overtaken by darkness, the party is forced to seek shelter in a cramped gully just below the summit. The five men are without water, little firewood can be found in the dark, the ground is uneven and rock-strewn,

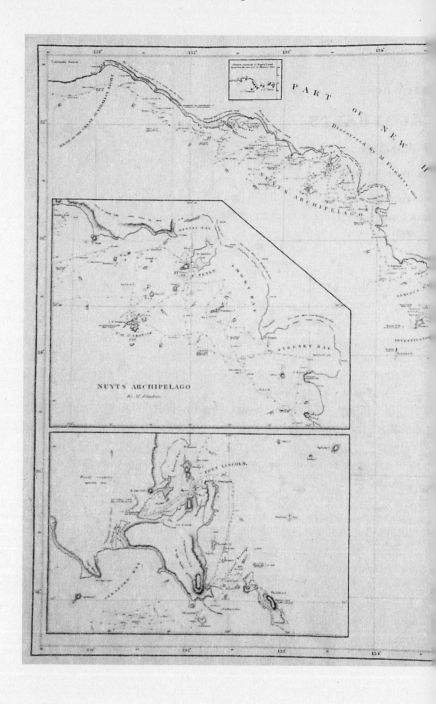

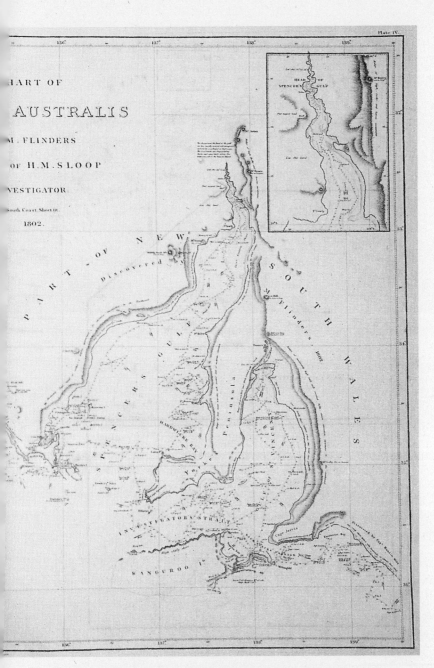

The Spencer Gulf region, 1802. Chart by Flinders.
[By courtesy of the Royal Geographical Society of South Australia]

*and the night air chilly at this altitude; sleepless, they count the hours
to sunrise.*

*At first light they resume the descent, stopping briefly while Westall
sketches the landscape to the west, lit by the first rays of the sun. Arriving
at the camp soon after seven, they find the servants beside a blazing fire,
a spring of clear fresh water nearby; a breakfast of salt beef and ship's
biscuit, washed down with cold water, has rarely been more welcome.*

Meanwhile, Flinders and Surgeon Bell took the cutter to explore the
country at the head of the gulf. Mangroves and shoal water made landing difficult, but at last they were able to get ashore near a high bluff
on the western side; after taking bearings, they returned to the boat
and rowed north along the shore, camping overnight among the mangroves. Next morning, on their way back to the ship, a flight of black
swans rose from the flats behind them; at the thought of fresh food
the men turned and rowed upstream, but although a hundred or more
swans were seen, there was 'not one that could not fly, and they all
escaped us'. Much time was lost in the hunt, and at six o'clock they
were still 9 miles from the ship. Missing their way among the mangroves and salt swamps in the dark, it was ten before they clambered
on board, hours after Brown and his footsore party had arrived.

The voyage resumed at dawn on the 13th, but within two hours the
ship grounded on a mud bank, 2 to 3 miles off the eastern shore; she
was hove off in half an hour with the aid of a kedge anchor. The shoals
continued southward, preventing Flinders from examining the barren-
looking eastern coast close to. On the 19th the coastline began trending
to the west, and towards evening they dropped anchor under a remarkable headland (Corny Point), well sheltered from the southerly winds.

Next morning, with the eastern headland lost to view behind, Flinders was able to obtain bearings on the land and islands on the western side of the gulf – proof that the survey of the great inlet was now
complete. He steered south, satisfied that

> for the general exactness of its form in the chart, I can answer with tol-
> erable confidence; having seen all that is laid down, and, as usual, taken
> every angle which enters into the construction.[20]

On his rough charts Flinders designated the gulf Great Inlet no. 12,
but later named it 'in honour of the respectable nobleman who pre-
sided at the Board of Admiralty when the voyage was planned', Earl

Spencer. The 'cliffy-pointed cape' at its eastern entrance became Cape Spencer, and three islands lying offshore the Althorpe Isles, after the earl's estate in Northamptonshire. The naturalists' cold and lonely vigil on the slopes of a waterless, 3000-foot peak was commemorated by naming it Mount Brown.

Dominion of the kangaroo

Off the Unknown Coast of New Holland, Sunday, 21 March 1802.

> *Blew a fresh Gale, with high land to the Southward, streaching to East &*
> *West as far as we could see on coming near the land we found ourselves*
> *much sheltered, with smooth water under the land stood to the Eastward,*
> *the whole country covered with wood, we passed two deep open Bays which*
> *run in to South or South West and anchored in the Evening well sheltered*
> *by high land near the East point … [Good, 1981, pp. 68-69]*

Flinders stood southwards from the gulf, finding the wind veering that way; land could be seen through the haze, but at a considerable distance. Fresh gales from the south-west – 'the hardest yet experienced on the coast of New Holland' – came on during the night, and he kept off and on until daylight, afraid to approach too close in the darkness. The winds moderated next day, and he ran east along the southern coast, looking for any signs to show 'whether or no [it was] a part of the main', but 'neither smokes, nor other marks of inhabitants, [were] perceived …) along seventy miles of its coast'.

The *Investigator* attempted to enter a large bay (Nepean Bay), with several coves within it promising good shelter, but 'the strength of the wind was such, that a headland forming the east side of the bay was fetched with difficulty'. She came to anchor at six o'clock, about a half mile off a small sandy beach. Though it was too late to go ashore, every glass in the ship was trained there, to see what might be observed:

> Several black lumps, like rocks, were pretended to have been seen in
> motion by some of the young gentlemen, which caused the force of their
> imaginations to be much admired; next morning, however, on going
> toward the shore, a number of dark-brown kanguroos [*sic*] were seen
> feeding upon a grass plat by the side of the wood; and our landing gave
> them no disturbance.[21]

The animals were extraordinarily tame, allowing themselves 'to be shot in the eyes with small shot, and in some cases to be knocked on the head with sticks'. Flinders carried a double-barrelled shotgun, and his companions had muskets; he killed ten himself, and the rest of his party made up the number to thirty-nine – 'the least of them weighing 69, and the largest 125 pounds'. The crew were put to work skinning and cleaning the carcasses:

> … and a delightful regale they afforded, after four months privation from almost any fresh provisions. Half a hundred weight of heads, fore quarters, and tails were stewed down into soup for dinner on this and the succeeding days; and as much steaks given, moreover, to both officers and men, as they could consume by day and night. In gratitude for so seasonable a supply, I named this southern land Kanguroo Island.[22]

When the butchery was finished, Flinders scrambled through the tangled undergrowth towards higher land to take his bearings, 'but the thickness and height of the wood prevented any thing else from being distinguished'. Fallen trees, most of them 'nearly of the same size and in the same progress towards decay', were so abundant that he thought a general conflagration many years before the most likely cause – the result of lightning, or perhaps the friction of two dead trees in a strong wind. 'Can this part of Terra Australis have been visited before, unknown to the world?' he wondered. 'La Pérouse was ordered to explore it, but there seems little probability that he ever passed Torres Strait.'

The men are kept busy replacing the topmasts, damaged in the recent storm, and rigging the new masts, stowing the booms afresh, gathering firewood for the ship's oven, and as always searching for fresh water. Robert Brown and his party stay ashore, examining the island's natural resources. Walking east along the shore, Peter Good finds several new plants, and comes to a spot 'where fresh water rises in tolerable plenty from the crevices of rock' (Hog Bay, Penneshaw). A watering party replenishes the ship's empty casks from this spring.

The officers go shooting, but the kangaroos have learnt from the previous day's massacre – the animals are now so shy that only a few are killed. Some emus are seen but escape the hunting party. In many places the soil appears rich and deep, 'and very capable of cultivation', but as everywhere they have visited along the southern coast there is a general want of water.

In the cabin that evening the captain philosophises, comparing the kangaroos' behaviour with that of the island's seals. Although they appear to dwell amicably together, he observes,

> the report of a gun fired at a kanguroo near the beach [frequently] brought out two or three bellowing seals from under bushes considerably further from the water side. The seal, indeed, seemed to be much the most discerning animal of the two; for its actions bespoke a knowledge of our not being kanguroos, whereas the kanguroo not unfrequently appeared to consider us to be seals.[23]

'March 24 in the morning, we got under way from Kanguroo Island, in order to take up the examination of the main coast at Cape Spencer, where it had been quitted in the evening of the 20th ...'. For three days the ship tacked between the island and the opposite coast, sounding the strait between:

> Of the two sides, that of Kanguroo Island is much the deepest; but there is no danger in any part to prevent a ship passing through the strait with perfect confidence; and the average width is twenty-three miles. It was named INVESTIGATOR'S STRAIT, after the ship.[24]

The soundings completed, *Investigator* entered a large inlet on the northern shore on the evening of the 27th. The soundings were regular, and increasing, and she continued until midnight before coming to anchor. The wind dropped next morning, leaving the ship becalmed, and Flinders gave orders to clean her; afterwards, it being Sunday, 'the ship's company cleaned themselves and were mustered and inspected as usual'.

Inland a ridge of mountains dominated the coastal plain. The country appeared well-timbered and fertile, and numerous smokes rising from the lower slopes 'bespoke this to be a part of the continent'. With winter drawing near, Flinders decided against sending a party to climb 'the loftiest of the hills', and continued the survey northwards.

Five days were spent examining the new gulf. He and Brown landed at the head – it too ended in almost dry mudflats, but finding a small channel they rowed to within a half mile of the shore. Struggling to it along a bank of mud and sand, they set off for a hummocky mount about 8 miles inland, but turned back before reaching it. The flats abounded with rays, so many that a boatload could have been caught had they taken a harpoon with them.

Investigator weathered Troubridge Shoal, at the entrance to the gulf, on 1 April, completing the examination of Inlet 14. When Flinders came to name this latest discovery it became, diplomatically, St Vincent's Gulf,

> in honour of the noble admiral who presided at the Board of Admiralty when I sailed from England, and continued to the voyage that countenance and protection of which Earl Spencer had set the example.[25]

The low headland at its eastern entrance he called Cape Jervis, neatly balancing Cape Spencer at the entrance to Spencer's Gulf. The peninsula separating them became Yorke Peninsula, after Charles Philip Yorke, First Lord at the time of Flinders's return.

Pelican Eden

Off Kangaroo Island, Thursday, 1 April 1802.

> *Having made myself acquainted with the shores of the continent up to Cape Jervis, it remained to pursue the discovery further eastward; but I wished to ascertain previously, whether any error had crept into the time keepers rates since leaving Kanguroo Island, and also to procure there a few more fresh meals for my ship's company. Our course was in consequence directed for the island, which was visible from aloft; but the winds being very feeble, we did not pass Kanguroo Head until eleven at night.*
> *[Flinders, 1814, vol. I, p. 181]*

Flinders anchored some 2 miles from his former anchorage, and about three-quarters of a mile offshore. Next day several parties went ashore, one to shoot more kangaroos for meat, another to cut wood, and the naturalists to explore along the beach. The kangaroos were less numerous here, and few were killed; the carcasses were brought off that evening, and the ship prepared to sail at dawn. No sooner was she under way, though, than Flinders found the timekeepers had run down once more – either Samuel or John Franklin, the senior midshipman, had neglected to wind them the previous day. Resignedly he dropped anchor again:

> ... as some time would be required to fix new rates, the ship was moored so soon as the flood tide made. I landed immediately, to commence the necessary observations, and a party was established on shore, abreast of the ship, to cut more wood for the holds. Lieutenant Fowler was sent in the launch to the eastward with a shooting party and such of the scientific

gentlemen as chose to accompany him; and there being skins wanted for the service of the rigging, he was directed to kill some seals.[26]

Making the best of the unwanted delay, Flinders decided to ascend a high sandhill visible at the head of Nepean Bay, 'from which alone there was any hope of obtaining a view into the interior of the island, all the other hills being thickly covered with wood'. Accompanied by Robert Brown, he set off in the cutter on 4 April.

At the south-west corner of the eastern cove, they find a small opening leading to a sizeable expanse of water, and by one of its branches approach within a mile or so of the hill. With no great difficulty they make their way through the wood at its base, and by one o'clock have reached the summit.

Expecting an uninterrupted view into the interior, Flinders is surprised to see instead the ocean less than $1\frac{1}{2}$ miles to the south – a large bight enclosed by two heads to the south-east and south-west, entirely exposed to the southerly winds, and backed by low craggy cliffs. At a great distance to the north-east Mount Lofty can be made out with his glass across Investigator's Strait. He names the eminence Prospect Hill.

Later they row up the eastern branch of the lagoon; it contains four small islands, one moderately high and woody, the others grassy and lower. It appears to be one vast pelican rookery: young birds, not yet able to fly, nest on the islets, while immense flocks of the older birds sit upon the shore. The islands must be their breeding places; 'not only so, but from the number of skeletons and bones there scattered, it should seem that they had for ages been selected for the closing scene of their existence'.

The two men camp for the night at the entrance to Pelican Lagoon. Warming themselves before a brushwood fire, and dining on oysters picked up on the shoals, they chat quietly about the day's excursion. Flinders, in philosophic mood, meditates on life and death in this pelican Eden:

> *Certainly none more likely to be free from disturbance of every kind could have been chosen, than these islets in a hidden lagoon of an uninhabited island, situate upon an unknown coast near the antipodes of Europe; nor can anything be more consonant to the feelings, if pelicans have any, than quietly to resign their breath, whilst surrounded by their progeny, and in the same spot where they first drew it. Alas, for the pelicans! Their golden age is past; but it has much exceeded in duration that of man.[27]*

181

On his return to the ship, Flinders was again faced with the day-to-day problems of command. Surgeon Bell came to tell him that Richard Daley, a seaman, 'having been simple enough to attack a large seal with a small stick', had received a serious leg wound and was probably lamed for life. Fowler reported that his shore parties had shot a few kangaroos, and had also obtained some seal skins. More fresh water had been collected from the spring on the beach east of Kangaroo Head – though not enough to supply the ship. The news on the timekeepers was good, indicating that 'the letting down had not affected the rates, and tended to give me confidence in their accuracy'. Everything was in order 'to prosecute the discovery beyond Cape Jervis', and at 2 p.m. on 6 April Flinders set sail for the eastern outlet of Investigator's Strait.

The encounter (the English version)

'Investigator *Exploring the Unknown South Coast of New Holland, between Kanguroo Id. and Bass's Strait. Interview with* Le Géographe.' *Captain's fair logbook, 9 April 1802:*

> The French expedition on discovery to New Holland, under Captain Baudin, had frequently furnished us with a topic of conversation, but when we first ascertained that it was a ship seen ahead, it was much doubted whether it was one of the French ships, or whether it was an English merchant ship examining along this coast for seals or whales. On going on board [on the afternoon of 8 April] I requested to see their passport which was shewn to me and I offered mine for inspection, but Captain Baudin put it back without looking at it. He informed me that after exploring the south and east parts of Van Diemen Land, he had come through Bass' Strait, and had explored the whole of the coast from thence to the place of our meeting; that he had had fine winds and weather since leaving the strait, but had not found any place in which he could anchor along the coast, there being no rivers, inlets, or places of shelter.
>
> I inquired concerning a large island in the western entrance of Bass' Strait, but he had not seen it and seemed much to doubt its existence. He had parted with the Naturaliste, his consort, in a gale of wind in the strait and had not since seen her. Captain Baudin was sufficiently communicative of his discoveries about Van Diemen Land and of his remarks upon my chart of Bass' Strait, many parts of which he condemned, but I was gratified to hear him say that the north side of Van Diemen Land was well laid down …

Captain Baudin was much more inquisitive this morning [9 April] concerning the Investigator *and her destination than before, having learned from the boat's crew that our business was discovery; and finding that we had examined the south coast of New Holland thus far, I thought he appeared to be somewhat mortified. I gave as much information to him of Kanguroo Island, the inlets no. 12 and no. 14 [the two Gulfs] and the Bay no. 10 [Port Lincoln] as was necessary to his obtaining water; and I offered to convey any information he might wish to the* Naturaliste, *in case of meeting with him; but he only requested me to say, that he should go to Port Jackson so soon as the bad weather set in.*

As Captain Baudin had an imperfect copy of my miscellaneous chart of Bass' Strait, I presented him this morning with a copy of the three charts lately published, and of the small memorandum attached to them. His charts, he said, were yet unfinished, but that when he came to Port Jackson, he should be able to make some return. Upon my requesting to know the name of the Commander of the Naturaliste, *before I went away, he 'apropos', begged to know mine; and finding it was the same as that of the author of the chart of Bass' Strait which he had been criticising, he expressed some surprize and congratulation; but I did not apprehend that my being here at this time, so far along the unknown part of the coast, gave him any great pleasure. I got information of a rock lying about two leagues off the coast, which has shoal water about. It was reported to lie in latitude 37.1 south and to be 22 leagues from the* Géographe's *situation yesterday noon [subsequently named Baudin's Rocks by Flinders].*[28]

Extracts from Peter Good's journal, 8 and 9 April 1802:

8 April. At 4 p.m. a Sail was seen to windward standing towards us, & soon came so near that we discovered French colours & a Union jack, about Sun down hailed and Captain & Mr Brown went on Board & learned it was the French National Frigate the Géographe, *Citizen Baudin commander. He informed them he had parted with his Consort Le Naturalist in Basses Streights, that he had lost a boat there with 8 men which he had sent in search of some Islands and had never returnd, they also learned he had lost many men at Timor by sickness among which was his Gardener who had been burried beside Nelson*[29] *& a monument erected to their memory – kept company all night.*

9 April. Our commander & Mr Brown went on board the French Frigate early and gained some farther information respecting their discoveries, that they had coasted around Van Diemans Land – and that he had

been on some part of the West Coast, and finding they were much in want
of wood & water directed him to where he could find both – at Captains
return we parted Company – the frenchman standing to Westward & we
to Eastward, light airs and cloudy all day in evening much lightning in
East & North East, some very heavy showers.[30]

'A new and useful discovery'

On board Investigator, *at sea, Friday, 9 April 1802.*

> *At the place where we tacked from the shore on the morning of the 8th,*
> *the high land of Cape Jervis had retreated from the water side, the coast*
> *was become low and sandy, and its trending was north-east; but after*
> *running four or five leagues in that direction, it curved round to the south-*
> *eastward, and thus formed a large bight or bay. The head of this bay was*
> *probably seen by Captain Baudin in the afternoon; and in consequence*
> *of our meeting here, I distinguish it by the name Encounter Bay. The suc-*
> *ceeding part of the coast having been first discovered by the French navi-*
> *gator, I shall make use of the names in describing it which he, or his coun-*
> *trymen have thought proper to apply. [Flinders, 1814, vol. I, pp. 194-195]*

Flinders and Baudin met about 6 nautical miles south-south-east of the
Murray mouth. From this direction and distance nothing could be seen
of the river mouth, and neither captain suspected its existence. Also
hidden behind the coast – 'a low sandy shore very uninteresting' – lay
the Coorong, a line of shallow saltwater lagoons curving southwards
for close to 80 miles behind the dunes.

The *Investigator* ran south-east along the coast for twelve days.
Autumn was well advanced, the weather was worsening, and Flinders
worried about making Port Jackson before the onset of the winter gales.
Knowing that Baudin had preceded him, he gave less attention to his
charting, and often continued his course during the night – 'a risky
undertaking in this uncharted region', says Geoffrey Ingleton. The former RAN officer believes that on the night of 12 to 13 April the ship passed
within a few hundred metres of Baudin's Rocks, of which the Frenchman had warned him, only narrowly escaping shipwreck. In the heavy
squalls and low visibility the rocks were not seen by the men on watch.

Bad weather on the 20th – 'a fresh gale from westward with frequent squalls and heavy showers of rain and hail' – obscured the coast,
and Flinders ran along it at a distance of up to 8 miles. During the

night, 'by favour of moonlight and a short cessation of rain', a high headland (Cape Otway) was seen on the lee some 3 to 6 miles off. Despite the danger to the masts, more sails were set to enable the ship to clear the rocky coast.

The weather eased in the morning, and the mainland coast was lost to view. Flinders suspected that the calmer seas were due to a large island to windward, whose presence had first been reported by a sealing captain in 1799. Unaware that it had since been surveyed by Lieutenant John Murray in the *Lady Nelson* (and named King's Island, after the governor), Flinders sailed south to search for it – 'more especially as it had escaped the notice of Captain Baudin'. Land was sighted on 22 April, and the following day the ship anchored off the island's north-east point.

The botanists spent a few hours ashore and, finding many new specimens, persuaded Flinders to allow them to return the next morning. They were so engrossed in their collecting that the captain was forced to fire two guns to recall them; he was anxious, he said

> to run over to the high land ... on the north side of Bass' Strait; and to trace as much of the coast from thence eastward, as the state of the weather and our remaining provisions could possibly allow.[31]

Cape Otway was sighted at five o'clock, and Flinders ran north-east along the coast. The country impressed him:

> The rising hills were covered with wood of a deep green foliage ... so that I judged this part of the coast to exceed in fertility all that had yet fallen under my observation.[32]

By noon on the 26th they were tracing the land round the head of a deep bight, about 5 miles offshore. Noticing a small opening on the west side of a rocky point, with water breaking across it, Flinders bore away to take a closer look:

> ... advancing a little westward the opening assumed a more interesting aspect ... A large extent of water presently became visible within side; and although the entrance seemed to be very narrow, and there were in it strong ripplings like breakers, I was induced to steer in at half past one; the ship being close upon a wind and every man ready for tacking at a moment's warning.[33]

The soundings were irregular, between 6 and 12 fathoms, until 4 miles inside the heads the water shoaled quickly to less than 3. Before the ship could come around, the flood tide set her on a mud bank, and she stuck fast. A boat was lowered to sound and, when deep water was found to the north-east, a kedge anchor was carried out and the ship worked up to it. With sails filled, she drew off into 6 to 10 fathoms – 'and it being then dark, we came to an anchor'.

At first Flinders supposed the harbour must be Western Port, although the narrow entrance by no means tallied with the description by his friend George Bass. Moreover, Baudin, who claimed to have sailed the coast in clear weather, had seen no inlet between Western Port and Encounter Bay, and he assumed the Frenchman had come to the same conclusion.

The exploration of the bay began at daybreak. Flinders went ashore with Brown and Westall, intending to climb a hill rising to about 1000 feet south-east of the anchorage (Arthur's Seat). The ascent was not difficult and they picnicked at the summit, lunching on oysters picked up at high-water mark, and enjoying an uninterrupted view of the expanse of water spread before them:

> even at this elevation its boundary to the northward could not be distinguished. The western shore extended from the entrance ten or eleven miles in a northward direction, to the extremity of what, from its appearance, I called Indented Head; beyond it was a wide branch of the port leading to the westward, and I suspected might have a communication with the sea; for it was almost incredible, that such a vast piece of water should not have a larger outlet than that through which we had come.[34]

Further along the ridge, the explorers were surprised to see to the east a second large stretch of water, 3 or 4 leagues off, and with an outlet to the sea. This must be Bass's Western Port, and Flinders and his party congratulated themselves 'on having made such a new and useful discovery'. However, their satisfaction proved premature. On arrival at Port Jackson Flinders learned that Lieutenant Murray in the *Lady Nelson* had again preceded him – by just ten weeks – and had named the great harbour, 'capable of receiving and sheltering a larger fleet of ships than ever yet went to sea', Port Phillip Bay, in honour of the colony's first governor.

Thursday 29 April. Leaving orders with Lieutenant Fowler to take the ship back to the entrance, Flinders sets off in the cutter with three days' provisions, to explore as much of the port as possible in that time. Much of the first day is spent sounding and surveying along the south-east shores of the bay, then towards evening he directs his boat's crew to row north-westwards towards Indented Head, 5 leagues off on the western side. With wind and tide against them it is hard work for the men, and it is past nine when they land and camp for the night.

In the morning, smoke is seen not far from the tent, rising from the ashes of a native camp-fire; the 'Indians' seem to have decamped. A small group show themselves a mile away but shyly keep their distance. To encourage them to come closer, Flinders hangs strips of red cloth, their favourite colour, about their camp. Shortly three men approach, and trustingly exchange their weapons for these and other trifling presents. As a gesture of goodwill, Flinders shoots a bird hovering near the boat, and offers it to them. To his surprise they run down to the water and accept the present without hesitation: 'their knowledge of firearms I then attributed to their having seen me shoot birds when unconscious of being observed', he writes, 'but it had probably been learned from Mr. Murray'. White men and black sit down to a shared meal of roast duck.

Saturday, 1 May. Accompanied by three seamen Flinders sets out for Station Peak, as he calls the highest part of the back hills. Their way crosses a low, flat plain; covered with small-bladed grass and slightly wooded, it presents ' great facility to a traveller', and by ten they arrive at the peak. From a large boulder atop the summit, Flinders surveys the whole extent of the port, stretching north and south for at least 30 miles, and 36 from west to east. Leaving the ship's name on a scroll of papers in a stone cairn at the peak, he and his men return to the tent at three o'clock most fatigued – they have walked 20 miles without finding a drop of water.

The crew row back to Indented Head, arriving at the camp site after dark. On the journey, lights are seen on shore, and two newly built huts are found, with fires still burning and food in a basket; the people have fled. Despite a close watch throughout the night, nothing is seen of the occupants until the boat pushes off in the morning. Seven men appear on a small hill nearby, and rush down to examine their huts; finding everything as they had left it – except for some water taken by the visitors to slake their thirst – they seem content, and for a short time follow the boat along the shore.

Investigator weighed anchor early on 3 May, and cleared the entrance with the ebb. The passage through Bass Strait was squally, but otherwise uneventful, and she entered Port Jackson on the 9th, anchoring in Sydney Cove at 3 p.m. 'There was not a single individual on board', Flinders wrote with pride,

> who was not upon deck working the ship into the harbour; and ... the officers and crew were, generally speaking, in better health than on the day we sailed from Spithead, and not in less good spirits.[35]

Notes

1. Flinders, 1814, vol. I, p. 47.
2. Good, 1981, p. 46.
3. Flinders, op. cit., pp. 100-101.
4. The whaler *Elligood* called at Cape Town in May 1801, her captain reporting that Captain Dixson and nine of her crew had died from scurvy. The plot may have been Dixson's grave.
5. Flinders, op. cit., p. 57.
6. Ibid., pp. 60-61.
7. Horner, 1995, p. 121.
8. Flinders, op. cit., p. 73-74.
9. Cape Barren Geese.
10. Ibid., pp. 93-94.
11. Ibid., p. 97.
12. Ibid., pp. 102-103.
13. Although most authorities attribute the name Point Bell to the surgeon, Flinders does not specify this. Possibly the name commemorates Surgeon Bell and Midshipman Bell.
14. Ibid., p. 134.
15. Austin, 1964, pp. 92-93.
16. Flinders, op. cit., p. 133.
17. Ibid., p. 139.
18. Ibid., p. 146.
19. Ibid., p. 156.
20. Ibid., p. 167.
21. Ibid., p. 169.
22. Ibid., p. 170.
23. Ibid., p. 172.
24. Ibid., p. 175.
25. Ibid., pp. 179-180.

26. Ibid., p. 182.
27. Ibid., pp. 183-184.
28. Cooper, 1952, pp. 66-67.
29. David Nelson, Bligh's gardener on the *Bounty*, died at Timor.
30. Good, 1981, pp. 72-73.
31. Flinders, op. cit., p. 208.
32. Ibid., pp. 209-210.
33. Ibid., p. 211.
34. Ibid., p. 213.
35. Ibid., p. 226.

CHAPTER SIX

⚓

Baudin in Tasmanian Waters, 1802

'To the extremity of the globe'

On board the Géographe, *Kupang Bay, Friday, 13 November 1801.*

> *We had been at Coupang 84 days, and our stay there, under all consid-erations, had been very fatal; we had lost a great deal of time; death had robbed us of several of our shipmates, and we were encumbered by a great number of sick on board each of our vessels. Such were the deplorable consequences of this long stay at Coupang; it even appeared very probable, that a farther residence in this island would have lost us all the remain-der of the crews of both ships. Which of us would not have thought at the time, that we quitted these destructive shores for ever. [Péron, 1809, p.137]*

'Impatient to leave the country where disease had dealt us such a blow', Baudin sailed before sunrise; but so weak was the wind that both vessels were soon becalmed. To their crews, both had the appearance of hos-pital ships, with a dozen men too sick to leave their beds on the *Natu-raliste*, and perhaps twenty on the *Géographe*.

In their airless quarters amidships on the lower deck, the sick men lay prostrate in the high humidity and oppressive heat. Within ten days their numbers had swelled to twenty-five on the flagship and eighteen on her consort, mostly with a vicious dysentery. 'The symptoms of this terrible illness [Baudin wrote in his journal] are so frightening that the moment one is struck down by it one feels dead already …'. The most dedicated efforts by the surgeons seemed in vain. Péron reported de-spairingly that

> whoever was attacked with any degree of violence by this terrible dis-ease infallibly died. It pursued us … to the extremity of the globe, and strewed the seas with our dead bodies.[1]

Sea funerals became a familiar ritual. The dead men were sewed into their hammocks, weighed with round-shot, and buried over the side. Baudin, so weakened by fever that he could barely stand, mourned the loss of some of his best men: the gardener's assistant, Sautier, noted for his 'zeal and diligence'; Corroyer, a helmsman, 'very steady and extremely active'; Poussin, 'a good seaman'; and two gunners, Mantel and Macon. Three days before his death the latter had attempted suicide by jumping overboard, but had been rescued by his messmates:

> When he was lifted back on board he said to us with great calmness:
> 'You have done me a great disservice, for I know well that I shan't recover ... You are only prolonging my sufferings for a few more days.'[2]

On 12 December the ships passed the tropic of Capricorn; entering these more temperate regions 'the heat became less oppressive, and the health of our sick seemed to mend as the thermometer sank'. For some men, though, the improvement came too late. The *Géographe*'s master sailmaker, 'a very respectable man ... much esteemed by our officers and crew', died about midnight the same day. On the 23rd, Hamelin signalled from the *Naturaliste* that Stanislas Levillain, the young zoologist who had accompanied Baudin to the West Indies, had succumbed to a second bout of dysentery.

The news of Levillain's death heightened Baudin's concern for René Maugé, his sole remaining colleague from the *Belle-Angélique* voyage. Seemingly recovered from dysentery, he too had had a relapse, and complained of acute pains in his lower abdomen.

> It seems I am destined on this expedition to be parted from my best friends, and to have not only the pain of seeing them die, but also my own reproaches, since it is only through friendship for me, and in order to accompany me, that they have joined it.[3]

Levillain was the eleventh, and last, to die on the two ships' passage south – eight from dysentery, two from fever, and one from a liver complaint. The remaining invalids convalesced slowly, although Maugé, Boullanger the geographer, and the mineralogist Depuch all gave Baudin cause for concern. On the *Naturaliste*, Milius, promoted commander at Timor, remained on the sick list with a chronic bilious colic that seemed resistant to all treatment.

They reached the latitude of Cape Leeuwin on 4 January. With the wind blowing a moderate gale from west-north-west, and the sea growing rough, Baudin gave up hope of sighting land in the heavy conditions. Signalling Hamelin to take advantage of the wind, he steered south-east for D'Entrecasteaux Channel in Van Diemen's Land.

Some 200 nautical miles northwards, in fresh squally weather, Matthew Flinders prepared to sail eastwards from King George Sound.

The ships run before the wind, taking full advantage of the westerlies to complete the passage from Cape Leeuwin in little more than a week. Mid-ocean they are joined by flocks of seabirds – albatrosses, gulls, and petrels 'of the kind that sailors do not like, most of them being convinced that their presence is a bad omen'. The seas are rough and swelling, the sky threatens; strong squalls and sharp gusts of wind fill the sails, straining the masts.

The Géographe *rolls incessantly, so much so that the oakum starts from the seams. The sick men's convalescence drags on. Maugé, white and haggard, withdraws to his cabin and refuses food. Boullanger, the geographer, unable to stand up to the ship's rolling, lies in his bunk, vomiting.*

The island's south-western coast is sighted through mist and rain at daybreak on the 13 January. Péron, always among the first on deck at a new landfall, watches spellbound as the ship follows the rugged south coast, doubles South-East Cape, and runs for shelter in Recherche Bay, the Naturaliste *in her wake:*

> *Every eye was now fixed on the land; we admired those lofty mountains, which nature has placed like so many ramparts of granite to oppose the stormy seas [which] extend as far as the frozen antarctic pole. We observed with admiration those large plains in the interior of the island, which rise in amphitheatres over the whole surface, and are covered with immense forests. The sea all this time was stormy and rough; the winds blew violently and in squalls from the S.W.; the temperature was cold, the sky thick, and long clouds of vapour gathered round the grey sides of the woods and mountains. This fog was succeeded by heavy rains, hail, and hoar frost; innumerable flights of boobies, gulls, cormorants, swallows &c. flew from the neighbouring rocks and encircled our ships, mingling their piercing cries with the noise of the angry waves; a long rank of white muzzled dolphins, with several large whales, played around us. In a word, every thing seemed to unite in giving a sort of solemnity to our arrival off these shores, and all proclaimed that we touched the extreme boundaries of the southern world.[4]*

The two vessels headed for Recherche Bay, Hamelin in the lead because his ship drew less water, taking soundings as he went. A large bank of rocks ahead, almost blotted out by the driving rain, forced him to veer to starboard. Unable to find a clear passage into the bay, he continued on into D'Entrecasteaux Channel, followed by the *Géographe*. At 4.30 p.m. the ships dropped anchor at the entrance to Great Taylor's Bay, east of Partridge Island.

D'Entrecasteaux Channel

At anchor in Great Cove, D'Entrecasteaux Channel, 14 January 1802.

> *To Citizen [Henri de] Freycinet, Lieutenant Commander.*
>
> *Citizen, I am putting you in command of my longboat to go and visit several places which, although thoroughly charted by General d'Entrecasteaux, may offer various objects of curiosity and Natural History …*
>
> *Upon leaving the ship, you will proceed to the Huon River. After entering its mouth, you will reach Port Cygnet, where you are to remain a little while in order to find some swans for us. I think that two days will be sufficient for the expedition, so you are to take only enough food for that time.*
>
> *There is no need for me to recommend you to be constantly and actively on the alert for the safety of the longboat and all those accompanying you. But if you should meet any natives, which is very likely, you are absolutely forbidden to commit a single act of hostility towards them, unless the safety of anyone in particular, or all in general, is at stake. According to what is known of their character, the people of this country do not appear to be savage, except when provoked. Therefore, you must influence them in our favour by kind deeds and presents …*
>
> *[signed] N.B.*
>
> *[Baudin, 1974, pp. 301-302]*

Freycinet's boat left for Port Cygnet at daybreak, with Péron, Lesueur, and Guichenot the gardener's boy on board; they carried hatchets, mirrors, knives, snuffboxes, beads and other trinkets for presenting to any natives they might meet. Baudin, accompanied by surgeon Lharidon and Bernier the astronomer, made for Partridge Island in the large dinghy. Captain Hamelin joined them as they passed astern of the *Naturaliste*; he too had sent several boat parties ashore.

The rain of the previous day had ceased, but conditions were far from pleasant; it took Baudin's men an hour and a half's strong rowing into the teeth of a south-westerly gale and against a strong-flowing current to reach the shore, barely a mile from the ships' anchorage.

Once they were on land, thick vegetation blocked the way, and Baudin and his companions returned to the shore, clambering over the granitic rocks lining the coast, at the foot of which the sea broke and boiled. A group of men came into view in the distance. On approaching closer they were recognised as Leschenault, the *Naturaliste*'s botanist, and two of her seamen, walking amicably with several natives. The latter, 'no doubt intimidated by our large number', ran off into the woods.

As Leschenault explained, he had no sooner landed than the natives came to meet him, their friendly gestures suggesting confidence and no harmful intent; moreover, they were unarmed. Mutual trust had quickly been established, and from then on they followed him everywhere. Leaving the botanist to continue his excursion, Baudin and his party made their way to a sandy beach where the drawnet might be thrown with advantage.

Before long a half-dozen dark shadows slipped from the shelter of the woods and came towards the party, showing not the slightest sign of distrust. Following Leschenault's example, 'we took them by the hand and embraced them, and gave them a few trifles as well'. Some of the party who had been eating bread or biscuit offered a piece to the natives, but none accepted the offering.

The Aborigines were of medium height and well-proportioned, except for weak and spindly legs. All were naked, apart from one wearing a skin that partly covered his back and shoulders, and were tattooed for about 6 inches below the shoulders, the lines running across and sometimes down.

> We could make nothing of their long speeches to us, but their signs were much more intelligible … Most of the things that we carried appeared to attract them greatly, and they would have been very pleased by our giving them our clothes … They are much paler in colour than the African negroes, and perhaps it is because they do not think themselves black enough that they daub various parts of their faces with charcoal … There was nothing unpleasant about these men. Their expression was one of liveliness and even gaiety, and their glance was quick. Their noses were rather flat and their mouths wide. They had beautiful white teeth that were well set and without a fault … They could have been from twenty-five to forty years old.[5]

Retracing their steps, the Frenchmen passed a second group of six or seven natives. These had clearly been fraternising with the sailors left at the boats, and were returning with their presents, carried slung around their necks. One wore a jacket given him by a sailor; another sported a piece of flag draped over his shoulders. The explorers had brought no dinner ashore, and Baudin ordered all boats to return to the ships.

With the wind behind them, Freycinet's party makes good progress up the Huon River to Port Cygnet. Lofty trees, crowded so close together it seems impossible to penetrate them, cover the shores on either side. Countless flights of brilliantly coloured parrots and cockatoos fill the highest branches, blue-collared tomtits flit through the shade below. Flotillas of black swans cruise majestically on the river's calm surface.

Freycinet steers in to land at the mouth of a small creek, which offers hope of a freshwater stream. Péron has scarcely set foot ashore when two natives appear at the top of a nearby hill. Responding to his friendly gestures, one of them bounds down the slope and into their midst. He is a young man, perhaps twenty-two to twenty-four years old, robust in build but with the reedy arms and legs characteristic of his race. His manner displays pleasure and surprise at the unexpected meeting.

> *M. Freycinet having embraced him, I followed his example, but the air of indifference with which he received this testimony of good will and friendship made us easily perceive that to him it had no meaning. What appeared at first to interest him most, was the whiteness of our skin, and doubtless wishing to ascertain whether the rest of our bodies was of the same colour, he successively opened our jackets and shirts, and expressed his astonishment by loud exclamations of surprise, and by very quick motions of his feet.[6]*

The longboat next attracts the native's attention and he leaps in, ignoring the rowers resting on their oars. Absorbed by something so new and unimagined, he examines in silence her ribs and planking, the rudder, the oars, the mast and sails. To amuse himself, one of the seamen offers him a glass bottle filled with arrack – part of the men's daily liquor allowance. Surprised at first by the glitter of the glass, the native soon loses interest and tosses the bottle overboard. The seamen's grins turn to shouts of alarm, and one of them dives into the water fully clad to retrieve it.

The *Géographe*'s two dinghies set out next morning to cast their nets off the sandy beach explored the previous day. Bernier the astronomer,

Petit the artist, and Maugé the zoologist went with them – the latter ignoring Baudin's pleas not to risk another relapse in the stormy conditions. Two armed marines accompanied the shore party – a sufficient defence, in the commandant's view, 'to ensure the safety of all those setting off'.

The sailors began their fishing, and a crowd of natives, about fifty in all, gathered at the beach to watch the nets being hauled in. Men, women, and children mingled happily with the Europeans, neither side showing any sign of mistrust or fear. With their work completed, the French lit a huge fire, and invited their visitors to sit down while the sailors shared part of their catch with them. Petit seized the chance to sketch several of the men – the women and children, less patient, would not stay still for more than a moment.

As the day wore on, relations between the two groups around the fire became warmer and more relaxed. Several women 'used various ways and means to draw some of the seamen aside, and the signs they made were too expressive for anyone to mistake their intentions'. Midshipman Maurouard, a strong and energetic young man, decided to while away the time by challenging a robust-looking native to a trial of strength.

> He began by clasping his wrist in the way that young men do in Europe, making signs to his adversary to pull as hard as possible. The other understood perfectly, but could not equal Maurouard's strength. After this first success, the two athletes grappled with each other to see which one could down his opponent. Maurouard was again the victor and threw the other over. As this scene took place on the sand, the fall could not be dangerous to either. Moreover, it all happened amidst much laughter, and the loser seemed happier with his tumble than the victor with his triumph.[7]

The fire died down, and the natives withdrew into the wood, loaded with presents and seemingly in excellent spirits. The French were preparing to board their boats when, without warning, a spear struck Maurouard on the neck, grazing the bone and cutting through the flesh at the base of the neck. It had been hurled with considerable force, for it passed through a lined woollen coat before piercing the skin.

The boat's crew, provoked by the cowardly assault, 'wanted to pursue the attackers and punish them as they deserved', but no-one had seen where the spear had come from. Further off, a small group of natives, all armed with spears, could be seen at the edge of the wood beyond the beach. They made no move to attack, and Bonnefoi, the boat commander,ordered his men to hold their fire; the crew embarked without further trouble.

Children of nature

It is for want of having made a proper distinction in our ideas, and seen how very far [the savage nations] already are from the state of nature, that so many writers have hastily concluded that man is naturally cruel, and requires civil institutions to make him more mild; whereas nothing is more gentle than man in his primitive state, as he is placed by nature at an equal distance from the stupidity of brutes, and the fatal ingenuity of civilised man. [Rousseau, Discourses in Inequality, *1755]*

At Port Cygnet the shore party commenced its explorations. Lesueur set off alone in search of forest animals and birds, while Péron and Henri de Freycinet made their way into the bush – the scientist hoping to make contact with the land's original inhabitants, the officer more concerned with finding fresh water. Climbing the hill behind the landing place, they met the young native's companion, an older man perhaps fifty years of age, and two women, whom they took to be his wife and daughter; all were naked.

The older people seemed friendly and well-disposed; 'notwithstanding some unequivocal signs of fear and disquiet, it was easy to discover kindness and candour' in their countenance. The younger woman was more timid than her parents, and hesitated a moment before approaching. She carried at her breast a small baby girl wrapped in a kangaroo skin; her breasts, already a trifle sunk, were otherwise well-formed, and abundantly supplied with milk. But it was her eyes that held Péron's attention:

[they] had an expression and fire which astonished us, and which we have never since observed in any other female of that nation. She appeared besides to be extremely fond of her little infant, and her care of it had all that kind and affectionate character which is acknowledged by everybody as the peculiar attribute of maternal love.[8]

The Frenchmen offered various presents to the family; 'but every thing which we offered them was received with an indifference that surprised us, and which we had often occasion to observe among individuals of the same country'.

Freycinet takes several of the seamen with him to look for a freshwater stream, while Péron stays with the savages, 'occupied in observing them, to describe their natural habits, and in endeavouring to collect some words of their idiom'. The young man, to all appearances the father of the baby girl, helps the remaining sailors to collect branches for a fire, for even though it is midsummer the thermometer stands at less than 10°Re.[9] He lights the fire with the aid of a torch he has concealed nearby, and it is soon blazing merrily.

The sailors warm their hands at the blaze; one of them takes off his fur gloves and puts them in his pocket. At the sight the young woman screams and jumps backward; it seems she has taken the gloves for real hands, or at least a sort of living skin, that can be peeled off and re-placed at will. The French laugh heartily at the mistake, but are less amused when the old man seizes a full bottle of arrack and attempts to carry it off:

> *As it contained a great part of our stock, we were obliged to make him restore it, at which he seemed to express a great deal of resentment, for he left us, together with his family, notwithstanding all that we could do to make them remain with us longer.[10]*

Freycinet and his party return, having found no trace of fresh water despite a long and laborious search. Lesueur has been more successful, bringing back a dozen or so birds, including several species of parrot and a blue-collared tomtit. They sit down to a frugal meal, prepared by the sailors over the fire's embers, before setting off to explore further along the coast.

A short distance away the explorers found a primitive hut, little better than a windbreak made of bark, arranged in a semicircle and propped up by some dry branches; before it lay the remains of a dying fire, surrounded by piles of seashells that gave off a nauseating smell. Two canoes were drawn up at the water's edge, each formed of three rolls of bark loosely held together by strips of the same material. There could be little doubt they belonged to the family they had just met; and a moment or two later the same group were seen approaching along the sandy beach.

> As soon as they observed us they shouted for joy, and mended their pace to join us. Their number was now increased by a young girl about sixteen

or seventeen years of age, a little boy of four or five years, and a little girl of three or four years. The most aged of them seemed to be the father and mother, the young man and his wife seemed also to be brother and sister, and we supposed the young girl to be also the sister of these last; the children we concluded might be the offspring of the young man and woman.[11]

The family were clearly returning from a fishing trip, for each was loaded with shellfish. Taking Freycinet by the arm, the old man led him towards the hut, gesturing to the others to follow. The fire was quickly rekindled, and after requesting their visitors to sit down (*médi, médi!*), the natives squatted on their heels and began their meal. The cookery was simple, the shellfish being baked in their shells and eaten without further preparation or seasoning. Given a portion to taste, the visitors pronounced it 'succulent and well-flavoured'.

> While our good Diemenese thus enjoyed their simple repast, the idea of treating them with a little music entered our heads, not so much to amuse them, as to see what effect our singing would have on our audience. We chose the hymn which was so unhappily prostituted during the revolution ['La Marseillaise'], but which is nevertheless so full of enthusiasm and spirit, and so likely on this occasion to produce effect.[12]

At first the music seems to confuse the natives, but after a moment's unease they listen attentively; they stop eating, and show their delight by so many bizarre gestures that the Frenchmen can barely refrain from laughing aloud. The young man in particular is beside himself, pulling his hair, scratching his head vigorously with both hands, throwing himself from side to side, and whooping repeatedly at the end of each chorus. Next, the French sing some light and tender melodies, but although the savages 'seem to comprehend the sense of these', it is obvious that sounds of this kind do not affect them.

The scene changes, and takes on a more familiar character. The young girl, whose name they learn is Ouré-Ouré, is remarkable for the softness of her features and the expression of her eyes, at once tender and spiritual. As naked as her parents, she is not in the least aware that her nudity can be considered indecent or immodest. Less robust than her sister and brother, she is also more lively and passionate. M. Freycinet is seated next to her, and appears to be the particular object of her regard. It is easy to see in this innocent child of nature that delicate shade, which gives to the most simple playfulness, a character of serious preference; coquetry itself seems to be called in to the assistance of the natural attractions of the sex.[13]

Ouré-Ouré shows them the basis for her make-up, and demonstrates how it is applied. Taking several pieces of burnt charcoal in her hands, she crushes them to a fine powder, which she rubs first on her forehead and then on her cheeks, making herself 'frightfully black'. The new adornment gives 'an added degreee of self-satisfaction and confidence to the expression of her countenance'. Thus, Péron concludes, 'a fondness for ornament and a sentiment of coquetry prevails in the hearts of the whole sex'.

The little children, meanwhile, have been imitating their parents' gestures and grimaces, their feet beating the ground in time to the rousing choruses of 'La Marseillaise'. Lively, merry and mischievous, they soon lose their reserve, becoming

> *as much at their ease with us, as if we had been long acquainted ... It is curious to find at the extremity of the globe, and in this unformed state of social intercourse, these amiable and affecting characters, which, among us, also distinguish the days of infancy.*[14]

It was late evening when the French got ready to leave. Freycinet took Ouré-Ouré's arm, Péron walked with the old man and Lesueur with the younger one, while a seaman led the boy. The rest of the family remained at the hut.

The path to the boat was full of brambles and thorny shrubs, 'and our poor savages, being naked, were much scratched by them'. Ouré-Ouré ignored the scratches to her legs and belly, totally absorbed by her chattering to Freycinet:

> Provoked at not being able to convey her ideas, or to understand him, she accompanied her discourse with so many winning gestures, and gracious smiles, that her coquetry was very expressive.[15]

Years later, Péron recalled their departure with mixed feelings:

> Thus ended our first interview with the inhabitants of Van Diemen's Land ... The confidence which [they] showed us, the affectionate testimonies of goodwill which we could not but understand, the sincerity of their demonstrations, the frankness of their manners, the affecting ingenuousness

of their caresses, all seemed to unite in developing the kindest and most interesting affection and friendship ...

I saw realised, with inexpressible pleasure, those charming descriptions of the happiness and simplicity of a state of nature, of which I had so often read, and enjoyed in idea. I was at the time far from conjecturing the many privations and miseries to which such a state is liable.[16]

Fire and water

At the entrance to North West Bay, D'Entrecasteaux Channel, Tuesday, 19 January 1802.

> *This short voyage between the two shores of the channel was very pleasant, and the prospects very picturesque ... but what [most] astonished us, was the multiplicity of fires which we perceived. In every direction immense columns of flame and smoke arose; all the opposite sides of the mountains, which form the bottom of the Port North-West, were burning for an extent of several leagues. [Péron, 1809, p. 187]*

The two ships moved further up the channel, coming to anchor in the sheltered waters of North West Bay on 22 January. Wooding and watering parties were sent ashore, a hospital tent was pitched close to the water's edge, and the astronomer Bernier set up his instruments nearby to observe Jupiter's passage across the sun. On board, crew members were employed below decks, clearing the holds for stowage of the water-casks and examining the spare sails; several of these were found to be damaged by salt water seeping through leaks in the deck seams.

The watering parties returned empty-handed, having failed to find any freshwater streams. Quite by chance, Engineer Ronsard made the discovery while hunting at the far end of the bay. For Milius, 'the fresh water was as delightful to us as the water of the New World must have seemed to Columbus and his companions'. Ronsard also brought back from his hunting two black swans, several teal, and a large pelican – a very handsome bird that would make 'a superb exhibit in the Natural History collection'.

Baudin discussed with Hamelin the surveys needed to complete Admiral d'Entrecasteaux's observations in 1793. Together, they agreed to send out two parties, one from the *Géographe* under Henri de Freycinet to follow the course of the North River (the Derwent) to its source, and the second from the *Naturaliste* to examine Tasman's 'Frederik

Hendrik Bay' – which D'Entrecasteaux's charts showed as a possible strait leading to the open sea. Both boats were to leave in the early hours of the morning.

Freycinet's party, Péron among them, left the ship before dawn. To the west, the slopes of Plateau Mountain (Mount Wellington) were shrouded in fog, which soon 'resolved into a very thick cold dew'. By noon they were abreast of the furthest point reached by D'Entrecasteaux's men, a great bluff on the eastern bank; just beyond it the boat ran aground on a large mud shoal, 'which, with all our strength, we tried in vain to get over'. They camped overnight, and next morning Freycinet set off on foot with a group of armed sailors to follow the river's course upstream. Pushing on through marshes and scrub for about 12 miles, he reached higher ground, from which he could trace the river's direction until it disappeared in the hills beyond.

Curious to see the interior of the country, Péron takes leave of his friends and sets off into the wilderness, alone and unarmed. The morning is fine but hazy, the air refreshing; pale shafts of sunlight slant through the early mist. In the shadow of a deep ravine, a natural rampart barring his way, he comes across a dozen or more primitive bark huts, similar to that of Ouré-Ouré's family; smoke curls lazily up from fires still burning before several of them. Doubtless they have just been abandoned by their occupants, frightened by the noise of his comrades' gunshots carried on the breeze.

Searching these humble shelters, Péron finds scattered among them the freshly discarded bones of kangaroos and birds; nearby lie several large flat stones, still warm and greasy, and obviously used as broiling plates. He picks up a number of knives and hatchets, mere fragments of granite, very fine and hard, and varying in size according to their purpose; 'with this substance the savages also make their weapons, which are a sort of tomahawk, and their pointed sagaies [spears]'.

Piercing cries coming from the bottom of a nearby gully cut short his examination. Hurrying away, he comes to a high hill, and makes out from its summit 'the whole course of the North River, which after making a great elbow, is lost among a lofty chain of mountains'. Beyond them lie even higher peaks, some of them seemingly covered with ice and snow. At four-thirty Péron returns to the boat, which has been left under the guard of a midshipman and some seamen. Freycinet and his weary party rejoin them three hours later.

Pierre Faure, the *Naturaliste*'s geographer, left with the second boat two hours after Freycinet's departure. His task, wrote Baudin, was 'to examine Tasman's Port Frederik Hendrik [and] to see if there were a practicable passage for our corvettes through the opening that it appears to have into Marion Bay' (on present-day maps, from Norfolk Bay through to Blackman Bay).

Faure's party was away for eleven days. From the anchorage in North West Bay, he headed east across Storm Bay for Tasman 'Island'. Following the coastline, he entered Norfolk Bay and, after closely examining its western and southern shores, came to a long narrow inlet at its eastern extremity. Here he made the discovery that he and Baudin considered his most important achievement:

> According to the report of Faure, the geographer, ... it seems that Abel Tasman Land has been wrongly made an island.[17] It is joined to Van Diemen's Land by an isthmus, which at no stage could be covered by the sea and which is about ninety paces in length and roughly the same across. Citizen Faure walked all over it and in every direction. On the chart that he has drawn up one can see not only Frederick Henry Bay, but also the isthmus connecting the so-called Tasman Island with Van Diemen's Land. There is most certainly no separating channel.[18]

Before returning to the ship, Faure also surveyed the northern coastline of Frederick Henry Bay, discovering the Carlton River estuary and Pitt Water:

> a piece of water that was rather shallow, but of great extent; [and] so sheltered, that it might at all times be a safe place of mooring for small vessels and boats.[19]

Faure's successful expedition did little to raise Baudin's spirits. The evening before, Dr Lharidon had come to tell him

> that Citizen Maugé had been in great discomfort during the day and he was beginning to despair of being able to save him, for his illness showed very dangerous symptoms.[20]

The news came as a heavy blow, and led Baudin to fear 'that what I had had an inkling of for a long time now might actually come to pass'.

From their camp on the North River, Freycinet and Péron again attempt a passage over the mud shoal. No more than 200 to 300 paces broad, beyond it lies deep water, enough to take them higher up the river to the base of the mountains. With wind and tide in their favour, they exhort the men, toiling knee-deep in water, their hands black with mud, to haul still harder; but after seven hours of 'excessive labour and fatigue' they are forced to withdraw, 'carrying with us the sad certainty, that the river could be of no advantage to navigation'.

Their geographical work thus at an end, the two explorers turn their attention to natural history and the study of the natives. It seems to them that the people in this part of the country are still more savage than elsewhere; though they have seen but a few individuals here and there, it has proved impossible to join them, 'as they all when they saw us, fled into the middle of the woods'.

Turning back downriver, they enter a small bay below the high bluff on the eastern shore (Mount Direction). The countryside presents a dreadful spectacle, a repeat of the scene facing the ships on their first arrival in North West Bay:

> *wherever we turned our eyes, we beheld the forests on fire: the savage inhabitants ... appeared to wish, even at this price, to drive us from their shores. They had retreated to a high mountain, which appeared like an enormous pyramid of flame and smoke; from this spot their shouts were distinctly heard, and the people who flocked to them seemed to be very numerous.[21]*

Leaving Midshipman Brue in charge of the boat and tents, Freycinet and Péron, with five men all well armed, advance towards the burning mountain.

The way through the smoking undergrowth is difficult and dangerous; many of the tallest trees have fallen in flames across their path. The closer they approach the mountaintop, the louder grow the cries of the savages. Braced for an attack, the French proceed; abruptly the shouts cease, and they are astonished to see the natives in flight, abandoning their huts and their weapons. They follow them for some time, but without success. At length, overcome with weariness, hunger and thirst, the little party make their way back to the tents, where they arrive at nightfall.

'Kaoué, Kaoué!'

Explorers with the purest and most honest intentions have often been led into error about the character of people by the behaviour they meet with.

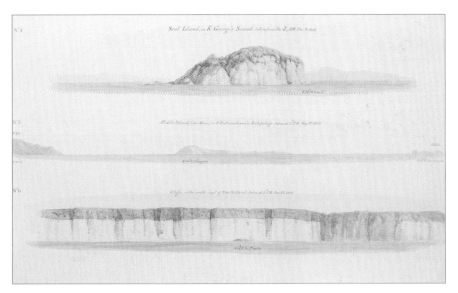

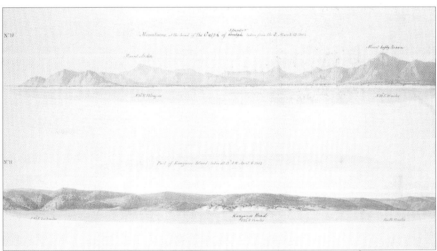

Top and above: 'Views on the South Coast of Australia', by William Westall, circa 1802. [By permission of the National Library of Australia]

Left: Plaque on The Bluff, Victor Harbor, South Australia. [Photo by A.J. Brown]

Top: *The* Investigator *(left) and* Géographe *in Encounter Bay, April 1802. [Painting by John Ford]*

Above: *Prospect Hill viewed across Pelican Lagoon, Kangaroo Island. [Photo by M. Hobbs]*

Left: *Cape Raoul, Tasman Peninsula, from the north-east. [Photo by Jack Ward]*

*They have inferred too lightly from the circumstances of their reception,
conclusions about the absolute and ordinary character of the men among
whom they have penetrated. They have failed to consider sufficiently that
their presence was bound to be a natural source of fear, defiance, and
reserve ... [Degérando, 1969 (1800)]*

While Freycinet and Faure were away, the remaining boats kept up a
regular shuttle between the ships and the shore. The longboats made
repeated trips to Ronsard's freshwater stream at the head of the bay;
here the water was collected in butts, then decanted into large stow-
age-casks in the holds. Other crewmen were set to fishing, while the
smaller boats ferried the officers and naturalists on hunting and scientific excursions – the latter in search of animal, botanical and mineral
specimens for the expedition's expanding collections. Though birds
and animals were scarce, wrote Baudin, 'we were luckier on the botanical side, for the gardener collected several things to plant, and I
found quite a number for the herbarium'.

Freed from their shipboard responsibilities, Baudin and Hamelin
made several excursions ashore, often together. They discovered a
mutual interest in natural history, though Hamelin acknowledged the
commandant to be 'an infinitely better judge than I am'. Noting the
disappearance almost overnight of both swans and pelicans from the
waters of the bay – 'undoubtedly because of the great number of shots
fired over the last three days' – they agreed to ban the crews of both
ships from taking guns ashore, 'and to limit their means of self-defence
to a sword'.

On one such trip, Hamelin caught sight of two natives crossing from
the mainland to Bruny Island in one of their flimsy bark canoes. Immediately on landing, he told Baudin, they had set fire to the growth
along the shore; 'no doubt they do this for fear of their footprints being seen in the grass'.

A group of sailors from the Géographe, *net-fishing in a bay at the northern end of Bruny Island, seek refuge ashore from a sudden gust of wind
and driving rain. They cook themselves a makeshift meal in the shelter
of a rocky outcrop, and as they are eating seven natives emerge from
the woods and stand around them, shouting and laughing. Though all
carry spears, their friendly gestures and happy mood make clear their
peaceful intentions.*

Reassured, the unarmed fishermen invite their visitors to share what is left of the meal. Ignoring the food, they display great interest in two rum bottles the men have brought with them. These are handed over without hesitation, one still containing a little rum; to the sailors' surprise the natives empty the liquid on the ground, so great is their delight at being given the bottles. With confidence firmly established on both sides, the two groups begin dancing around the fire.

Soon tiring of this diversion, all sit down together, and the natives turn their attention to the men's pockets, running their hands over anything made of metal – particularly the buttons on the sailors' coats. Their pleading is so transparent that several of the Frenchmen cut off their buttons and present them to their visitors; the 'unfortunates' have nothing to offer in exchange. For their part the sailors, surprised at the absence of women in the group, make it clear 'by unmistakable gestures ... what sort of desires they are referring to'. The natives respond with boisterous laughter.

After about two hours' good fellowship, the sailors return to their fishing. The embarkation goes without a hitch, and they part from the natives 'the best friends in the world'. Their catch, without being plentiful, is 'none the less big enough for all the crew to share in it'.

Fires blazed all night at the landing place, encouraging the French to return next morning. However, when Baudin, accompanied by Hamelin, Milius and other officers, went ashore, the group of natives awaiting them showed none of the friendliness of the previous day. Despite accepting some glass beads as presents, they remained wary, their gestures leaving Baudin in little doubt

> of their not being pleased to see us, for they asked us to go back to our boat. Among the different words that they spoke, one, '*Kaoué, Kaoué*', kept recurring, and was even said in a tone of command. In the language of the negroes of Mozambique and the African coast, it means 'Go away', but I cannot say if it has the same meaning with the natives of Van Diemen's Land.[22]

Shadowed all the while by three armed natives, Baudin and his companions continued inland to search for plant specimens. The forest seemed a wasteland, studded with half-burnt trees, and to all appearance abandoned by wildlife. In a three-hour trek they saw a solitary kangaroo; not even birds disrupted the silence. Baudin brought back just one new plant, a heather with tiny indigo-coloured flowers.

On their way back the officers invited their trackers to join them. After much hesitation they did so, laying down their spears behind a tree, and gesturing to the white men to put away their weapons. Never, wrote Milius, 'had [the natives] shown such a lack of trust as on this particular day; it was only necessary to make a single movement for them to show uneasiness'.

By degrees they relaxed, 'and fitted into our midst with as much familiarity as though we had all been comrades'. They were especially curious about the explorers' appearance beneath their clothes, examining 'every part of our bodies, with the exception of one organ, which out of modesty we did not wish to expose to their inquisitive looks'. They also repeated, clearly and with little difficulty, various French words which were said to them – only those containing *R*s and *S*s seemed a problem.

The French spent two hours or so with the natives, and before reaching the boats 'smothered them with presents and endearments'. They were about to embark when, without warning, 'these ingrates we had given presents to and whom we had just embraced', attacked them with a hail of stones. Baudin was hit on the hip by a heavy stone as he bent to examine a shell. He immediately aimed his musket at his assailant, so frightening the man 'that he put his head down and ran for his life'.

While their companions safely embarked, Baudin and Hamelin advanced towards the stone-throwers, now joined by a dozen others, all carrying spears. At the sight of the officers' guns the natives split up and vanished among the trees.

Péron, returned from his excursion to the North River, also lands on North Bruny Island, with a view to observing the manners and customs of the 'miserable hordes' of these parts. Walking along the beach with surgeon Bellefin and Midshipman Heirisson, he notices a group of about twenty 'savages' in the distance. The latter withdraw into the shelter of the trees, and the three Frenchmen content themselves 'with calling after them, showing them several things as presents, and waving handkerchiefs'.

Hesitantly, the natives emerge from hiding, and are now seen to be women, with not one male among them. They seem to have been fishing, for each carries a large bag fastened around the forehead with a sort of string, and hung low on the back. These hold numerous crabs and other shellfish, and are quite heavy; 'we pitied them sincerely for having such burdens to carry'. In no time the women shed their reserve, and with 'all the natural vivacity of their character' ply their visitors with questions, 'seeming often to criticise our appearance and to laugh heartily at our expense'. Bellefin, fancying his abilities as a performer, entertains them with a song and dance, and one of the older women, apparently their leader, responds in like manner, mimicking his actions and his tone of voice, much to the delight of her companions.

Apart from a few with kangaroo skins draped over their shoulders, the women are naked:

> *but without seeming to think at all of their nudity, they so varied their attitudes and postures, that it would be difficult to give any just idea of all that this interview presented of the whimsical and picturesque.[23]*

Two or three young girls, no more than fifteen or sixteen years old, with an agreeable form and pleasant features, stand out; these girls have

> *something ingenuous in the expression of their countenance, something soft and tender in their manners, as if the most amiable qualities of the mind were always, even among the most savage hordes of the human species, the more particular appendages of youth and beauty.[24]*

The older women, by contrast, mostly appear unhappy and depressed, 'signs which misery and servitude always print on the faces of those who are compelled to bear the yoke'. Many are covered with sores, 'the sad consequence of the ill-treatment received from their ferocious husbands'. One only among them preserves a degree of confidence and displays a lively and merry temper – the singer who imitated M. Bellefin in such a truly original manner. Her name is Arra-Maida.

Coming up to Péron, the same woman takes charcoal from her bag and rubs it into his cheeks; Heirisson receives the same adornment.

> *Thus it appears that the fairness of skin, of which Europeans are so vain, is an absolute defect, and a sort of deformity, which in these distant climates must yield the palm of beauty to the blackness of coal, or the colour of red ochre.[25]*

Several of the women take Heirisson into the trees to satisfy their curiosity – they are inquisitive, not only to see the white men's chests, 'but also to find out if they resemble the native men in form and function'.

The women accompanied Péron and his companions back to the beach, only to find their menfolk already there:

> At this unexpected rencontre, all the unfortunate females who had followed us seemed greatly terrified; and their savage husbands gave them such looks of rage and anger, as were not at all likely to reassure them. After having deposited the fish they had brought at the feet of these men, who immediately divided them among themselves, these humiliated wives placed themselves in a group behind their husbands ... and during the remainder of our stay, these unfortunate women did not dare to speak or smile, or even lift up their eyes from the ground ...[26]

Baudin and Hamelin, with a dozen sailors, were also at the landing-place. They had been fishing with the drawnet, watched by a score of natives, mostly children, whose 'capers and exclamations' showed their delight at seeing fish caught in the net. At first the children had been very timid, but on seeing the sailors haul in the net they had rushed to help, imitating the men's actions and repeating intelligibly everything they said. They soon

> seemed to forget that we were strangers and played games with our sailors, leaping up on to their necks as though they were on the best of terms with them. They showed a lively sense of enjoyment and an irrepressible friskiness by a never-ending display of antics and feats of suppleness ... and played a thousand tricks on our men to get them to chase after them; many times they challenged us to a race. Our sailors readily joined in all their games, which seemed to greatly please the children's mothers.[27]

These games were interrupted by the arrival of the warriors who had made the attack the previous day. They approached without fear or mistrust, and their presence seemed to inspire respect in the other natives, even the children becoming 'models of silence'. Though one man tried to take Hamelin's sword as he was climbing into his boat, the embarkation was otherwise uneventful. The whole party was back on board before dusk.

The tombs of Maria Island

At anchor in North West Bay, D'Entrecasteaux Channel, Friday, 5 February 1802.

> *In the evening ... the disk of the sun at its setting appeared of the most beautiful and bright red colour; the wind was then N.E., but in the course*

> *of the night it changed to the north, and blew in impetuous squalls, which
> lasted till ten or eleven o'clock on the following morning. These squalls
> were so violent, [and were] attended with such an extraordinary degree
> of heat … that it was scarcely possible to breathe even in the open air; the
> wind seemed like the heat from a furnace, and immediately all the sur-
> face of the sea appeared to smoke; an immense quantity of water spread
> throughout the atmosphere, and during the rest of the day we were as if
> plunged in a bath of hot vapour. [Péron, 1809, p. 202]*

The expedition made ready to leave the anchorage on 5 February, but
searing northerlies, followed by several days of sultry mist, oppressive
humidity, and flat calm, confined the vessels to the bay for more than
a week. Below decks, conditions became unbearable. Visiting René
Maugé in his tiny, airless cabin, Baudin found his sick friend 'so weak
and listless as to have difficulty in moving', and despaired of his recovery. The ship's doctors had warned him this was no longer possible,
but he had not given up all hope. It was Maugé's acceptance of his
fate that convinced Baudin: 'for while I was with him, he spoke of
nothing but his imminent death, and his regrets for the friends whose
advice he had disregarded by undertaking the voyage'.

A light breeze blowing from the east, along with fine weather and a
favourable current, enabled the ships to set sail for the east coast of
Van Diemen's Land on 17 February. They doubled Cape Raoul – 'peaked
in every part with jutting projections of prisms' – and Cape Pillar, also
covered with immense basalt columns, next morning, and in the afternoon dropped anchor in Oyster Bay, on the western side of Maria Island
(named by Tasman after the wife of his patron, Governor Anthony Van
Diemen of Batavia). Here Dr Lharidon came to Baudin with the news
he had long been dreading – his friend was sinking fast, and would
probably not last the night. Though expected, the doctor's report 'so
upset [me] that I was ill', Baudin confided to his journal.

Meanwhile the expedition's work had to continue. Four boat-parties
were organised and sent out next day to explore the little-known
southern part of the east coast. Boullanger and Maurouard, with Péron
also on board, were to circumnavigate and survey Maria Island. Faure
and Bailly were entrusted with the examination of Schouten's Islands
to the north, while the Freycinet brothers, in separate boats, were to
explore the mainland coast to the north and south of the island.

Boullanger sails from Oyster Bay for the island's southernmost point (later named Cape Péron). From here to Cape Maurouard, towering cliffs of granite guard the south-east shores from the turbulent waves; sheer against the wide blue arch of the sky, the massive ramparts, spangled with bright-red and brimstone-coloured splotches of lichen, mimic the battlements of some ancient fortress. At two o'clock the boat steers into a great bay on the eastern side (Riedlé Bay), and the crew land on the isthmus separating it from Oyster Bay to the west.

While his companions busy themselves with their surveys, Péron sets off 'to make observations on the island and its productions, the nature of the soil, the temperature, and the inhabitants'. Struggling through the undergrowth, so thick in parts he is unable to force a passage, he comes upon a path trodden flat by the natives. It leads to an open grassed space, in the centre of which, shaded by ancient casuarinas, stands a large cone-shaped structure, roughly formed of eucalyptus bark, and supported by four long poles driven into the ground. These intersect wigwam-fashion, and are bound together by long wisps of bark to form a kind of dome.

Removing several pieces of bark, Péron peers into the cone's interior. The upper part is empty, but at the bottom lies a flattened mound of fine grass, 'disposed in concentric layers' and held down by eight wooden hoops, the ends of which are buried in the earth and weighted by slabs of granite. Beneath the grass a heap of white ashes can be made out; sifting carefully through its midst, he draws forth something solid, and shudders to find himself holding 'the jaw-bone of a man, to which there yet adhered some remains of flesh'. Probing further, he finds more bones charred by fire, easily identified as belonging to a human vertebra, leg, and shoulder; they crumble into powder at his touch. They have been placed with care in the bottom of a circular hole, some 16 or 18 inches in diameter and 8 to 10 inches deep, before being covered with grass.

Next day Péron finds another monument. It is older than the first, the poles supporting the bark have fallen to the ground, the grass covering the ashes is decomposed, but the bones and other remains are disposed in much the same way. His eye catches a memorable detail— the underside of several of the larger pieces of bark bears rudely engraved characters, similar to the tattoos on the natives' arms.

The discovery of the tombs appears to provide proof of the practice of ritual cremation by the Diemenese:

[It] agrees both with the general habits of these people and their particular situation. Fire in these countries ... seems to be esteemed as something very superior to all other objects of nature, and these primitive ideas have not a little contributed to the idea of burning the dead ... The necessary materials were always ready; it required neither reflection nor labour; no tool was requisite; the execution was quick and easy to be done; it prevented both the putrefaction and the consequent infection; a few bones was all that remained after this duty was performed, and the ashes of the fire were sufficient alone to cover them. The whole ceremony required only a few hours, and the preparations had also a tendency to make the ceremony more solemn and sacred.[28]

Boullanger and his boat's crew have not quite completed their circumnavigation of the island when, at 9 a.m. on 21 February, they hear cannon shots at regular intervals from the ships in Oyster Bay. It can only be the ceremonial salute for René Maugé.

Maugé had clung to life for two days and nights. Baudin visited the zoologist's sickbed every few hours, and was filled with pain at his dying words: 'I was too devoted to you and scorned my friends' advice. But at least remember me in return for the sacrifice that I have made for you.' Maugé's death occurred just before midnight on 20 February. At daybreak the ship's yards were cock-billed (tilted so that they lay at an angle to the deck) as a sign of mourning, the flag half-masted, and a boat sent to the *Naturaliste* to inform Hamelin of their loss. All officers and naturalists were invited to acccompany the commandant's boat ashore at nine o'clock for the funeral ceremony. As the dead man's body was lowered into the *Géographe*'s boat, the ship's guns thundered a salute, followed by a second at the halfway point to the shore, and a third as the funeral party landed.

René Maugé was buried with full military honours on the southern point of Oyster Bay, where the sailors had cleared a site for his grave between two eucalypts and two casuarinas. For the commandant, his friend's death was more than a grievous personal blow; it was, he wrote, an

irreparable loss [for the expedition]; alone, he did more than all the scientists put together. Occupied solely with his work, he thought of nothing but performing his duties well, and I was never [forced] to remonstrate with him on this head.[29]

Péron shared in the general grief. He had worked closely with the older man, and valued his colleague's collecting and embalming skills. His death, he noted, 'was universally regretted by all on board. He was deservedly esteemed for many good qualities, and his zeal for the success of our expedition'. It was probably his doing that the zoologist's burial place appeared on Freycinet's charts as Point Maugé.

Filling the gaps

At the anchorage in Oyster Bay, Sunday, 21 February 1802.

> *Today, Citizen Commander, imbued like me with the importance of the study of man and equally distressed to see it so neglected in your expedition, you have imposed upon me the obligation to occupy myself with it in a more particular manner. You wish me to devote all my attention to establish a comprehensive relationship with the natives, to occupy myself with their language, to gather information about their customs, their practices and their habits, all of which deserve to hold the attention and researches of the philosopher. ['Anthropological Observations by F. Péron on the natives of Maria Island', quoted in Plomley, 1983, pp. 83-84]*

Following his friend's burial, Baudin spent the day ashore with Hamelin and other members of the staff, leading them on a search for fresh water in a wooded ravine; they had little luck, finding only a patch of stagnant water in a muddy hole. Afterwards they encountered a small group of islanders, spending several hours hours peacefully with them but learning nothing of consequence. At four o'clock, still grieving at Maugé's death, Baudin returned to the *Géographe* with Hamelin.

Baudin was in his cabin when Boullanger returned from his circumnavigation of the island. The geographer presented his rough chart of the coastline, but it was Péron's report that lifted the commandant's spirits – the tombs he had so providentially discovered were 'the most skilfully and carefully-made things that we have seen [and] infinitely superior to anything else we know of belonging to the natives'. He ordered the naturalist to return to the site with Nicolas Petit to sketch the structures: after Péron's detailed description, 'Citizen Petit's drawings will fill in any gaps that may be left'.

Péron and Petit set out in a pousse-pied, *or punt, with three sailors, their only protection a faulty musket concealed in Petit's baggage – the commandant has absolutely forbidden any arms to be taken ashore. On landing they head for a large fire burning in the scrub some distance behind the beach. The Frenchmen approach with caution, signalling to the dozen or so savages around the fire that they come as friends. 'Médi, médi,' respond the latter, inviting their visitors to sit down, at the same time laying down their spears and clubs. Rouget, the coxswain, who carries the musket, does likewise, but keeps it close to hand.*

The two groups observe each other at close quarters. The islanders begin the familiar ritual of examining their visitors' chests and other body parts, persisting with 'so much warmth and obstinacy' that Péron asks a young sailor to satisfy their curiosity. He shows himself to be formed like themselves, exhibiting such striking proof of his virility that the natives utter 'a cry of joy and acclamation that perfectly stunned us'.

Péron is no less observant of the Aborigines. Most are young men, aged from sixteen to twenty-five; two or three are in their thirties, while one, the oldest, seems at least fifty years of age – he alone wears a kangaroo skin. They vary in height from about 5 feet 2 inches (157 centimetres) to 5 feet 6 inches (168 centimetres). Their faces mirror their quickly changing passions:

> *Fierce and ferocious in their menaces, they appear at once suspicious, restless, and perfidious. In their joy, the figure displays a convulsion that has the appearance of madness; among the aged there is an expression that is at once sad, sullen, and severe; but in general, among all these people, there is to be noticed at some moments an insincerity and ferocity, which cannot escape an attentive observer, and which but too well corresponds with their character.[30]*

The mutual examination completed, Petit diverts the natives with conjuring tricks. Rouget stills their gleeful cries – producing a long pin from his pocket, he jabs it hard into his flesh, with no sign of pain and without producing a drop of blood. They look at each other in silence, then with one accord shout like madmen. Péron, unluckily for himself, has given similar pins to several natives; one of them, 'wishing to satisfy himself whether I possessed the like insensibility', creeps behind him and stabs him in the calf with the pin, 'so dextrously and decidedly that I could not help crying out'.

Baudin expectantly awaited Péron's return to the ship; gone were his usual strictures on the naturalist's dilettante approach to his work. 'Citizen Péron,' he wrote, 'as a student of anthropology, will no doubt give a very detailed account of the structure of the tomb, as well as his ideas on the origin of this custom [cremation] among these peoples.'

Péron's report[31] covered far more than the tombs and native burial customs. Seeking to profit from their meeting with the islanders, he and Petit made the most of their good humour. While the artist sketched some of the younger natives, Péron engaged the rest in conversation, 'giving my attention exclusively to those whose intelligence was more ready and whose replies were more definite'. Building on the words he had previously collected in the d'Entrecasteaux Channel region, he rapidly extended his vocabulary:

I obtained successively replies to the words *to yawn, to burn oneself, to urinate, to defecate, to belch, to laugh, to cry, to whistle, to blow, to slap, to tie, to untie, to wrestle, to tear, to strangle, to have a bellyache, to have an erection,* etc. … I easily obtained the words for our *dinghy,* our *glass beads,* their *waddy,* their *spear,* their *lice* (because they did not eat them), *to kill lice,* etc. … They seemed to me to be very intelligent and they easily grasped the meaning of all my gestures and seemed to understand both their object and their purpose … they laughed heartily when, attempting to repeat [their words], I made mistakes or pronounced them very badly.[32]

One observation struck him as having particular interest – the Aborigines seemed to have no idea of the acts of kissing or caressing. When, to make clear what he was asking, he brought his face close to theirs so as to kiss them, 'they all had that look of surprise and uneasiness which an unknown act always arouses in us'; and when on kissing them he asked '*gouanarana?*' ('what do you call that?'), they replied '*nidego*' ('I don't know'). Caressing was no less strange to them, for when he made 'the gestures peculiar to this action', their surprise showed they had no conception of it.

Consequently, these two actions, kissing and caressing, which are so full of charm and which seem to us so natural, appear equally unknown to savage natural man. This double source of the most lively enjoyment, of the most exquisite sensations of love and sensual delight do not exist for them …[33]

As the day wears on, the mood of the 'savages' changes; by degrees they become more uneasy and fearful. Petit has trouble completing his sketches, and Péron's questions go unanswered. Only their awe of Rouget's musket seems to hold them back; they tease him to fire at some birds on a nearby tree, but the coxswain, knowing the weapon is old and faulty, declines. His refusal serves as 'another cause of suspicion and disquiet'.

Péron's brightly coloured jacket attracts one of the group. While the naturalist's attention is elsewhere, the man makes a grab for it, at the same time pointing his spear at the Frenchman, as if to say, 'Give it to me, or I will kill you.' Péron affects to treat his menaces as a joke, but quietly points at Rouget, who has levelled his musket at the attacker's chest:

> I added one single word of his own language ('mata', death). He understood me, and laid down his weapon with as much indifference as if he had done nothing to offend me.[34]

Even as Péron congratulates himself on his narrow escape, another native slips unobserved behind him. This man covets the large gold rings in the naturalist's ears; stealthily he puts a finger through one of the rings, and jerks it 'so violently, that he would certainly have torn my ear, if, fortunately, the ring had not opened'.

As the French prepare to move off the savages snatch up their weapons and advance, shouting in unison. With Rouget in the rear, his musket pointed at the leaders, Péron and his companions retreat with caution, walking backwards towards the boat; the threat of the firearm keeps the natives at a distance. The latter follow the boat along the beach, still shouting, but disappear into the forest when they see other boats dragging for oysters.

While Péron and his companions were occupied on Maria Island, the other boat parties were 'exploring the nearest parts of Diemen's Land, and the isles adjoining'. Louis de Freycinet was the first to return, on the evening of 22 February. He had surveyed the east coast from Cape Bernier northwards to Cape Bougainville, and on to Point Bailly, opposite Schouten Island; his chief discovery was the large inlet now known as Prosser Bay.

His brother, Henri, and Pierre Faure, the geographer, returned from their separate excursions on the 26th. Henri de Freycinet's orders were to examine the coast south from Cape Bernier (named after the astronomer who accompanied him). In eight days he surveyed Marion Bay and today's Blackman Bay, and proved, by walking right round

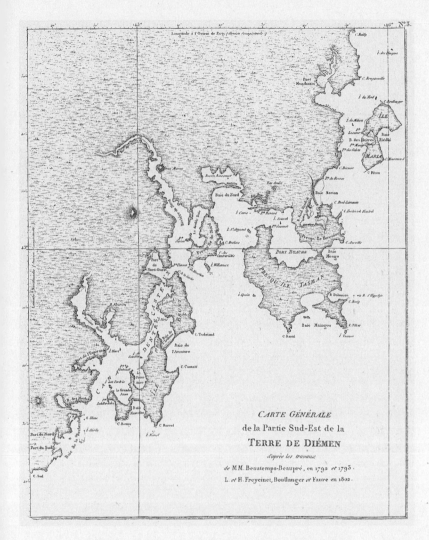

South-eastern Terre de Diémen (Van Diemen's Land), 1802.
Chart by Freycinet, Boullanger and Faure, also using charts by
Beautemps-Beaupré from 1792 and 1793.
[By courtesy of Scotch College, Hawthorn]

the latter, that the existing Dutch, French, and British maps of the area were all wrong. His survey, added to Faure's earlier discoveries, established beyond doubt that 'Tasman Island' was in fact two joined peninsulas; the northern one became Forestier Peninsula.

Faure had been given the most important task – to examine the 'Schouten Islands' of Tasman and Furneaux and the adjoining mainland coast. He found there was but one island, and those apparently lying to the north were part of a long, mountainous peninsula (today's Freycinet Peninsula). He followed the coast as far north as the 42nd parallel, then, returning south, surveyed the shores of Great Oyster Bay inside the peninsula.

Listening to the reports of his boat commanders, and examining their charts, Baudin had reason to be pleased with their discoveries. Faure and the elder Freycinet had corrected significant errors on his predecessors' maps, and filled the gaps. The mistakes of Tasman and Furneaux were understandable: 'this error frequently occurs at sea when one is rather a long way off' and cannot distinguish the details of the coast. As for Péron's report, the naturalist seemed once more to have allowed his 'fertile imagination' free rein. There was, however, gold amid the dross – not least his descriptions of the tombs and of the natives' vocabulary. Petit's drawings, as expected, filled in the gaps left by Péron's imaginative speculations.

On 27 February the ships sailed from Maria Island in fine weather and set their course for Bass Strait and the Unknown Coast. The winds were consistently from the north, 'and too contrary to our course for [us] to be happy with it'. Next day mist and fog shrouded the horizon.

Lost at sea

Off the east coast of Van Diemen's Land, Saturday, 6 March 1802.

> *During the whole of the [previous] eight days we were almost continually enveloped in a thick fog and moist atmosphere, so that our two ships could scarcely see each other, and several times we were obliged to make the necessary signals to the* Naturaliste *with the guns. All our decks ran with water, even in the day, and during the night the more condensed mists dissolved into such a penetrating moisture, that nothing could escape its power ... This deplorable state of the atmosphere much increased the suffering of those who were sick. [Péron, 1809, p. 238]*

Fine, clear weather returned on 6 March. Baudin dispatched Midshipman Maurouard and the geographer Boullanger – so short-sighted he could 'only take bearings and angles with his nose on the ground' – in the large dinghy to make a nearer survey of the coast than was possible from the deck of the *Géographe*. They were instructed to stay in sight of the ship, and to return before nightfall; they carried two days' rations.

By midday the dinghy was lost to view. Towards evening the *Géographe* stood in towards the land to pick up the crew, but without result. By eight o'clock, with night closing in, Baudin brought to under topsails, putting up lights to indicate his position; orders were given to fire a rocket at hourly intervals, and a shot from the swivel gun every half-hour. At nine o'clock the *Naturaliste* passed astern, and he signalled Hamelin to emulate *Géographe*'s efforts, 'for a greater number would be more clearly seen or heard'. Again it was in vain.

For three days the ships searched for the lost boat. The *Géographe* closed the land to within $1\frac{1}{2}$ miles, sometimes less, close enough 'to make out a person on the beach, had anyone been there'. Even Péron acknowledged how dangerous it was 'to steer with a large vessel close along the bendings of a wild and unknown shore'. On the second night, a blazing fire on shore reassured them that the boat's crew were alive – especially when it was kept burning throughout the night. In the morning, Bonnefoi was sent in the longboat to investigate, but found nothing. His anxiety for the men's safety brought on a crisis in Baudin's health; unwell since leaving Timor, he was gripped by acute colic pains, and could not leave his bunk.

To further complicate matters, all contact with the *Naturaliste* was lost as the weather worsened once more. Temporarily handing over command to Henri de Freycinet, Baudin instructed him to seek the views of the ships' officers and the petty officers on continuing the search. Both groups advised searching south of Cape Tourville, where the dinghy was last seen.

On the 10th, a small schooner-rigged ship was seen ahead; she proved to be the *Endeavour* from Port Jackson, on a sealing expedition to Maria Island. Her captain, brought aboard in the *Géographe*'s boat, had welcome news. He had met the *Naturaliste* further north, and passed on a message from Hamelin that he was bound for Banks Strait (off the north-east coast), and would await Baudin's arrival at Waterhouse Island. The sealer knew the Furneaux Group well, and recommended some good anchorages where wood and water could be found.

Called together a second time, the ship's staff and petty officers resolved to call off the search and sail northwards; it was, they thought, 'a moral certainty' that the dinghy 'had unhappily been lost at sea'. Baudin himself, in constant torment with his stomach pains, was not so sure; he still clung to the hope that it might be found at the entrance to the strait:

> it is true that the men in it will have suffered greatly through lack of food and possibly water, but in short, if they have had courage, they will still be alive.[35]

Unfavourable winds drove the *Géographe* away from the coast, and a week passed before she ran in to Banks Strait, at the north-eastern tip of Van Diemen's Land. The horizon ahead was covered with black storm-clouds, but 'since we were well advanced, there was nothing to be gained by retracing our course'. Buffeted by wind and rain, they continued through the strait, reaching Waterhouse Island on 18 March. The *Naturaliste* was not there, and despite several days' searching, here and in the Furneaux Group to the north, Baudin could find no trace of his consort. His frustration growing with every wasted day, he broke off the search on the 24th and headed across Bass Strait for Wilson's Promontory on the mainland coast.

At nightfall on 9 March – the very day that the Géographe *resumed the search for the missing boat south of Cape Tourville – Maurouard, Boullanger, and their men are picked up by the English brig* Harrington, *Captain Campbell, at the entrance to Banks Strait.*

The *Naturaliste* in Bass Strait

On board the Naturaliste, *Wednesday, 10 March 1802.*

> *'The horror of our situation,' said M. Boullanger, 'may be conceived; the small portion of provisions and water which we had received as our day's allowance on leaving the ship, was exhausted, and we were sinking under fatigue, from being incapable of sleeping, and soaked with sea water. On the 8th, however, we caught a number of cormorants, and had the happiness to discover the Isle Maurouard, where we found fresh water, and passed the night. On the 9th ... we proceeded along the shore, till we came in sight of the Furneaux Isles, when we fell in with an English brig, the* Harrington, *commanded by Captain Campbell.' [Péron, 1809, p. 264]*

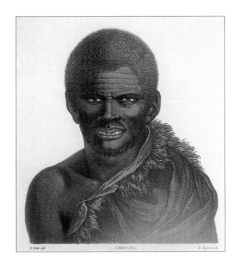

Top: *Aboriginal canoes on the Schouten Islands, off the east coast of Tasmania, by C.-A. Lesueur. [Collection Lesueur, Muséum d'Histoire Naturelle, Le Havre]*

Above: *An Aboriginal camp in Van Diemen's Land, by N.-M. Petit. [Collection Lesueur, Muséum d'Histoire Naturelle, Le Havre]*

Right: *Parabéri, a man of Bruny Island, by Petit. [Collection Lesueur, Muséum d'Histoire Naturelle, Le Havre]*

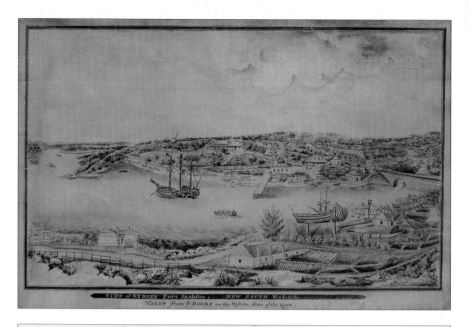

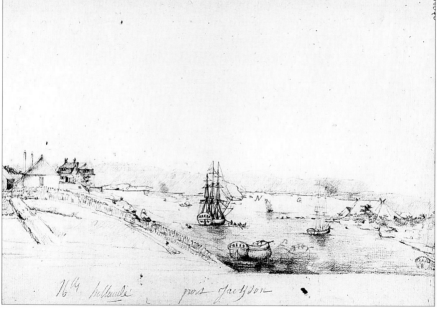

Top: *View of Sydney, Port
Jackson, taken from The Rocks,
by J.W. Lancashire, 1803.
[Dixson Galleries, State Library
of NSW]*

Above: *'Part of Sydney Cove,
Port Jackson', by C.-A. Lesueur,
1802. The observatory tents to
the right are on Cattle Point
(the site of the Opera House on
today's Bennelong Point).
[Collection Lesueur, Muséum
d'Histoire Naturelle, Le Havre]*

Hamelin had lost sight of the *Géographe* during rough weather on the night of 8 March. At dawn he stood in for the land, hoping to meet up with his consort. Instead, he encountered a small English sealer, the *Endeavour*, bound for Maria Island; her captain had seen nothing of the *Géographe* or of Boullanger's boat. Another sail was sighted two days later, in the morning. She too was an English brig, the *Harrington*; she approached and put a boat down:

> What was our surprise, on recognising the long-boat of the *Géographe* [Louis de Freycinet told Péron at Port Jackson], with Messrs. Boullanger, Maurouard, and the eight sailors [with] them ... We learned from M. Boullanger, having in vain on the evening of the 6th endeavoured to rejoin the *Géographe*, M. Maurouard and he had resolved to keep close to the shore; that the 7th had been occupied by them in coasting along it, and continuing their geographical observations; but being forced to pass another night in the open air, they had suffered much from the cold and rain ...[36]

Captain Campbell, Boullanger told Hamelin, had received them kindly, and given them every assistance; he was bound for Port Jackson, and offered to land the boat party there. Boullanger, however, declined, having learnt from the English skipper that his lookouts had seen a large ship southwards. Confident this must be either the *Géographe* or the *Naturaliste*, Boullanger and Maurouard decided to await the vessel's arrival in Banks Strait. They were preparing to leave the *Harrington* when the *Naturaliste* was sighted from the masthead.

Information provided by Captain Campbell, coupled with Boullanger's account (which left little doubt that the *Géographe* was still to the south), determined Hamelin to wait for Baudin in Banks Strait. Rather than sailing at once for the rendezvous at Waterhouse Island, west of the strait, he anchored at Swan Island, on its southern side. He put the waiting time to good use, dispatching two boat-parties on hydrographic surveys – Boullanger to complete his survey of the east coast, and Faure north to the Furneaux Group.

As ill luck would have it, the weather again worsened while the boats were away. Gale-force winds snapped two of the *Naturaliste*'s anchor cables, forcing Hamelin to sail before they returned. He made for a more sheltered anchorage at Waterhouse Island, where the boat parties rejoined him a week or so later.

Faure, Leschenault, and Bailly the mineralogist land on Preservation Island in the Furneaux Group, their boat threading its way through rock-strewn breakers into the mouth of a sandy creek; the wrecks of ships scattered along the coast bear witness to the frequent storms in the strait. Much of the island's surface is covered with a meagre stratum of top-soil, barely enough to nourish a few slight shrubs and tussock grass, and difficult and tiring to walk on. Honeycombed with the nests and burrows of innumerable white and blue fairy penguins, the ground is 'so hollow, that the feet of the passenger sink knee-deep at almost every step'.

The timid birds make little attempt to defend themselves from the men who hunt them. During the day they remain squatting in their burrows; but once night arrives,

> *they rush in crowds towards the shore to search for fish and other ani-mals, on which they feed; and from these excursions they do not return till break of day; to keep their nests from cold, they cover them with dry leaves and feathers; in these nests they rear their young, till they are strong to walk to the shore and feed themselves.*[37]

At nightfall the sailors collect bundles of dead wood and dried grass, and the party warm themselves around the fires. For supper they feast on the flesh of penguins broiled on the live coals; the taste is somewhat oily, not unlike that of red herring. The heat and light attract the birds, for during the night they approach the fires in great numbers, though many are burnt.

One sailor, wrapped in a blanket and lying close to the fire, is so tor-mented by the birds that he cannot sleep. Repeatedly they attempt to creep under his covering and share the warmth; enraged, he rings the necks of several, but the others 'return incessantly to the charge'.

With his survey parties back on board, Hamelin sailed from Waterhouse Island on 18 March, expecting to find the *Géographe* off the east coast of Van Diemen's Land. Unknowingly, they passed each other in Banks Strait, invisible in the morning mist. For a week Hamelin tacked along the coast northwards from Maria Island, buffeted by a succession of vio-lent squalls. Convinced at length that Baudin must have passed him un-seen, he returned a second time to the rendezvous at Waterhouse Island.

The *Géographe* was not there, nor did his shore parties find any sign of her presence on the island. As a last resort, Louis de Freycinet was sent to look for her at Port Dalrymple (site of Launceston), with Faure also on board to check the accuracy of Flinders's chart of the port. Returning on 7 April, they reported that its main features were correctly set out in the English chart. Navigation was made difficult by sandbanks and rocks, but there were many coves where ships could anchor safely; the land seemed rich, and the vegetation vigorous. They could find no trace of the *Géographe*.

Tired of playing hide-and-seek, Hamelin crossed Bass Strait to resume his hydrographic work on the mainland coast, at Wilson's Promontory. For ten days the *Naturaliste*'s boats explored the rugged coastline, from the promontory westwards to Western Port, discovered by George Bass in 1798; their charting was meticulous. At last, on 18 April, his supplies almost exhausted, his crew sickly and in need of rest, Hamelin broke off the survey and made for the English settlement at Port Jackson – where, he had little doubt, the *Géographe* was already in harbour.

Napoleon's Land

On board the Géographe *in Bass Strait, Sunday, 28 March 1802.*

> *From the 21st to the 26th of March we experienced one of the strongest gales that we had ever encountered in these seas; several of our sails were carried away by the squalls, and we had nearly been lost in the night of the 21st on the Isles of Furneaux. To escape this misfortune we were obliged, notwithstanding the force of the tempest, to carry all our sails, and in the course of the morning of the 22nd, we succeeded in clearing the strait by the pass between Kent's group and the promontory. At nine o'clock in the morning we doubled the pyramid, an enormous rock, which at a distance has the appearance of a gothic ruin; then standing right in to the west of the isles of Kent's group, we succeeded in doubling them, though with much difficulty ... These terrific granitic rocks presented a majestic and dreadful spectacle, naked and barren; the roaring waves broke against them with such noise and force as seemed to threaten every instant to bury them under torrents of foam ... [Péron, 1809, pp. 241-242]*

The storm had blown itself out by the morning of 27 March, and, says Péron, 'we hastened to enter the straits again, as we were impatient to make the south-west coast of New Holland'. By midday on the 28th, Wilson's Promontory was in view, and next day the *Géographe* lay off Western Port – the starting point for their exploration of *Terre Napoléon* (Napoleon's Land).

[D'Entrecasteaux] not having gone beyond the isles St. Pierre and St. Francis, which form the eastern boundary of Nuyt's Land, and the English not having carried their researches towards the south farther than Port Western, it consequently follows, that the whole of the coast laying between … was entirely unknown at the time we arrived at these shores; and as the point in question was nothing less than to resolve by this exploration the problem of New Holland being a continent, and to discover if there was any large river belonging to that continent, we all felt an additional degree of zeal and courage.[38]

Unaware of Flinders's voyage, and of Grant's and Murray's recent discoveries, Baudin pressed on to the west. Bypassing Western Port and rounding Cape Schank, the *Géographe* entered the large bight on which Port Phillip lies; 'we continued along the coast', wrote Baudin, 'and at three o' clock reached the cape that seemed to form the furthest point of the land in view to the West'. The coast, he told Flinders at Encounter Bay, '[had] no place in which he could anchor, there being no rivers, inlets, or places of shelter'.

Five years later, in his history of the voyage, Péron gave a more creative account, better suited to the record of a voyage of exploration:

On the eastern coast of this bay … is a port, of which every winding may be perceived from the mast-head; we described it by the name of Debut Port; but having afterwards learned that it had been already more particularly reconnoitred by the English brig the *Lady Nelson*, and that at that time it was named Port Philip [sic], we preserved this name with so much more pleasure, as it reminded us of the founder of a colony in which we received such generous and effective assistance.[39]

After the gales in Bass Strait, conditions became ideal for navigation. 'The sky was serene and clear, the sea smooth, and the wind favourable; so many advantages gave us an opportunity of ranging the coast very near,' added Péron, 'and thus our observations were made with the utmost precision.' The *Géographe* followed the coast westwards, the staff charting capes and headlands as they passed. On 2 April, from a league out to sea, they observed two mountain peaks inland (Mount Schank and Mount Gambier).

From this point they became true explorers – the first Europeans to sail this coast. The shoreline trended north-west; mostly it was a monotonous succession of high dunes, inspiring only gloom and disappointment. For Péron, this part of New Holland is still more terrific than those we have hitherto described. An immense surf broke the whole length of this shore, making a dreadful noise … the most hideous sterility is seen in every direction, and there is not the least appearance of even the smallest stream of fresh water.[40]

On the 5th, 4 or 5 miles from land, the lookouts sounded the alarm for a line of reefs ahead. On drawing closer, it was seen to be a vast school of dolphins:

of such great length, that at first, we thought them to be an immense chain of reefs; but their swift progress soon convinced us of our error, and we then began to think of making war on them.[41]

In a few minutes the sailors had killed nine animals with their harpoons, 'such a large quantity of fish [seeming] like a gift from Heaven'.

The meeting with the dolphins proved to be the prelude to a more momentous encounter, for 'we had but just finished our fishing, when signal was made at the mast head, of a sail being in sight'.

La rencontre (the French version)

Aboard Le Géographe, *at sea, Thursday, 8 April 1802:*

In the afternoon we continued along a stretch of coast consisting entirely of sand-hills. But towards three o'clock we began to see some high terrain which looked as if it must be pleasant. Shortly after, we sighted a ship which we thought at first could only be Le Naturaliste, *for we were far from thinking that there would be any other Europeans in this region and at this time of year. Nevertheless, we were greatly mistaken, for as we drew near her, we realised from her masts and size that she was not our consort. Finally, at five o'clock, when we were both able to see each other clearly, this ship made a signal which we did not understand and so did not answer. She then ran up the English flag and shortened sail. We, for our part, hoisted the national flag, and I braced sharp up to draw alongside her.*

As they spoke us first, they asked what the ship was. I replied that she was French. Then they asked if Captain Baudin was her commander. I was very surprised, not only at the question, but at hearing myself named as well. When I said yes, the English ship brought to. Seeing her make ready to send a boat across, I likewise brought to to wait for it. The English captain, Mr. Flinders … came aboard, expressed great satisfaction at this agreeable meeting, but was extremely reserved on all other matters.

As soon as I learnt his name, I paid him my compliments and told him of the pleasure that I had in making his acquaintance, &c. I informed him of all that we had done up till then in the way of geographical work. As it was already late, Mr. Flinders said that if I were willing to stand off and on till dawn, he would return the following day and give me various pieces of information concerning the coast that he had examined from Cape Leeuwin as far as here. I was very gratified by his proposal and we agreed to remain together during the night. The weather was very fine.

9 April. Seeing that Mr. Flinders was making ready to come aboard again, ... I hove to once more to await him. He arrived at half past six, accompanied by the same person as on the earlier occasion [Robert Brown]. As he was much less reserved on this second visit than before, he told me that his ship was the Investigator *and that he had left Europe about eight months after I had. He also told me that he had begun his exploration of the coast of New Holland at Cape Leeuwin. He had visited the Isles of St. Peter and St. Francis, as well as all the coast up to the point of our meeting. In addition, he informed me of the layout of a port that he had discovered on an island. This latter was only 15 or 20 leagues from where we were, and he had named it Kangaroo Island because of the great numbers of that animal that he found there ... He stayed six weeks there and so had time to examine it well.*

Before we separated, Mr. Flinders gave me several charts published by Arrowsmith since our departure. As I told him of the accident that had befallen my dinghy and asked him to give it all the help he could if he should chance to meet it, he told me of a similar misfortune that had happened to him, for he had lost eight men and a boat on Kangaroo Island. His companion ship had also been separated from him during the equinoctial gale, part of which I had weathered in Bass Strait and the remainder outside. Upon leaving, Mr. Flinders said that he was going to make for the strait and try to find some land which was said to exist between the Hunter Group and the place they have named Western Port. We parted at eight o'clock, each wishing the other a safe voyage. [Baudin, 1974, pp. 379-380]

Extract from François Péron's A Voyage of Discovery to the Southern Hemisphere:[42]

8 April 1802. We had but just finished our fishing, when signal was made ... of a sail being in sight. At first, every one thought it was the Naturaliste, *and our joy was general; but as we came nearer the ship, we soon perceived that it was not our consort. As she made for us with all sail set, she was presently under our stern, when she hoisted the English flag, and we at the same time hoisted French colours, and lay to, in imitation of their example. The English captain now hailed us, and asked if we were not one of the two ships which had sailed from France to make discoveries in the Southern Hemisphere? On our answering in the affirmative, he hoisted out a boat and came on board us; we now found that he was captain Flinders, the same who had already made the circumnavigation of Diemen's Land; that the name of his ship was the* Investigator; *and that it was then eight months since he had sailed from Europe, with the intention of reconnoitring the whole coast of New Holland and the archipelagos in the South Seas ...*

Deadly survey

On board the Géographe, *at sea, Saturday, 17 April 1802.*

> *As we still had plenty of work to do on the coast of New Holland, and had only enough water for another two months, I judged it right to begin cutting down on it in good time; that is, instead of giving each man two and a half bottles of water, I gave him no more than two. Undoubtedly it was not a very great reduction, for we were still allowing more than the regulations for long voyages prescribed. Nevertheless, it produced malcontents – not amongst the sailors, but in another quarter. Be that as it may, it will in no way alter my decision on the matter, for I am convinced that people can well do without tea and coffee twice daily. So, satisfied or not, I was the first to conform – and with a good grace. [Baudin, 1974, p. 386]*

After separating from Flinders on 9 April, Baudin continued westwards, determined to finally resolve whether New Holland was indeed an entire continent, or whether a vast strait bisected it from north to south. Frustrated by the weakness of the winds, the *Géographe* did not reach the first anchorage recommended by Flinders – Antechamber Bay, on the east coast of Kangaroo Island – until the evening of the following day. The ship's anchor dragged in the strong current, and the crew spent the night tacking back and forth in the channel between the island and the mainland coast. Next morning, Baudin searched unsuccessfully for Flinders's second anchorage – at Kangaroo Head – then crossed the strait to examine Gulf St Vincent.

In the shallow inshore waters of the gulf, he feared to take the ship in close enough for an accurate survey to be made. The men stayed on duty throughout the night as the *Géographe* tacked from one shoal to another, the leadsman calling the depth. For Baudin it was the *Golfe de la Mauvaise*, 'because of the fatigue it caused the whole crew'. (On Freycinet's 1808 chart of *Terre Napoléon* it appeared as *Golfe Josephine*; its larger neighbour to the west was named *Golfe Bonaparte*.)

Conditions were no better in the western gulf. 'Twice we attempted to penetrate to the farther part of Golfe Bonaparte,' wrote Péron, 'and twice we … nearly perished.' For four days the ship was trapped by continuous gales from the south-west, unable to beat out of the gulf. Most dreadful was the night of the 19 April:

> Impetuous winds from the W.S.W. blew with terrible violence, the sky was covered with thick black clouds; torrents of cold rain, like melted snow, fell, accompanied by flashes of terrific lightning; the ground swells were so violent and so sudden, that we were obliged to tack continually till day appeared.[43]

Baudin lacked the naturalist's talent for hyperbole, noting only that 'the rain-bearing squalls were very cold, and sometimes the water was like half-melted snow'. While he and the crew spent the night constantly on deck, the officers, except for those who changed watch,

> passed it just as peacefully in their beds as if the ship had been absolutely secure … I was not in the least surprised by it and left them in complete peace.[44]

It has been two months since they left Maria Island. The water allowance has been cut by the captain's order; after eight weeks in a cask it is foul and polluted, barely fit to drink. For brandy and wine, 'so indispensable to European seamen', has been substituted half a bottle of inferior rum, made in the Isle de France, and drunk there only by black slaves. The biscuit is riddled with weevils, the salt provisions tainted; the smell and taste are so offensive that the famished seamen often throw their allowance overboard, even in the captain's presence. 'The officers and naturalists, reduced to the same allowance as the seamen, [suffer] the same afflictions of body and mind.'

The surgeon's stock of antiscorbutics is long used up. The sick list grows by the day— men with swollen gums and fallen teeth, with ulcerous sores, with purple blotches disfiguring their skin, and all with a dreadful lethargy from which nothing can rouse them. Dysentery, the bloody flux, also wreaks havoc, claiming one life in the large gulf, another off the west coast. Half the crew is disabled, unfit to work the ship; only two helmsmen still have the strength to take the wheel.

Replacements for the sick helmsmen must be found among the crew. Henri de Freycinet, second-in-command and senior deck officer, requests authority to order the master carpenter and second caulker to take a turn at the helm. It is not orders that are needed in such circumstances, suggests Baudin, but rather 'a simple request made with good grace', because the work is so different from the men's normal duties. 'Such a proceeding,' bristles Freycinet, 'with regard to men so inferior [in rank], could not be undertaken by an officer like [my]self.'

Baudin settles the problem his way, ordering the midshipmen to take the helm for an hour and a half during their watch: 'this was not an arduous task ... and could only help in their education'. To his amazement the young gentlemen regard it as 'dishonouring', and scheme to evade it. Midshipman Bougainville refuses point-blank to steer, pleading sickness 'to disguise his disobedience'; Midshipman Brue claims that the regulations exempt him from such duty. Midshipman Charles Baudin alone accepts the work willingly, 'thus drawing upon himself many jests on the part of his friends'; five days later he is promoted to midshipman first class. As for the others, the captain prohibits them 'from any kind of duty on board, saying that I would do without them for steering quite as easily as I had done for other things'.

On 8 May the *Géographe* sighted Cape Adieu – Baudin's objective, the point at which d'Entrecasteaux had called off his eastward survey a decade earlier. Belatedly, French navigators had realised Fleurieu's dream, the reconnaissance of the entire southern coast of New Holland. In the event, fate and the British had conspired to deny them the credit for first discovery of the Unknown Coast.

Baudin could console himself with the thought that, notwithstanding the hostility of his staff, he had carried out his instructions faithfully, and had confirmed, as Flinders had said, that no great strait divided the continent. Now, with the men exhausted and in urgent need of fresh food and water, he heeded his officers' pleas:

... the weakness of my crew, which now consisted of only thirty men for the handling of the ship, our pressing need for firewood, the shortness of the days … all decided me to abandon the coast for D'Entrecasteaux Channel, where the anchorage is good, and from there to proceed to Port Jackson … Everyone expressed satisfaction [at this change of course], and truly, we were all very much in need of a little rest.[45]

The ship anchored in Adventure Bay, on the eastern shore of Bruny Island, on 20 May. Supplies of wood and fresh water were soon replenished, and a fishing party brought back a plentiful catch for the famished crew. Officers and men alike looked forward to a swift passage to Port Jackson, where they could enjoy all the comforts of a European settlement promised by Captain Flinders.

Notes

1. Péron, 1809, p. 136.
2. Baudin, 1974, p. 271.
3. Horner, 1987, p. 191.
4. Péron, op. cit., p. 171.
5. Baudin, op. cit., p. 303.
6. Péron, op. cit., p. 173.
7. Baudin, op. cit., p. 305.
8. Péron, op. cit., p. 175.
9. Péron was using the Réaumur temperature scale (°Re), established in 1730 by French naturalist René-Antoine Ferchault de Réaumur. On this scale, at normal atmospheric pressure the freezing point of water is 0°Re, and the boiling point is 80°. Thus 10°Re is 12.5°C, or 54.5°F. Although use of the Réaumur scale was once widespread, it had all but disappeared by the late 20th century.
10. Ibid., p. 176.
11. Ibid., p. 177.
12. Ibid.
13. Ibid., p. 178.
14. Ibid., p. 179.
15. Ibid., p. 180.
16. Ibid., pp. 180-181.
17. Though this conclusion – that the 'Island' is a peninsula – is of course correct, it was based on faulty information. With supplies running low, Faure made only a quick running survey of the western shore of Forestier Peninsula; consequently he missed the second (northern) isthmus

separating Norfolk and Blackman bays, which was the primary purpose of his mission.

18. Baudin, op. cit., p. 325.
19. Péron, op. cit., p. 201.
20. Baudin, op. cit., pp. 324-325.
21. Péron, op. cit., p. 191.
22. Baudin, op. cit., p. 320.
23. Péron, op. cit., p. 196-197.
24. Ibid., p. 197.
25. Ibid., p. 198.
26. Ibid., pp. 199-200.
27. Milius, 1987, p. 35. (Passage translated by J. Treloar.)
28. Péron, op. cit., p. 211.
29. Baudin, op. cit., p. 340.
30. Péron, op. cit., pp. 217-218.
31. An incomplete copy, held in the Muséum d'Histoire Naturelle at Le Havre, is reprinted in Plomley, 1983.
32. Plomley, op. cit., pp. 86-87.
33. Ibid., p. 87.
34. Ibid., p. 89.
35. Baudin, op. cit., p. 359.
36. Péron, op. cit., p. 263.
37. Ibid., p. 275.
38. Ibid., p. 244.
39. Ibid., pp. 244-245. As Philip Edwards (1994) astutely observes, 'all voyage narratives are self-serving, and to watch ... the development of a narrative is to see the record being adjusted, massaged and manipulated in the writer's interests'.
40. Péron, op. cit., p. 248.
41. Ibid., p. 250.
42. Ibid.
43. Ibid., p. 255.
44. Baudin, op. cit., p. 388.
45. Ibid., p. 401.

⚓

CHAPTER SEVEN

Port Jackson Interlude,
April to November 1802

PRELUDE, 1788

'All Europeans are countrymen'

Botany Bay, Saturday, 2 February 1788.

> *At 2 in the morning, Lieutenant Dawes of the marines and myself sett off in a cutter for Botany Bay to visit Monsieur De La perouse [sic] on the part of Governor Phillip and to offer him whatever he might have occasion for. We got down to the harbour's mouth at daylight; finding a light air from the southward, we were obliged to row all the way and arrived on board the* Boussole *at 11 o'clock in the morning, where we were received with the greatest politeness and attention by Monsieur De La perouse and his officers. After delivering my message to him, he returned his thanks to the Governor for his attention to him and made the same offers which he had received, and added that, as he should be in France in 15 months, and having stores enough on board for 3 years, he should be happy to oblige Mr. Phillip with any that he might want. Monsieur De La perouse informed me that a number of the convicts had been to him and offered to enter, but he had dismissed them with threats and gave them a days provision to carry them back to the settlement ... As the wind came on to blow fresh from the northward, I yielded to the sollicitations [sic] of the French Commodore and consented to dine with him ... and return to Port Jackson next morning. [Lieutenant P.G. King, 'Remarks and Journal kept on the Expedition ...', in Cobley, 1987]*

'At such a distance from home,' the Comte de La Pérouse writes to Paris from Botany Bay, 'all Europeans are countrymen.' The young naval officer sent by Governor Phillip to formally welcome the French ships to the British colony shares the sentiment. Lieutenant Philip Gidley King,

RN, admires French culture, speaks the language well, and is mercifully free of the xenophobia common among many senior officers – notably the future Admiral Nelson, who loathes the French, habitually calls them villains, and makes no secret of his belief that 'one Englishman is worth three Frenchmen'.

King readily accepts the commodore's invitation to dine with his staff on board the flagship; the meal is enjoyable, French wine flows freely, and information is as freely exchanged. He returns to Governor Phillip full of praise for the work of the French scientists and for La Pérouse. The commodore's parting remark, he reports, was remarkably generous: 'Enfin, Monsieur Cook a tant fait qu'il ne m'a rien laisse à faire, que d'admirer ses oeuvres!' ('Captain Cook has done so much that he has left me nothing to do, except to admire his work!')

PORT JACKSON, 1802

Arrivals I – Emmanuel Hamelin

Port Jackson, April 1802.

> *This being a period when war raged in all its fury between Great Britain and France, Captain Hamelin conceived that he should not be allowed to put into the port ... But his alarm was soon dispelled; for the English received him instantly, with that charming generosity which the height of European civilization can alone explain, and is alone capable of producing. The most distinguished houses in the colony were thrown open for our crew, and during the whole time we remained here, we experienced that delicate and affectionate hospitality, which is equally honourable to those who confer it, and those who are its objects. [Péron, 1809, p. 270]*

Hamelin sighted Cape Howe on 21 April – thirty-two years to the day, he noted, 'since the immortal Cook observed 37°58'S. in sight of the same shore'. The *Naturaliste* arrived off Port Jackson Heads three days later, but a fresh south-westerly prevented her making the entrance before nightfall. Hamelin stood off all night, always keeping in view the beacon light burning on South Head. Below decks, the cramped and ill-lit quarters of the weary crew buzzed with fevered rumours and speculation – some doubted whether the British would allow them into the port, while others looked forward hungrily to tasting the delights of civilisation after so many months at sea.

Alerted by the lookouts on South Head, the signal battery's gun announced the French ship's arrival. His Excellency the Governor, Captain P.G. King, RN, ordered the port pilot to leave at dawn to guide her safely through the heads. Hamelin, meanwhile, lacking any charts of the port, hoisted out a boat to go in search of a pilot.

Making for the entrance the boat meets the pilot's barge heading out to sea, and is taken in tow. Unluckily the boat capsizes while tacking, and her crew are thrown into the choppy seas; turning to go to their aid, the barge runs over the boat's bow, staving it in. Hastily the Naturaliste *launches a second boat, with Midshipman Moreau in command, to go to the rescue, and within minutes the men in the water are all pulled aboard. The sinking craft is towed to the shore by Moreau's boat, while the pilot boards the ship.*

Landing in a remote bay, Moreau sets about repairing the damaged boat. His men and those he has rescued are without food or water, but the skipper of an English boat fishing nearby offers to sail to the ship and bring back some provisions. Hamelin sends rations for one day only, observing that hunger will bring them back sooner. His 'barbarous behaviour' outrages Midshipman Heirisson, for one: 'the indignation felt by the ship's company is more eloquent than anything I can say'.

Three days elapse before the castaways return, Moreau reporting that the men are well, the only casualty being one seaman with a broken shoulder. They have endured hunger and thirst with great courage, despite their hard work in saving the boat. Some natives and the English fishermen have shared a few fish with them, which has kept up their strength. He requests they be given the rations due to them, nine meals in all, for the time they were ashore, but Hamelin again refuses.

The *Naturaliste* passed through the heads at three-thirty on the 25th, and moored in 10 fathoms inside the entrance. The south wind was too strong for her to be towed up to the harbour, and in the rough seas, Hamelin put down a second anchor – a timely precaution, for one of the cables broke during the night. Waiting for the winds to moderate, he sent Commander Milius to Sydney Town with a letter for the governor, informing him of their needs and the reasons for their call.

King, no doubt recalling his reception by La Pérouse in 1788, received Milius warmly. Any concerns that their countries were at war were put aside as the French officer related their desperate shortage of fresh water and provisions, and the spread of scurvy among the crew. The governor's response was spontaneous and practical; the hospital would be readied at once to receive the French sick, and he sent boatloads of vegetables from his garden to the *Naturaliste* at her moorings. Wheat was supplied from the colonists' reserve stocks, and several oxen were killed to succour the famished crew. To Hamelin he wrote: 'I have much pleasure in assuring you that everything will be done to supply your wants, as far as the colony is capable of doing so.'

The French vessel was permitted to anchor at the entrance to Sydney Cove. As she moved towards the anchorage, the officers and scientists saw spread before them a well-wooded and fertile country, dotted with fine-looking country houses. Sydney Town, no more than a large village, lay at the base of two low hills with a stream running between; and the rows of little houses, wooden and single-storeyed and mostly set in a small square of garden, had 'the appearance of dominoes spread across a greensward'.

People from every corner of the settlement crowded the harbourside to see the enemy ship, a group of red-coated officers and their ladies prominent in the front ranks. Soldiers of the New South Wales Corps paraded under arms, and as the anchor dropped the military band struck up, the ladies waved, and the crowd erupted in a great burst of cheering. 'We had not expected,' wrote Milius, 'to find such signs of luxury and civilisation at the birth of a colony.'

Governor King, resplendent in full dress, epaulettes and gold braid glinting in the sun, and attended by the civil and military officers of the colony, boards HM Armed Brig Lady Nelson *to visit the* Naturaliste *at her moorings. A gangway is run across between the two vessels, and the French greet their visitors with three rousing cries of* 'Vive la République!' *King and his retinue remain on board for an hour; before he leaves the governor invites Captain Hamelin and his staff to a ball and supper to be held in their honour at Government House.*

The two-storeyed whitewashed house, simple but elegant in the Italian style, its gardens descending to the harbour shore, provides a perfect setting for the event. All the distinguished people of the colony attend, national rivalries are forgotten, and the festivities continue late into the night. It is near dawn when the guests of honour make their farewells and return to the ship.

Next day Milius calls upon the Governor's Lady to express his thanks for her hospitality; graciously she invites him to ride with her on an excursion through the environs of Sydney. He is surprised by the number of horses and coaches on the roads, the well-built houses of the colonial gentry, and the cultivated fields, which remind him of his native Gironde. Mrs King points out the hospital where the French sick are recuperating; the foundations of the colony's first permanent church; and the Female Orphan School, a great interest of hers, in which are educated 'in the principles of religion, morality, and virtue' young girls whose parents are too poor or too depraved to provide due parental care.

On their return Mrs. King and Milius call on the lieutenant governor, Colonel William Paterson, and his wife. After lunch the colonel shows off with great pride a superb garden at the rear of his house. Here Paterson – a distinguished naturalist and member of the Royal Society of London, a correspondent of Sir Joseph Banks – cultivates many useful plants and vegetables collected from Europe, Africa, and Asia. He overwhelms the Frenchman 'with the most tactful attention'.

While Milius and his fellow officers treated 'every day like a holiday', there was little respite for the crew. Hamelin worked them from dawn to dusk, determined to see the ship reprovisioned and ready for sea before winter prevented a return to Isle de France. On 5 May, with watering completed, he begged the governor

> to add to your many favours by letting me have one bower anchor, 200 bushels of wheat, a few gallons of brandy, and from 8 to 10 cwt. of potatoes ... Provided with these articles I intend to sail on Monday or Tuesday next.[1]

King gallantly replied: 'that as I do not consider the trifles of which you speak worthy of thanks, I am much obliged to you for allowing them to be received'.

Returning sealers brought back reports of fresh wreckage found in Bass Strait, raising fears for the safety of the *Géographe* and her crew. Hamelin and his officers were relieved when a large vessel appeared off the coast on 8 May, everyone thinking it must be their missing consort. Though she proved to be the British ship *Investigator* – also on a voyage of discovery – their minds were set at rest when Captain Flinders came aboard to tell them of his meeting with the commandant on the south coast. All looked forward to an early reunion with their lost comrades.

Arrivals II – Matthew Flinders

Port Jackson, Friday, 21 May 1802. Governor King reports to His Grace the Duke of Portland, HM *Secretary of State:*

> *Previous to the* Naturaliste*'s sailing from hence, I was highly gratified by the arrival of His Majesty's ship* Investigator *on the 9th, and was still more pleased to find that Captain Flinders had surveyed the S.W. coast to within six degrees of Basses Strait before he met the Géographe, which it appears had passed through the straits after parting company with the* Naturaliste, *and that it was the Commodore's intention to come here for refreshments, in consequence of which the captain of the* Naturaliste *intends cruising off the coast till Mons'r Baudin arrives.* [Historical Records of New South Wales, *vol. IV, 1800-1802, p. 763]*

Flinders had completed the passage from Port Phillip Bay in less than a week. From the pilot who boarded the ship at the entrance to Port Jackson he learnt that the *Naturaliste* was already in the port, along with the *Lady Nelson*, the *Porpoise*, and some whalers, but 'no ships had been from England later than us'. He hastened ashore to present his papers from the Admiralty to Governor King, the colony's senior naval officer.

King greeted the younger man warmly – he had been lieutenant governor of Norfolk Island at the time of Flinders's visit in the *Reliance* in 1796 – and listened with growing elation to his account of his discoveries on the south coast. In his turn he brought the explorer up to date with the recent discoveries by lieutenants Grant and Murray in the south – in particular that of Port Phillip Bay by Murray.

The two men celebrated Great Britain's success – their country could now claim the prior discovery of the Unknown Coast, from Cape Nuyts to Port Phillip Bay, excepting the stretch of 50 leagues of barren shore between Encounter Bay and Grant's Cape Banks;[2] at the last moment the French had been forestalled. Though disappointed by the news that Murray had beaten him to the discovery of the great bay, Flinders was buoyed by King's promise that the brig *Lady Nelson* would assist the *Investigator* on his northern survey.

'On the morning after our arrival,' writes Flinders,

> *we warped to a convenient situation near [Cattle] point, and sent on shore the tents, the sail-makers and sails, and the cooper with all the empty casks. Next day the observatory was set up, and the time keepers and other astronomical instruments placed there under the care of Lieutenant Flinders; who, with Mr. Franklin his assistant, was to make the necessary observations and superintend the various duties carrying on at the same place; and a small detachment of marines was landed for the protection of the tents.*[3]

The seamen are employed in stripping and rerigging the masts, checking the stores, and preparing the holds to receive a fresh supply of provisions and water. The greenhouse, shipped at Sheerness for the plant collection, is brought up from the hold for erection on deck. Finding it too heavy for the ship's upperworks, Flinders sets the carpenters to work reducing its size and height. Four convict carpenters are sent aboard to help with alterations to the quarterdeck bulwarks, necessary because they obstruct the captain's observations from the deck when the ship heels over in an offshore wind. Impatient to resume the survey, Flinders drives the men hard, allowing little time for recreation.

Resentment brews in the crew's quarters; after ten months almost continuously at sea they have been counting the days to their next shore leave. Five sailors brought before the captain for drunkenness and mutinous talk are flogged. Others, less rebellious or more circumspect, keep their thoughts to themselves. Replacements are found for the men lost at Cape Catastrophe; with the governor's permission nine convicts, several experienced seamen among them, are taken aboard, with the promise of a free pardon on their return. John Aken, mate of the Hercules, *is appointed master in Thistle's place.*

Brown and his team took lodgings ashore – 'for convenience of Drying & collecting without being detained for Boats etc.', according to Peter Good. On an excursion into the interior, he and and the naturalist fell in with the French botanist, Leschenault de La Tour, and together they collected many fine plants. Brown was impressed by the young Frenchman, describing him as 'an acute observer' in a long letter to Sir Joseph on 30 May. He also briefed his patron on their meetings with the French discovery ships – the *Géographe* at sea and the *Naturaliste* in port:

> Of their past transactions I have learn'd a little; of their future plan of operations, nothing to be depended on. My information comes mostly

from the botanist and mineralogist of *Le Naturaliste*. From the Isle of France they made the south-west cape of N. Holland, and appear to have run along the greater part of the west coast, from which they proceeded to Timor; from thence to their arrival at Van Dieman's Land I have learn'd nothing of their proceedings. Van Dieman's Land they seem to have minutely examin'd, especially towards its southern extremity; they do not seem to have very accurately surveyed Basses Strait. Cap'n Baudin and the *Géographe* we met after having pass'd through it. He had neither been in Port Dalrymple or Western Port, nor had he discover'd Port Phillip, or even King's Island … [he] had not once anchor'd on the south coast.[4]

The *Naturaliste* had already sailed when Brown wrote. Hamelin left Port Jackson on 17 May, ostensibly to meet with the *Géographe* on the south coast – though few of his officers doubted that he intended to make directly for Isle de France. 'It proved to us,' Lieutenant St Cricq noted in his journal, 'that Captain Hamelin was looking only for separation … [and did not wish] to have to consult anyone but himself.'

The whaler *Speedy* left for England a few days later. Along with Brown's letter and a consignment of seeds for Banks, she carried Flinders's account of his south-coast discoveries for the Admiralty. Unfortunately his charts, eagerly awaited in London by his superiors, 'being unfinished, were obliged to be deferred to a future opportunity'. Governor King forwarded with his dispatches a rare trophy for Sir Joseph – 'a New Hollander's head' preserved in spirits. The head was that of the outlaw Pemulway, recently killed in a skirmish with settlers; according to Peter Good, 'he was shot by the Master of the *Nelson* brig, that was shooting in the Woods'. His severed head had been presented to the governor, reportedly as a peace-offering at the request of other Aborigines.

Friday, 4 June 1802. The observance of His Majesty's birthday interrupts the unremitting work on the ship. Before leaving for Government House, Matthew pens a letter to Ann:

> This is a great day in all distant British settlements, and we are preparing to celebrate it with due magnificence. The ship is covered with colours, and every man is about to put on his best apparel and to make himself merry. We go through the form of waiting on His Excellency the Governor at his levée, to pay our compliments to him as the representative of majesty; after which a dinner and ball are given to the colony, at which not less than 52 gentlemen and ladies will be present.[5]

While officers and scientific staff celebrate at Government House, the rest of the ship's company make merry with a special issue of roast beef and an extra ration of grog – both equally welcome. They have not tasted fresh meat since their arrival – in the captain's opinion the market price is exorbitant, and the governor has allowed them none on the public account. For this festive occasion Flinders has exchanged 'an equal quantity of salt meat [for] a quarter of beef for my people'. Later, men with wages still unspent make their way to the jumble of tenements, taverns, and brothels at The Rocks. In the time-honoured custom of HM *ships, many 'temporary wives' slip aboard to share the bunks and hammocks of their sailor friends.*

Sunday, 13 June. The transport Coromandel *arrives in the port, carrying along with her human cargo twelve months' stores from the Navy Office for the* Investigator; *the ship can now be adequately equipped for the northern survey. She also brings two letters from Ann – the first for almost a year. As Matthew reads he becomes more and more dismayed.*

Ann has been ill, so ill that an operation was needed to save her sight. Worse still, she questions the sincerity of his love, in preferring to follow his career rather than be with her. (Matthew remembers his pledge to Sir Joseph: 'I will give up the wife for the voyage'.) He writes her a letter from the heart:

> *My dearest friend, thou adducest my leaving thee to follow the call of my profession, as a poor proof of my affection for thee. Dost thou not know, my beloved, that we could have barely existed in England? ... It was only upon the certainty of obtaining an employment, the produce of which would be adequate to thy support as well as my own, that I dared to follow the wishes of my heart and press thee to be mine. Heaven knows with what sincerity and warmth of affection I have loved thee – how anxiously I look forward to the time when I may return to thee, and how earnestly I labour that the delight of our meeting may be no more clouded with the fear of a long parting. Do not then, my beloved, adduce the following of the dictates of necessity as my crime ... Adieu, my beloved, be happy, and ever believe me to be affectionately and sincerely thy anxious husband.[6]*

Arrivals III – Nicolas Baudin

Aboard the Géographe *off Port Jackson, Thursday, 17 June 1802.*

From the 7th to the 15th of June, the stormy weather was incessant. On the night of the 14th we had much thunder and hail, and the lightning

was so vivid and constant, that we were almost blinded with its refulgence.
At length, on the 17th, the man at the mast-head sung out, that a ship
was in sight and preparing to board us, and shortly the vessel was along-
side. [Péron, 1809, p. 261]

The ship was a whaler bound for New Zealand. Her captain informed
Baudin that the *Investigator* was berthed at Sydney Cove; that Captain
Hamelin had been in the port, but had since sailed; that Boullanger
and his men, picked up off Van Diemen's Land by an English vessel,
were with him; and that he and his men were 'expected with the utmost anxiety in the Colony, [where they] should be sure of meeting
with every possible accommodation'. Most welcome of all was the captain's news that 'only a few days before he sailed, [word] had been
received of the conclusion of peace between England and France'.[7]

With the winds contrary and most of the crew disabled with dysentery or scurvy, Baudin tacked fruitlessly against the headwinds, reluc-
tant to attempt the hazardous passage through the heads. The voyage
from Van Diemen's Land had been nightmarish, due in large part to
his obstinacy. Unaware of Boullanger's charting of the island's east coast,
Baudin had insisted on surveying it himself; in consequence, wrote
Louis de Freycinet, the ship 'stayed on the coast for thirteen days, vainly
occupied in verifying some geographical positions already fixed by our
preceding operations'.

Most of this time was wasted fighting wind and tide, which either
held the *Géographe* far out to sea or threatened to drive her onto the
coast. In the sick bay, filled with the stench of decaying bodies, men
lay inert in their own filth, openly complaining that the commandant
'wanted to bring about their death at sea'. One sailor, a helmsman,
jumped overboard to escape twelve strokes of the lash (ordered to
quieten an attack of hysteria), but was hauled from the sea. The baker
was among those who fell ill, suddenly cutting short the supply of bread
for the invalids. To ease their distress Baudin gave them 'one of my
three remaining pigs ... They were given enough for three meals and
the rest was salted for their future use'.

Conditions worsened on the passage to Port Jackson; daily the number of sick increased, and when the heads were sighted on 17 June
'there were only four men [from each watch] able to remain on deck'.
For three days the *Géographe* cruised off the entrance, until Governor
King, guessing the ship's distress from her manoeuvres, sent a pilot

boat with additional men to help bring her into port. Its appearance, wrote Lieutenant Ronsard, was greeted 'with universal joy', but even then it was nearly dark before they anchored inside the heads. Next morning *Investigator* and other vessels sent their boats to help tow the French ship to an anchorage in Neutral Bay.

Ronsard left with the pilot to pay a formal call on the governor, and to request permission for the expedition to remain at Port Jackson 'for some time, as we all want a little rest, having been at sea for nine consecutive months'. King, hospitable as ever, invited his visitor to join his wife and himself for supper, then watched benevolently as the Frenchman ate enough for four. He dismissed Ronsard's excuses for his appetite, saying in excellent French 'that having gone round the world himself, he had more than once been in a similar position'. When Ronsard returned to the ship he took with him baskets of fresh food from King's garden for the commandant and his fellow officers. He also carried a letter from the governor for Baudin, assuring him that although their countries were now at peace,

> yet a continuance of the war would have made no difference in my re-
> ception of your ship, and affording every relief and assistance in my power;
> and altho' you will not find abundant supplies of what are most requisite
> ... to those coming off so long a voyage, yet I offer you a sincere wel-
> come. I am much concerned [he adds] to find from Mons'r Ronsard that
> your ship's company are so dreadfully afflicted with the scurvy. I have
> sent the Naval Officer with every assistance to get the ship into a safe
> anchorage. I beg you would give yourself no concern about saluting.
> When I have the honour of seeing you we will then concert means for
> the relief of your sick.[8]

Once more the hospital was thrown open to the French, and more than twenty men were admitted. Two died, but Péron was astonished at the speed with which the rest, 'even those on the brink of the grave', recovered their health:

> We were lost in wonder at the magical effect of the country and the vege-
> tables upon a disorder, to counteract which all the medicines on board,
> all the most active operations and energetic attentions, had proved fruitless.[9]

Baudin and Flinders exchanged visits to each other's ships. On 24 June the French captain was piped on board *Investigator*, Flinders taking pains to show the commandant 'that attention that was due to his employment and to his rank'. In the great cabin, Flinders spread out his own manuscript chart of the South Coast, pointing out to his visitor 'that part first explored by him, and distinctly marked as his discovery'. Baudin, unaware of Grant's discoveries, was surprised and disappointed to find his portion 'so much less than he had supposed', but did not object.

Next day, Flinders returned the call, hoping to see the latest French charts, but Baudin excused himself, saying they were not constructed on board the ship; his chart-maker, Charles Boullanger, was away with Hamelin in the *Naturaliste*, and they would not be drafted until later. They talked instead of such problems as determining geographic positions from a ship at sea. It was agreed between them that the French could erect their observatory tents alongside those of the *Investigator* at Cattle Point (Bennelong Point), and that their observations would be made available to Flinders for comparison.

Two days later the *Naturaliste* reappeared off the heads, requesting once more the colony's hospitality. 'The friendly and generous manner with which you received me two months ago,' Hamelin wrote to King, 'leaves no doubt in my mind as to the result of this request.' He brought back with him an English deserter, found on board after his departure; he had planned to hand the soldier over to the first British ship they met, but as he could now return the man to his own officers, he begged the governor to grant him a pardon: 'I ask it from you as a favour, which will further entitle you to my gratitude.'

The *Naturaliste*'s reappearance surprised the French camp as much as the British. Baudin had been taken aback by Hamelin's perplexing decision to leave the port rather than await his arrival there, but it seems there were no reproaches when the two captains met. Milius wrote:

> The *Naturaliste*, having sailed several hundred leagues towards Isle de France, had been forced to return to Port Jackson, to take on more food, not having enough on board to continue the voyage! M. Baudin was not angry at this reverse, and the reunion was pleasant for us all.[10]

After so many months at sea, the French ships were in need of a thorough refit. Damage to their upperworks, sails, and rigging had to be made good, and where necessary replaced. The *Géographe* was careened at Cattle Point, close to the *Investigator*, to allow her copper sheathing to be repaired. Emptied of stores and provisions, the *Naturaliste* was found to be overrun by rats; they had eaten sails and cables as well as wheat and rice – even the fair copy of Hamelin's journal had been gnawed. For five days *Naturaliste* was fumigated with sulphur, gunpowder, and arsenic, and baits were strewn everywhere, after which, 'more than a thousand rats were found dead, and as many more killed running ashore, but there were still some aboard'.

The officers find billets ashore, and lose no opportunity to take part in the colony's social life as invitations flow in from the civil and military officers and their ladies. Lieutenant Henri de Freycinet, attending a function at Government House, finds himself conversing with Captain Flinders, and for a moment allows his frustrations to show: 'Captain, if we had not been kept so long picking up shells and catching butterflies at Van Diemen's Land, you would not have discovered the South Coast before us'.

Though the ships are 'in no forwardness for sailing', Baudin makes plans for the next stage of his expedition. When they are ready for sea, he informs the governor, the Naturaliste *will return to France 'with the very extensive collections in every Branch of Natural History made on the different Coasts of this Country'. He obtains King's consent to purchase a colonial-built schooner of some 30 tons, then on the stocks, and names her the* Casuarina, *after the she-oak timber used in her construction. With her shallow draught, she is far better suited for inshore surveys than the larger ships. The price is negotiable – the merchant and shipbuilder James Underwood settles for £50 and 150 gallons of rum.[11]*

The *Lady Nelson* and her second mate

At Port Jackson, May to July 1802.

> *In consequence of the directions given by His Majesty's principal secretary of state for the colonies, the* Lady Nelson, *a brig of 60 tons, commanded by acting-Lieutenant John Murray, was placed under my orders as a tender to the* Investigator. *This vessel was fitted with three sliding keels ... When the sliding keels were up, the [brig] drew no more than six feet of water; and was therefore peculiarly adapted for going up rivers, or other shallow places which it might be dangerous or impossible for the ship to enter.* [Flinders, 1814, vol. I, p. 232]

HM Armed Surveying Brig *Lady Nelson*, commanded by Lieutenant James Grant, had arrived in the colony in December 1800. At Cape

Town on the outward voyage Grant had received dispatches from the Admiralty enclosing a copy of Flinders's first chart of Bass Strait, and ordering him to survey the strait from west to east. He was instructed to enter any large rivers, and 'take possession in His Majesty's name, with the consent of the inhabitants, if any'.

Making his landfall near Cape Banks on 3 December, Grant sailed east along the unexplored south coast, naming Cape Northumberland, Mount Gambier, Mount Schank (after the brig's designer, Captain Schank), Portland Bay, Cape Otway, and other landmarks. On 8 December he passed the opening to what appeared to be a large bay, which he named after Governor King but did not enter – thus losing the chance to claim for himself discovery of the present Port Phillip Bay. Short of provisions, he decided to run for Port Jackson along the coast previously charted by Bass and Flinders.

Governor King placed the brig on the colony's naval establishment, with a complement of seventeen officers and crew. In 1801 he twice despatched her south to complete the charting of Bass Strait. On the first voyage, Grant surveyed the western side of Wilson's Promontory and Western Port before gales forced him back to Sydney. There he resigned, admitting to King that he lacked experience in nautical surveying, and sought leave to return to England. Lieutenant John Murray took over the command, and sailed for the strait on 12 November; on board as second mate was a young Danish seaman, Jorgen Jorgenson, enlisted under the name of John Johnson.

Murray was out for eighteen weeks, surveying the Kent Group, King Island, and Western Port before earning his niche in history as the discoverer of Port Phillip Bay: 'it is a most noble Sheet of Water larger even than Western Port', he noted in his log, 'with many fine Coves and entrances in it and the appearance and Probability of Rivers'. On 8 March he raised the colours and took possession of the land 'in the Name of His Sacred Majesty George the Third of Great Britain and Ireland'. Following Grant's example, Murray named the great bay 'In honour of Governor King, under whose orders I act', but back in Sydney the recipient of the honour modestly persuaded him to rename it after his own former chief, Rear Admiral Arthur Phillip, now living quietly in retirement in England.

Murray left Port Phillip on 11 March – a fortnight later and he might well have encountered Baudin in Bass Strait – and entered Port Jackson on the 23rd. The *Lady Nelson* was still there when Flinders arrived in

May, but she sailed for the Hawkesbury to load grain on 12 June, not returning until 5 July. To help Murray with his surveying and watch-keeping, Flinders lent him Midshipman Denis Lacy, while Henry Hacking, the Port Jackson pilot, joined the brig as first mate. The young Dane 'John Johnson' kept his post as second mate.

A competent seaman with eight years of experience in British ships (including a spell as a pressed man on one of HM ships in action against the French), Jorgenson had come to the colony aboard the brig *Harbinger* in January 1801. Governor King, as senior naval officer, Flinders and Murray were very likely aware of his true identity, but chose not to ask questions. Junior officers with his experience and ability were hard to come by on the station, and there was nothing to be gained by following regulations.

On a visit to the observatory tents on Cattle Point, Captain Baudin falls into conversation with an English seaman – it may be that Flinders brings about their meeting, for the young man is John Johnson, the Lady Nelson*'s second mate. He speaks French well, and most probably has instructions to fraternise with Baudin and his officers so as to discover the real purpose of their visit.*

Baudin soon learns the young man's true identity. The commandant has many Danish friends, and his brother Augustin sails under the Danish flag. As the two men talk, Johnson tells, with pride and pain, the story of the fierce naval battle of Copenhagen the previous year – he has learnt the details from the London papers carried by the transports Canada, Minorca, *and* Nile. *They rejoiced in yet another victory won by Admiral Nelson – this one, though, fought against the battle fleet of a neutral country, and against unmasted ships anchored along the city's foreshore. At the battle's height, Nelson ignored the signal to withdraw, placing his telescope against his blind eye so as not to see it. Johnson concludes the tale with Nelson's comment on his bloody victory: 'I have been in a hundred and five engagements, but that of today is the most terrible of all.' Clearly the Dane feels the irony of his position – second mate on a British vessel named after the wife of the admiral responsible for his country's humiliation.*

His listener, though, has other concerns than the war at sea. Johnson finds Baudin a man with an intense desire to distinguish himself 'by doing something that no man before him had ever accomplished'. Baudin intends to make an excursion further into the interior of New South

Wales than any Englishman had been before, and induces the Dane to accompany him as a guide. By boat and on foot, they journey together a considerable distance up the Hawkesbury River, a marked tree or the remains of a rough hut serving as evidence that other Europeans have preceded them.

The younger man grows increasingly impatient at the French captain's insistence on continuing his journey 'with so futile an object as that of returning to Paris and boasting that he had been where no other traveller had stood before'. At length, spying a large white rock ahead, he runs forward and stands upon its summit, exulting that no European has passed this point. Baudin marches some twenty paces further, and turns back 'with his ambition fully satisfied'.

Writing from Sydney to Professor Jussieu at the Paris Museum, Baudin claims that on an excursion to the interior he has reached 'the furthest points known to the English', and encloses some seeds collected on the expedition.[12]

Naturalists at large I – The British

'At the Hawksburry & Richmond Hill', Tuesday, 1 June 1802.

> *In the morning sallied out into the immense woods alone the fine morning induced me to go a considerable distance when the day becoming obscured and at noon set in a heavy rain, the thickness of the woods and flatness of the Country, I having no Compass lost my direction, and with great difficulty I reached the house I had set out from before 10 at night very much fatigued, having been about 10 hours in a heavy rain which made the long grass & thick brush very troublesome to walk in and I had walked a considerable time in a morass knee deep ... [Good, 1981, p. 79]*

'In Mr. Good,' Robert Brown wrote to Banks, 'I have a most valuable assistant; a more active man in his department could hardly have been met with.' For three weeks after their arrival at Port Jackson, the gardener was, by his own account, 'constantly employed collecting when the weather would permit & in wet weather preserving those collected'; as well, he was required to sort the seeds gathered on the south coast, and prepare these for dispatch to England by the *Speedy*. Then, probably at Brown's direction, on 30 May he set out for the 'Hawksburry' on a lone excursion.

His exhausting trek through the woods near Richmond Hill brought Good back to the banks of the Hawkesbury River. From there he travelled downstream to the Green Hills (Windsor), and made his way overland to Constitution Hill (near Toongabbie). Again he reached his destination late at night and 'much fatigued, the roads being very bad in one place for near two miles being knee deep from the heavy rains'. He returned to Sydney disappointed that 'in all this distance [50 miles] I did not find such a variety of new plants as I expected, however I found several sufficient to fill all my boxes'. Two weeks later he was back on the same road, this time in company with Brown, the artists Bauer and Westall, and their servants, on a more ambitious excursion to Parramatta, the Hawkesbury, and the foothills of the Blue Mountains.

Thursday, 17 June. Brown and his party set out to walk to Parramatta. The road from Sydney Town, the longest yet built in the colony, cuts through vast eucalypt forests like 'an immense avenue of foliage and verdure'. Though unpaved, it is in good condition and wide enough for much of its length to allow three carriages to drive abreast. Extensive clearings, studded with the stumps of newly felled trees, line the route; herds of fat cattle and numberless sheep graze in the fields, all forms of poultry fill the farmyards of the settlers' huts.

They reach Parramatta towards evening. The town, planned by Governor Phillip to be his capital, lies on the banks of the Parramatta River, and can be reached by small vessels from Sydney. It contains a hospital, a prison, a factory for female convicts, and a barracks. The governor's country residence on the summit of Rose Hill overlooks the settlement; the fine botanical garden within the grounds collects 'the most remarkable of the indigenous plants intended to enrich the royal gardens of Kew'.

George Caley, the garden's superintendent and a collector for Sir Joseph Banks, is waiting for them; Brown has instructions to 'look over his collection' and make sure he is 'doing justice' to his employer. Caley joins his visitors on day excursions to Jerusalem Rocks (North Rocks) and Castle Hill, but little of interest is collected. Brown finds his bluff-spoken colleague 'loquacious' and his names for plants 'mostly bad Greek', but feels he should be encouraged as a collector.

Eager to reach the foothills of the Blue Mountains – beyond which 'all distance is a blank' – the naturalists press on to the Hawkesbury by way of Toongabbie and the Seven Hills. It is midwinter, and though the

*weather by day is fine and clear, at night sharp hoarfrosts and thick
fogs blanket the countryside. Their destination is Commissary John
Palmer's farmhouse at Green Hills, where their arrival is expected – a
warm fire blazes in the grate and a roast dinner is in the oven.*

Woken not long after two in the morning of 22 June, Brown and his
companions made their way through the still darkness to the riverbank,
the hoar-tipped grass crackling beneath their feet. Fog lay heavy on
the water as they boarded the waiting boat and set off upriver, and did
not lift until long after daybreak. They passed the junction with the
Grose River (where the Hawkesbury becomes the Nepean), but soon
found their progress blocked by an extensive gravel bar in the river.
Brown decided to leave the boat and make for the mountains on foot:

> [We] walkd about 3 miles ... Saw the Grose twice its banks then steep &
> rocky but not perpendicular its channel considerably contracted Got upon
> the first hummocks of inconsiderable height The ridge next these wooded
> to the top not high – flat topped rocky & rather steep but apparently not
> impassable ... A few plants found.[13]

They were back at the farmhouse by midnight, and rose before dawn
to sail downstream for Portland Head. On the two excursions, Good wrote,

> we sailed on the River Between 30 and 40 miles for about 20 miles of
> which distance the land is cleared and Cultivated on both sides and is
> probably as fertile & productive in wheat & maiz as any in the world ...
> The River about Portland Head is very romantic consisting of low hills
> abrupt Cliffs of Rock and fine fertile vallies through which the River winds
> Majestically bold & deep with the most luxuriant Vegetation on its shores.[14]

On the 24th, Good set out from Green Hills on foot, 'and after taking some refreshment at Toon Gabie [*sic*] arrived at Parramatta about
8 p.m.'; he reached Sydney the following afternoon. Bauer was not
with him, writing to his brother later that 'I returned [to Port Jackson]
with the *Lady Nelson* on the Hawkesbury River'. Brown's diary is silent
on the matter, but most probably he sailed with Bauer and Westall,
taking the opportunity to enjoy some hunting with Palmer and his
friends and to see new country from the brig's deck.

John Murray, the brig's master, was already under orders to join Flinders on the northern survey. Though the purpose of his trip to the
Hawkesbury (from 12 June to 5 July) was to load grain from the

scattered river settlements, it seems likely that Flinders arranged for him to provide logistical support for Brown's excursion. If so, it was perhaps one of the brig's boats, crewed by ticket-of-leave convicts, that carried the naturalists on their river excursions.

Naturalists at large II – The French

At Port Jackson, June to November 1802.

> *Our scientific researches met with every encouragement; a guard of Eng-lish soldiers were appointed expressly to protect our observatory, which we placed on the north point of the eastern bank of Sydney Cove. The whole of the country was open to the excursions of our naturalists, and we were even permitted to wear our arms, as were the persons of our suite; while guides and interpreters were furnished us for our longest journeys. In short, the English government behaved to us with such generosity, that they ac-quired our warmest gratitude. [Péron, 1809, pp. 278-279]*

Robert Brown does not seem to have sought out François Péron in Sydney, but in a small town of some 2500 inhabitants (many of them convicts) there can be no doubt that the two men met socially before the *Investigator* sailed – on board the ship at Flinders's reception for the French officers and scientists; at one of the governor's functions at Government House; or at Lieutenant Governor Paterson's house on George Street, with its botanical garden. Brown and Péron were wel-come visitors, and the latter, who had lodgings nearby, became a regular guest – in fact, he wrote later, Paterson received him 'as a son'. Through the colonel he got to know the colony's leading men, among them Commissary John Palmer, the Reverend Samuel Marsden, Surgeon D'Arcy Wentworth, and others.

It was Paterson who organised an excursion to Parramatta for Péron and his colleague Dr Bellefin, surgeon of the *Naturaliste*. He assigned a sergeant of the New South Wales Corps as their guide, 'with orders to obtain for us such facilities as we might require', wrote letters of introduction to the local dignitaries, and arranged accommodation for them at the Freemason's Arms, run by James Larra, a Frenchman of Jewish descent.

The highway from Sydney to Parramatta crosses Liberty Plains, an area where many former convicts have taken up land grants and cleared large tracts of ancient forest. The settlers' slab huts look out on grass-lands where 'the bulls frisk about with a vigour equal, or even superior to that of the cold meadows of Ireland, while the cow, more fecund, gives a greater quantity of milk ...'. To the Frenchmen the situation of these 'banditti, so long the terror of their mother country' has been 'the subject of astonishment and contemplation' since their arrival – per-haps, muses Péron, 'there never was a more worthy object of study pre-sented to the philosopher' than the unique experiment of building a new society with the outcasts of the old.

The sergeant introduces Péron and Bellefin to several of the small-holders. These 'former robbers, rogues and pickpockets, criminals of every kind, the disgrace and odium of their country', give them the warmest welcome. 'All these unfortunate wretches,' marvels the natu-ralist, 'have become, by the most inconceivable metamorphosis, labori-ous cultivators, and happy and peaceable members of their commu-nity.' The reformation of their womenfolk is no less remarkable; former prostitutes 'have imperceptibly been brought to a regular mode of life, and now form intelligent and laborious mothers of families'.

Credit for the transformation rests, in Péron's view, with the British government – 'never was the influence of social institutions proved in a manner more striking and honourable to the distant country in ques-tion'. The majority of the convicts, he writes, 'having expiated their crimes by a hard period of slavery, have been restored to the rank which they held amongst their fellow men'. For the rest of his short life he advo-cates that France should follow Britain's example and set up a penal colony overseas.

James Larra, their host at Parramatta, provided a further example of the redemptive powers of transportation. A thief and forger before being sentenced at the Old Bailey in London, he arrived in the colony with the Second Fleet in 1790. He was pardoned, received a land grant, married a woman 'of his own nation and religion', and through suc-cessful speculation had become one of the richest landowners in the district. The value of Colonel Paterson's recommendation on behalf of Péron and Bellefin soon became apparent:

> during the six days we remained at Parramatta, we were served with an elegance, and even a luxury, which we could not suppose obtainable on these shores. The best wines ... always covered our tables; we were served

on plate, and the decanters and glasses were of the purest flint; nor were the eatables inferior to the liquors. Always anxious to anticipate [our] wishes, Mr. Larra caused us to be served in the French style; and this act of politeness was the more easy to him, because amongst the convicts who acted as his domestics was an excellent French cook, a native of Paris, [and] two others of our countrymen.[15]

Péron and Bellefin were taken on a tour of the small town. At one end of the main street stood the barracks, brick-built in the shape of a horseshoe and enclosing a gravelled parade ground. At the hospital they were greeted by the principal physician, Dr D'Arcy Wentworth, before moving on to inspect the female factory and the public school for the young girls of the colony – an innovation by the governor to protect the girls' morals and teach them 'the useful arts appropriate to their sex'.

Crowning Rose Hill, at the western end of the town, was Government House, known as The Crescent. The Frenchmen were more interested in the gardens spreading down the hill's eastern slope; it was from this spot, wrote Péron, 'that England has acquired most of her treasures in the vegetable kingdom; and which has enabled the English botanists to publish many important volumes'. George Caley, alerted to their coming by Paterson, conducted the visitors around the garden. The two Frenchmen knew little enough English, and their host no French, so they attempted to converse in Latin; unfortunately the bluff Yorkshireman's Latin 'was so different from his visitors' in the pronunciation as to be unintelligible to each other'.

They spent several days collecting in the district, and 'succeeded in procuring a variety of animals as beautiful as they were various' – insects, butterflies ('mostly of the grandest colours that can be conceived'), lizards, frogs and toads, worms, shells, and fish. So many new objects came before Péron that he 'was obliged to sacrifice particulars, in order to admit the general and simple enumeration which it was necessary for me to make'.

The two travellers returned to Sydney on one of the small vessels trading between Parramatta and the port. Calling at the riverside farm of Lieutenant Cox, the colony's paymaster, at Kissing Point, they enjoyed a leisurely meal with Cox and his family before re-embarking, 'and in a few hours got to Sydney Town'.

Détente cordiale

At Port Jackson, winter 1802.

> *During his stay in Sydney Captain Flinders often had us to dinner aboard his vessel. He seemed to us a very distinguished officer and one from whom we could learn a great deal. He had already made several voyages along these coasts, [and] we were indebted to him for various pieces of information that would be very useful for the next part of our voyage ... Captain Flinders appeared confident that on his return he would be able to provide news about the wreck of the unfortunate Monsieur De La Pérouse. [Milius, 1987, p. 48]*

Recent arrivals from England confirmed the signing of the Peace of Amiens in March. In mid-July Flinders entertained Baudin and Hamelin, with Péron and other French officers, in the great cabin; they came aboard with Lieutenant Governor Paterson, and 'were received under a salute of eleven guns', befitting the colonel's office. The news of peace enlivened the party 'and rendered our meeting more particularly agreeable', as glasses were raised to the health of the first consul and the longevity of King George. Flinders listened attentively as Baudin talked of his plans for the next stage of his voyage, but privately doubted the Frenchman's prediction that they would meet again off the northern coast:

> I understood that he meant to return to the South Coast, and after completing its examination, to proceed northward, and enter the Gulph with the north-west monsoon; but it appeared to me very probable, that the western winds on the South Coast would detain him too long to admit of reaching the Gulph of Carpentaria at the time specified, or at any time before the south-east monsoon would set in against him.[16]

Prior to his departure on 22 July, Flinders left two copies of his charts of the south coast with the governor – one copy of each to be sent to the Admiralty at the first opportunity, and the remainder to be kept until his return, 'or until [King] should hear of the loss of the *Investigator*, when it was also to be sent to the Admiralty'. Before the first set could be dispatched, however, two events threatened to disrupt the *détente cordiale* that King and Flinders had fostered with their French colleagues. The trouble began with the arrival of the *Atlas* transport from Cork, before the *Investigator* sailed.

Governor King considers he is duty bound to cleanse the colony of vice and corruption. The root of the evil is the infamous liquor traffic controlled by the officers of the New South Wales Corps, the nemesis of Governor Hunter, his predecessor. Determined to tackle the problem at its source, King withdraws their monopoly rights and reduces the quantity of liquor that can be landed – moves which (in Manning Clark's words) 'did not please those vultures who had been enriching themselves at the expense and the existence of their fellow creatures'.[17]

Conditions on board the Atlas *firm the governor's resolve. The ship is a floating hell, a charnel-house – sixty-three of the 151 male convicts on board, and two of the twenty-eight females, have died on the voyage; many of the survivors, too weak to walk or work, have to be carried ashore. An inquiry points to greed and disease as the dual causes of the tragedy – crammed like cattle in insanitary quarters 'tween decks to make room for 'the spirits and private trade' that the master planned to sell in the colony, the men had succumbed to virulent fevers and dysentery. King forbids any of the 2000 gallons of spirits aboard to be landed for sale.*

The French ships are exempted from the ban. Commodore Baudin is allowed to purchase 800 gallons of brandy, King accepting his word of honour 'that nothing with respect to spirits should happen that could anyways differ from the allowed rules and customs observed by the officers of the colony'. The latter, disappointed in their expectations of a large profit from the cargo, openly complain of the governor's partiality for the French, and bide their time.

For the French, 22 September – *1 Vendémiaire* in the republican calendar, and the tenth anniversary of the founding of the Republic – was New Year's Day. The ships dressed to celebrate the occasion, but to the disgust of the younger officers no salutes were fired, nor did the crews receive the customary extra allowance of liquor. Henri de Freycinet wrote in his journal:

> it seems to me that this memorable day has been very poorly commemorated by us; what could be a more glorious time, or closer to the heart, for a French republican. Ah! I shall never forget the obligations that bind me as a soldier to my beloved homeland! Whatever the circumstances in which I find myself, I shall always do my duty.[18]

The governor, mindful of the diplomatic niceties, sent orders to British ships in the harbour 'to hoist their colours in compliment to the French flag'. All complied, until Captain Campbell of the *Harrington* (rescuer of Boullanger and his boat's crew) observed that His Majesty's flag occupied an inferior position on the *Géographe*'s upper masts to the national flags of Spain and the United States, and hauled down his colours. When the Naval Officer, Surgeon Harris of the New South Wales Corps, investigated, Campbell told him bluntly that 'the English flag must be placed higher before he would again hoist his ensign'.

National honour as well as protocol was involved, and the incident escalated. Harris, seeing for himself that the *Géographe* flew the Union Flag at her main yardarm instead of at the masthead, lodged a formal protest with Baudin. The commandant in turn demanded an explanation for the supposed slight from his officers. Ronsard and Freycinet pointed out that the flag's placement on the yardarm was 'only occupied in France by the national flag, and it was by excess of deference that we gave it to the English colours' – clearly implying that their captain should have known this.

King was satisfied, but Baudin, though not without blame for the mistake, did not let the matter rest, writing a sharp letter to Harris complaining of his behaviour, and sending a copy to the governor. Called on to defend his conduct, the Naval Officer argued that the matter would have been settled amicably at the outset if Baudin had explained the custom observed by the French when dressing their ships. Incensed at the 'animadversions of my conduct', Harris went on to inform King of the rumours circulating about the French officers trading in spirituous liquors – these were so widespread, he said,

> that the officers of the New South Wales Corps made many reflections on the Commodore and [his] officers being allowed to purchase spirits from the *Atlas*, whilst they could not be allowed any from that ship.[19]

The governor, not surprisingly, took the allegation as an attack on his conduct and character by his own officers; moreover, if true, the French commodore had broken his word. Sending for Baudin, King rebuked him 'upon the impropriety of his officers' conduct and his deceiving me, if privy to any such transaction'. The affair, he warned, could 'ultimately become the subject of representation between His Majesty and the French Republic'.

Baudin, as astonished by the allegation as the governor, assures him that he will ascertain who could be those of [my] officers … who had dared to disobey your orders and mine in a manner so contrary to the laws of honour of our Navy.[20]

He is soon convinced that the accusation is false – his officers deny having sold spirits from the Atlas; some have traded rum from their daily ration for vegetables, but this is a routine transaction in the colony. 'I give you my word of honour,' he informs King, 'that not one pint of the 800 gallons of brandy … from the Atlas has been landed, the whole of that quantity being kept for consumption at sea.' However, he adds, every matter which attacks an officer's reputation is very delicate:

> be it out of thoughtlessness or wickedness, all my officers and myself are compromised in this affair … I am waiting for the reparation which is due to outraged honour, for you cannot doubt that, were my officers to ignore the reciprocal regards which men owe to one another, I would compel them to submit to them.[21]

King's investigations reveal Captain Anthony Fenn Kemp to be the troublemaker behind the affair. Seeking to discredit the governor in public, and force him to reinstate the officers' trading rights, Kemp had called a former convict named Chapman to testify on the parade ground that he, Chapman, had purchased 'eight bottles of spirits from the first lieutenant of the Géographe [Henri de Freycinet], for which he had paid him 5s. per bottle'. When questioned, Lieutenant Freycinet 'gave his word of honour, and everything that was dear to him as an officer, that no transaction of the kind alluded to had ever taken place'. King writes approvingly,

> I place the fullest confidence [in his account], and his being innocent of the foul and unsupported charge brought against him by a miscreant whose villainy was roused and put in action by being disgraced.[22]

As for Captain Kemp, Governor King informs Colonel Paterson his conduct 'most certainly has occasioned the present misunderstanding between the Commodore and the French officers, with myself and every other officer in the garrison'. He does not doubt 'but you will cause that justice to be done which the laws of honour and hospitality … demands'.

The affair owed as much to colonial politics as to any deep-seated animosities between the two groups of officers. Nonetheless, it soured relationships for a while, until Paterson and his fellow officers prevailed

on Kemp to send a formal (though transparently insincere) apology
to Baudin:

> I beg you, sir, and the French officers under your command, will be fully
> assured how much I am concerned that any occurrence brought forward
> by me should be considered as done with a view of injuring their honor,
> as it is so totally different from my wishes and so unconnected with my
> ideas of them ... I am further requested by my brother officers to say that
> the officers on board the French ships will be considered by them in the
> same estimation as they were on their arrival in the colony.[23]

There the matter rested, officially at least, though several of Baudin's
junior officers refused to allow Kemp to escape so lightly. Their revenge
was ingenious, and owed much to Petit's skills as an artist. A caricature of the gallant captain, sporting a huge pair of stag's antlers on his
head, and with a cumbersome padlock fastening his sword-hilt, was
circulated among the officers of the garrison, to great acclaim. The 'lively
and inviting' Mrs Kemp was rumoured to have a number of admirers,
who went unchallenged by her husband – nor did he make any public complaint about the caricature.

Neither King nor Baudin allowed the unfortunate events to interfere
with their growing friendship. The two men met almost every day,
socially or on business. They were of similar age (King was four years
younger), rank, and social background, each had a genuine interest in
botany; King matched his fluent French against Baudin's somewhat
fractured English. Above all, they shared the loneliness of command,
each recognising and sympathising with the other's problems with
intrigue, envy and insubordination among his staff, which must have
seemed as intractable as his own.

Baudin was often a welcome guest at Government House, where
he charmed Mrs King with his polite attentions and his concern for
her charitable work with the Orphan School. At the governor's table
he made the acquaintance of the colony's notables – among them Judge
Advocate Richard Atkins, ranked second to King himself in the civil
administration. The fifty-five-year-old Atkins lacked legal training, but
was cultured and well-educated, and despite an addiction to the bottle was 'a perfect gentleman when himself'. He read and spoke French,
his residence was a few minutes walk from Government House, and it
would not be surprising if Baudin sometimes joined him for an evening's
conversation over a convivial drink.

'Evidences of future grandeur'

Timor, May 1803. Commandant Baudin reports to the Minister of Marine in Paris:

> *I should warn you that the colony of Port Jackson well merits the attention of the Government, and even of the other European powers, especially that of Spain. People in France and elsewhere are far from being able to imagine how large and prosperous the English have been able to make this colony in the space of fourteen years – a colony whose size and prosperity can only increase further each year through the efforts of the Government. [Horner, 1987, p. 273]*

The progress made by the colony since 1788 amazed Baudin and his staff. Péron went further than his captain when reporting to General Decaen at Isle de France:

> this colony, which people in Europe still believe to be relegated to the muddy marshes of Botany Bay, is daily absorbing more and more of the interior of the continent. Cities are being erected, which [though] in their infancy, present evidences of future grandeur.[24]

Milius did not doubt that 'if the English government continues to encourage agriculture and industry, Port Jackson will grow into [their] main commercial centre in the south seas'.

The woodlands around Parramatta and the abundance of 'a fine and sweet-scented grass of excellent quality' combined, in Péron's view, 'to form an immense pasturage … proper for rearing all kinds of cattle'; in thirty years the colony 'would be covered by innumerable herds'. The sheep were even more fertile – quoting Captain Macarthur, he reported that 'in a few years millions of these animals might be raised without any other expense than the payment of the shepherds'. Their fleece was remarkable for its fineness and its silky quality, and considered equal to that of Spain.

The South Sea fisheries had also opened up new sources of wealth and power to the English. The whale fisheries were 'extraordinarily lucrative', as were the Bass Strait islands with their countless seal colonies. Convict-manned ships sailed monthly from Port Jackson to hunt the great herds of seals and sea lions on their shores, and returned loaded with barrels of oil 'infinitely superior to whale oil', and with thousands of skins for the China trade – 'the sale of a shipload in that country is as rapid as it is lucrative'.

The convicts' contribution to the growth and prosperity of the colony fascinates the French. They note the contrast between the cowed and sullen work-gangs shuffling through the streets of Sydney, and the flourishing businesses operated by emancipist traders such as Henry Kable, Simeon Lord and James Underwood. It is Underwood who negotiates the sale of the Casuarina *– under construction in his shipyard – with Baudin.*

Most tradesmen – carpenters, masons, blacksmiths, bakers and the like – are convicts who have completed their sentence, or have the governor's permission to follow their trades instead of being employed on public works. Even men with no trade, if they behave well, may be granted a smallholding to clear and cultivate. However, those who do not conform, or disobey their masters, incur further punishment: 'they are sentenced to be put in chains, to further deportation [to Norfolk Island], and are given the most loathsome tasks'. It seems to Milius that the rewards offered on the one hand – and the punishment meted out on the other – 'would necessarily bring about an improvement in these people'; but, he observes,

> *the vices they have brought with them to the colony seem to remain with them ... If the English government continues to populate the colony with undesirable characters, it is very much to be feared that these people may end up declaring a state of revolt ...*[25]

The Irish convicts are another matter. Most of these self-proclaimed 'martyrs of liberty', veterans of the 1798 rising in Ireland, claim to have served under General Humbert's flag at Killala.[26] *Forty or more come to the ships seeking asylum; they are returned to the colonial authorities. To hear them talk, all manner of weapons (pikes, muskets, blunderbusses, pistols) are being stockpiled to overthrow the British oppressor. They are an embarrassment to the French. 'We took the wise precaution,' says Milius, 'of having nothing to do with these wretched people, and I must say ... that in this respect the British governor will have no reason to reproach the officers of our expedition.'*

From Sydney Town the western mountains had the appearance of 'a bluish curtain, rising but little above the horizon'. At Hawkesbury, only 8 or 10 miles distant, they dominated the landscape, closing in the horizon to the west and north-west; 'no fracture or peak [broke] their uniformity', and from north to south they formed an immense and impregnable barrier.

Many attempts had been made to cross the Blue Mountains, Colonel Paterson told Péron, but none had succeeded; what sort of country lay beyond was as little known to the Port Jackson natives as to the Europeans. Despite the general ignorance, all forward-looking people in the colony recognised the necessity of finding a pass, because the colony's future progress depended on opening up new lands to the west. Paterson himself had made a failed attempt in 1793; although well-equipped and supplied, his party like its predecesors had been forced to turn back: 'frightful precipices everywhere presented themselves; and no sooner had one summit been escaladed, than others appeared still more barren and difficult of access'.

A more recent attempt, made by the explorer George Bass in June 1798, had also come to naught. Péron heard the full story from Bass himself:

> With his arms and feet protected by iron crotchets, Mr. Bass several times escaladed horrible perpendicular mountains; being often stopped by precipices, he caused himself to be let down by ropes into their abysses. But even his resolution was of no avail, and after fifteen days of fatigue and unparalleled danger he returned to Sydney, confirming by his own failure all that had been asserted of the impossibility of going beyond those extraordinary ramparts.[27]

Discouraged by so many 'sacrifices and useless efforts', the colonial government made no further assault on the mountains until October 1802, when a new expedition set out under Ensign Francis Barrallier, the governor's aide-de-camp and a Frenchman by birth. Péron volunteered to join it, but to his disappointment King 'did not think himself justified in so far extending his complaisance'. Barrallier, however, proved no more successful than Bass; his men and supplies exhausted, 'unable to procure any beasts for the subsistence of [his] troops, except some snakes', he too was forced to retreat.

Péron, as it happened, had no time to brood over the lost opportunity. For three weeks he and Lesueur were occupied day and night preparing their collections for return to France on the *Naturaliste*. 'It may be imagined what we had to undergo,' he wrote, 'when it is known that we arranged in the most methodical manner more than 40,000 animals of all descriptions, collected ... during a period of two years.' They filled thirty-three large packing cases with these collections – 'more valuable and numerous than any voyagers had ever sent to Europe', they excited the admiration 'of all the learned Englishmen in the colony, particularly of the celebrated naturalist Colonel Paterson'.

The naturalists having safely stored their collections, Paterson invited them to join him on one last excursion before they sailed. After a comfortable stay at the governor's residence at Parramatta, they moved on to Castle Hill, the colony's most recent settlement. The tiny township contained a dozen houses,

> but already were to be distinguished on the neighbouring hills vast tracts of cultivated land, while several handsome farms were settled in the vallies. Six hundred convicts were employed in felling trees ... and in twenty quarters might be seen rising immense volumes of flame and smoke, produced by the burning of new concessions.[28]

Paterson took them to visit a nearby farm owned by a French émigré, the Baron de la Clampe, at one time a colonel in the French Army in India. Forced by the Revolution to seek asylum in England, and refusing to bear arms against his countrymen, he had sought permission to settle in New South Wales; living the life of a recluse, he had dedicated his life to agriculture.

They find the baron stripped to the waist, working at the head of a gang of six convicts provided by the government. Mortified at being discovered so poorly dressed, he rushes into the house 'to make himself decent' before inviting the visitors inside. When Péron is introduced the baron embraces him warmly: 'Ah, Monsieur, how goes our dear France?' He listens overjoyed as the naturalist relates 'all the prodigies which a great man had performed for the happiness of our nation', then offers up 'his vows to heaven for the happiness and preservation of the First Consul'.

After a frugal meal, their host conducts the party round his farm, showing off flourishing plantations of cotton and coffee trees. 'If I am not mistaken,' he tells Péron with pride,

> *in a short time I shall have created for this colony two branches of commerce and exportation equally valuable; I have only this means of acquitting myself of a sacred debt of gratitude towards the people who received me in the time of my misfortune – a wish so agreeable to my ideas of delicacy and patriotism.[29]*

'A neighbourhood of voluntary spies'[30]

Paris, 7 Vendémiaire, an 9 [29 September 1800]. 'The Minister of Marine's Instructions to Citizen Baudin, Post-Captain and Commander-in-Chief ...':

> *I shall not dwell further, Citizen, on what solely concerns your inner conduct. Since you are sailing under the flag of truce, and since the sole aim of your labour is the perfecting of the sciences, you must observe the most complete neutrality and not give rise to a single doubt as to your exactitude in confining yourself to the object of your mission, such as it is announced in the passports obtained for you ... [Baudin, 1974, p. 8]*

Port North-West, Isle de France, 20 Frimaire, an 12 [12 December 1803]. Citizen Péron's report to Captain General Decaen:

> *Always vigilant in regard to whatever may humiliate the eternal rival of our nation, the First Consul ... decided upon our expedition. His real object was such that it was indispensable to conceal it from the Governments of Europe, and especially from the Cabinet of St. James. We must have their unanimous consent; and that we might obtain this, it was necessary that, strangers in appearance to all political designs, we should occupy ourselves only with natural history collections ... It was far from being the case, however, that our true purpose had to be confined to that class of work. [Scott, 1914, pp. 437-438]*

London, 1810. Sir John Barrow, Second Secretary to the Admiralty, reviews Péron's Voyage de découvertes aux terres australes *(Paris, 1807).*

> *The perusal of M. Péron's book has convinced us that [the French] application was grounded on false pretences, and that the passport was fraudently obtained; that there never was any intention to send these vessels on a voyage of discovery round the world as stated, but that the sole object of it was to ascertain the real state of New Holland; to discover what our colonists were doing, and what was left for the French to do, on this great continent, in the event of a peace; to find some port in the neighbourhood of our settlements [and] rear the standard of Buonoparte, then First Consul, on the first convenient spot; and finally, that the only circumnavigation intended in this* voyage d'espionage, *was that of Australia. [Quarterly Review, vol. IV, no. VII, August 1810]*

Explorers or spies? Baudin and Flinders in turn came under suspicion – at Isle de France in 1803 the Englishman and his crew discovered to their cost how fine a line separated factual data-gathering from military intelligence. Though it is doubtful whether Baudin himself approved any espionage at Port Jackson, some of his subordinates undoubtedly looked on themselves as 'voluntary spies' for the Republic.

The British government, sceptical of the geopolitical motives behind Baudin's voyage, nevertheless granted a passport for the French ships. Among the king's ministers, wrote Sir Joseph Banks to Professor Jussieu at Paris, he had found 'every proper disposition to promote as all men of education ought to do the increase and improvement of human knowledge by whatever nation it may be undertaken'. Simultaneously, dispatches were sent to Port Jackson warning Governor King of the French voyage, and making clear the government's opposition to any rival settlement in the region.

King did not overlook the probability that the French contemplated 'a settlement on the N.W. coast [of Bass Strait], which I cannot help thinking is a principal object of their researches'. As his friendship with Baudin warmed, he raised his concerns directly with the Frenchman. 'I communicated it to Mr. Baudin,' he reported to Banks, 'who informed me that he knew of no idea that the French had of settling on any part or side of this continent.' King accepted his friend's word, but drew up his own plans for a British outpost to protect the passage through the strait.

At Isle de France, a few days before Flinders's arrival in mid-December 1803, François Péron prepares a secret report on the convict colony for Captain General Charles Decaen, commander-in-chief of French forces in the Indian Ocean. Baudin is dead, his successor Captain Milius is about to sail for France, and the general requires information on the British settlement at Port Jackson – a legitimate military target now that the two countries are again at war. Péron writes in haste, for the Géographe must leave within the week:

> It would be very easy for me, Citizen Captain General, to demonstrate to you that all our natural history researches, extolled with so much ostentation by the Government, were merely a pretext for its enterprise, and were intended to assure for it the most general and complete success. So that our expedition ... was in its principle, in its purpose, in its organisation, one of those brilliant and important conceptions which ought to make our present Government for ever illustrious. Why was it that, after having

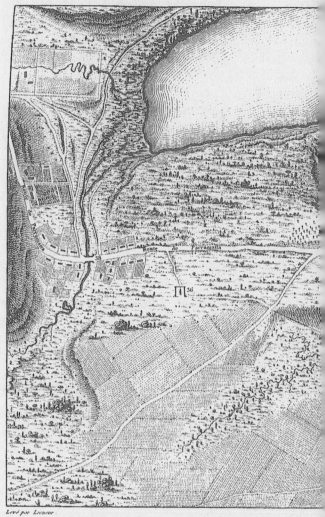

Levé par Lesueur.

NOUVELLE - I:

1. *Ruisseau.* 2. *Batterie du Pavillon des Signaux*
3. *Bâtimens de l'Hôpital.*
4. *Hôpital transporté d'Europe.*
5. *Magasin de M.ʳ Campbell.*
6. *Chantier de Construction.* 7. *Chaloupe de M. Bass*
8. *Calle de l'Hôpital.* 9. *Prison.*
10. *Magasin des Liqueurs fortes et des Salaisons*
11. *Place d'Armes.*

12. *Maison du Lieutenant-gouverneur-g*
13. *Jardin du Lieutenant-gouverneur-ge*
14. *Maison d'Éducation publique.*
15. *Magasin de Grains et Légumes Se*
16. *Casernes des Soldats.* 17. *Place des*
18. *Logemens des Officiers.* 19. *Magasin*
20. *Église.* 21. *Moulins à vent.* 22. *Pont*
23. *Batterie.* 24. *Saline.*

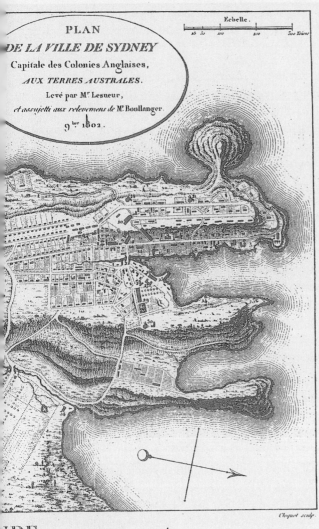

PLAN
DE LA VILLE DE SYDNEY
Capitale des Colonies Anglaises,
AUX TERRES AUSTRALES.
Levé par M. Lesueur,
et assujetti aux relevemens de M. Boullanger
9.bre 1802.

Echelle.

NDE : Colonies Anglaises.

du Gouvernement.
en général de Meubles, d'Instrumens &c.
in d'Habillemens, de Cordages &c.
. Public.
et Jardin du Gouvernement.
et Boulangerie du Gouvernement.
merie du Gouvernement et de la
de Sydney.

32. Habitation ...
33. Briqueterie ... de M. Palmer.
34. Chantier ...
35. 1.re Potence établie sur la Nouv.le Hollande (Ruinée)
36. Potence en activité. 37. Cimetière.
38. Village de Brick - field où se trouvent plusieurs fabriques de Tuiles, de Poteries, de Faiencece, &c.

Plan of Sydney, 1802, by Lesueur.
[By courtesy of Scotch College, Hawthorn]

265

done so much for the success of these designs, the execution of them was confided to a man utterly unfitted in all possible respects to conduct them to their proper issue?[31]

During their stay at Port Jackson, Péron assures the general, he 'neglected no opportunity' to collect information and ideas of interest to the French government. As a scientist he could ask questions 'which would have been indiscreet on the part of another, especially on military matters'. He was acquainted with all the principal people of the colony, 'and all of them have furnished me with information as valuable as it is new'.

Decaen is familiar with the posturings of self-appointed spies anxious to ingratiate themselves with higher authority. There is little of military value in Péron's report to hold his attention – save for brief details of the garrison, a warning of British designs on Spanish South America, and, intriguingly, the prospect of an Irish uprising 'if ever the government of our country should formulate the project of taking or destroying [this colony]'. Péron has no doubt that 'at the mere mention of the French name every Irish arm would be raised'; upon the Naturaliste's first arrival in the port, the Irish in the colony had at first taken her for a warship, 'and commenced to flock together; and if their mistake had not been so rapidly dispelled, a general rising would have taken place among them'.

Among the Irish prisoners at Port Jackson were several gentlemen – including Sir Henry Brown Hayes, 'General' Joseph Holt (a landowner and rebel leader in the 1798 revolt), and William Maum, a young teacher of Latin and Greek – who maintained their households under nominal surveillance by the colonial authorities.

Transported for a political offence, Maum was only twenty-one when he arrived in the colony in January 1800. Suspecting him of being involved in the Irish conspiracy later that year, Governor King termed him 'one of those incendiaries employed in promoting discord and fermenting Litigations'. In 1804, after the Irish uprising at Castle Hill, he exiled him to Norfolk Island, observing that 'his principles and Conduct have changed as little as the others, Nor can Time or place have any Effect on such depraved Characters'.

A few months later, King unexpectedly received from the island's commandant an unsigned declaration from Maum[32] 'calculated to confirm his worst suspicions' of the motives for the French visit in 1802.

Perhaps the young Irishman hoped by this means to procure his return to the mainland; he refused to sign his name 'for fear of its being discovered'. 'When the French vessels were in Port Jackson,' Maum wrote,

> two of the officers belonging to the *Naturaliste* came to see me, and in a seeming friendly manner informed me that they were desired by a friend of mine to apply to me for information for their Log for this colony, &c., ...and appointed a day to wait on me for that purpose. I wrote out in the Latin language such information as I judged most suitable for a Log, such as the Colonial Government, how Justice was administered, the number of Settlements, and such information as that.

The officers, 'one of whom was the captain', returned a day or two later, Maum went on, but seemed 'chagrined at what I wrote, as they expected some more useful intelligence'. They required information on

> the population of the colony – their general opinions – how they endured slavery – how many English – how were *their* sentiments – the Irishmen's sentiments he was assured of – and particularly whether the English in general were seamen or soldiers, who had been transported for any offence after their discharges, as they may be easily worked on.

The peace would not last, they told Maum, and as soon as the first consul learned of 'the great number of persons in this colony who were desirous of Innovation, and a change of condition', he would launch an attack – 'first taking Sydney, and procuring all the prisoners to enter the French service'. The outcome was not in doubt:

> one or two vessels could totally destroy the colony, as by keeping out until dusk they could get into Botany Bay, and take the garrison in their beds ... the stores, provisions, &c., independently of the political advantages to their nation, by obtaining seamen and soldiers, would amply compensate for the enterprise ...

Before leaving, the Frenchmen offered to take Maum with them to Mauritius, where he would be given rank in the Republican service; 'if such an expedition was ever formed [there], he could return with it to Sydney'. Maum had replied

> that I was under peculiar engagements which I could not think of breaking, and that I suffered many inexpressible hardships in Ireland in consequence of the French government's deceitful promises to that Island.

The situation now was completely changed, the officers assured him: 'the present Government of France would amply atone for the villainies that lost Ireland to them'.[33] Finally, wrote Maum, 'they asked me Holt's character; I gave him [that] of a murderer and robber, with which they seemed rather pleased than otherwise'.

King's reaction to Maum's claims is not recorded. He does not appear to have sent a copy to London, nor did he mention them in the strongly worded letter he wrote to Captain General Decaen in 1805 protesting the general's treatment of Captain Flinders 'in every respect as a spy, except in not being executed as one'.

'Port Jackson should be destroyed as soon as possible. Today we could destroy it easily; we shall not be able to do so in 25 years' time.' Decaen discounts Péron's urging; for all its wealth of detail, the naturalist-spy's report omits the military specifics needed to plan an attack on the settlement. However, a second document on his desk proves more useful. Though brief and unsigned, it offers precise information for an invading force. Despite the port's strong natural defences, 'the English have neglected every means of defence, and its conquest would be easy to accomplish'. It would be possible, the writer (almost certainly Milius)[34] explains,

> *to make a descent through Broken Bay, or even through the port of Sydney itself; but in the latter case it would be necessary to avoid disembarking troops on the right side of the entrance, on account of the arm of the sea [Middle Harbour]. That indentation presents as an obstacle a great fosse, defended by a battery of ten or twelve guns, firing from 18 to 24-pound balls. The left shore of the harbour is undefended, and is ... more accessible. The town is dominated by its outlying portions to such an extent, that it might be hoped to reduce the barracks in a little time. There is no battery, and a main road leads to the port of Sydney.[35]*

The captain general has no time to spare. His aide-de-camp and brother-in-law, Major Barois, is returning with Captain Milius to France on board the Géographe. To protect her passport, Barois is travelling as the ship's geographer, but carries with him secret dispatches for personal delivery to First Consul Bonaparte. The reports on Port Jackson are among them.

A clandestine passenger

Aboard the Géographe, *Port Jackson, Wednesday, 17 November 1802.*

> *At about 11 at night an English girl named Mary Bickaith [Beckwith]*
> *appeared on board in men's clothing. I had known her during my stay at*
> *Port Jackson, and she had more than once asked me to obtain Governor*
> *King's permission for her to return to England as Mr. Thomson's assist-*
> *ant.[36] I had promised to interest myself in her case and indeed spoke of it*
> *to the Governor. He would not have refused me this request had it not been*
> *contrary to the general instructions concerning deported persons, but he*
> *told me that if she wanted to leave, no inquiry would be made about her.*
> *She was therefore taken aboard the* Naturaliste *on the day before depar-*
> *ture; but as she is unable to re-enter England without authentic permis-*
> *sion from the Governor, I have embarked her to set her down somewhere*
> *in the Moluccas. Her youth will soon be noticed there and will find her*
> *some happy fate. [Baudin, 1974, p. 425]*

By mid-November the expedition was ready to depart – the ships provisioned, the repairs complete, and the men mostly restored to health.
Those whose fitness remained in doubt joined the *Naturaliste* for the
homeward voyage to France – soon followed by several of the *Géogra-*
phe's midshipmen, classed by the commandant among the 'useless men'.

During their five-month stay, the officers and scientists sought comfortable lodgings ashore. Baudin's residence is not known – most probably it was a rented house; possibly he lodged with one of the civil or
military officers, or in a merchant's household. It must have been not
far from Government House, for he met King almost every day. He
was also in frequent contact with Commissary Palmer and merchant
traders such as Simeon Lord and James Underwood, whose offices and
warehouses bordered Sydney Cove. By virtue of his rank he was surely
a frequent guest at the dinner tables of the colonial notables.

Surviving records fail to explain how, when, or where Post Captain
Baudin first came to know the convict girl Mary Beckwith. That he
should go so far as to risk his reputation and career by embarking her
on his ship – and do so, moreover, with the active complicity of his
second-in-command, Captain Hamelin, and the tacit if unwritten agreement of Governor King – suggests that there may well be more to Mary's
story than at first seems apparent. One needs to leaven the few known
facts with a little imagination.

Very likely his acquaintance with Mary came about through his contacts with colonial officials or merchants; it may be that she became a servant in his household. 'I had known her during my stay at Port Jackson', he wrote – adding that she came from a good family, and had voluntarily accompanied her mother, transported to the colony for stealing a length of muslin, rather than be placed in a convent. 'She is seventeen,' he noted, 'and is more interesting on account of her behaviour than for her prettiness.'

Mary Beckwith had arrived in the colony with her mother,[37] also Mary, on the *Nile* transport in December 1801. Sentenced to death at the Old Bailey, London, in June 1800 for stealing three pieces of printed calico, 46 yards in length and valued at £5, from a draper's shop in The Strand, the two Beckwiths had been reprieved and given transportation for life. The jury had not questioned the shop assistant's evidence that he had taken the calico pieces from under the girl's gown – she had concealed them, he testified, while standing 'about a yard and a half from where I was, not further'. Their verdict was unanimous – guilty as charged, but with a recommendation of mercy for the girl, on account of her youth. Mary was then fourteen.

The *Nile* carried only female prisoners, ninety-six in all, from various gaols in Britain. Upon arrival, the ship and two companion transports, *Canada* and *Minorca*, were inspected by the governor and his senior officials. King was delighted with the condition of the convicts, male and female – there had been just one death between the three ships on the six-month voyage. 'All the convicts were landed in health,' he reported to London, 'and were by far the best conditioned of any that have ever arrived here, being fit for immediate labour, which is not yet the case with many who came by former ships.'

Following their inspection, the governor and his entourage proceed to select the female convicts they consider 'most agreeable in their person' for domestic duties in their households. Mary is now sixteen, a healthy girl in the bloom of youth. Her mother is thirty-five, of respectable appearance, and a cut above the harlots and petty thieves making up the majority of the Nile*'s human cargo. When their companion Margaret Catchpole, four years older than Mary senior, is chosen by Commissary Palmer as his cook, it is more than likely that one or both of the Beckwiths should be chosen by one of his colleagues for his service.*

Judge-Advocate Richard Atkins is among the officers attending the shipboard inspection. His wife Elizabeth lives in Ireland, his Irish convict mistress Catherine Haggerty (Kitty) has left him to return to England at his expense, leaving their daughter Teresa in his care, and he is in need of a housekeeper. Whether Atkins chooses the Beckwiths as household servants at this time is not known. A few years later, however, Mary Beckwith senior is his housekeeper; after Elizabeth's death in 1809 she returns with him to England and eventually becomes his wife.

In October 1802, a month before the younger Mary sails on the Géographe, Atkins's domestic arrangements are unsettled by his wife's arrival from Ireland. Mrs Atkins's departure from her homeland after an eleven-year separation from her husband may, one suspects, be linked to Catherine Haggerty's return.[38] Atkins's influential family (with links to the Cecils) had financed his 'exile' to the colony a decade earlier, and were doubtless scandalised by this latest example of his feckless behaviour; they may well have put pressure on his wife to rejoin him and regularise the situation.

Mrs Elizabeth Atkins has 'a very fine figure' and a lineage to match. Coming from the Irish branch of the old Antrobus family, she should be an ornament to colonial society, limited as it is. Yet, says John Grant, a visitor to the household in 1804,

> *she has never been introduced into polite Circles here, which is singular as the Judge-Advocate's Lawful Wife – but the truth is, her pride prevents it, the origin of the first Women in the Colony is so obscure.[39]*

Grant adds that on arrival she found her husband 'living with a woman, as is customary with all the Gentlemen here … by whom he has 2 Girls whom Mrs. Atkins adopts as her own'. He fails to mention whether the girls' mother stays on as housekeeper.

One of the most surprising aspects of the Beckwith affair is Governor King's response to Baudin's request: 'he told me that if she wanted to leave no inquiry would be made about her'. This directly contradicted official policy; no convict could be granted a full pardon within twelve months of arrival in the colony, and none could be permitted to leave without one. Apparently King had second thoughts about the expediency of thus evading his responsibilities. The following year he refused a request from a fellow RN officer, Captain Colnett of HMS *Glatton*, to grant a free pardon to a young convict woman who had shared his cabin on the voyage to Port Jackson. Why was the usually punctilious King prepared to bend the rules in Mary Beckwith's case? His friendship with Baudin scarcely seems sufficient reason.

Other factors that perhaps influenced King's decision can only be guessed at. In the absence of any firm evidence, it is tempting to see Richard Atkins's hand at work. He was the governor's legal adviser, second to King in the civil hierarchy, and his family links in England still carried considerable weight; more particularly, the possibility of a personal motive on Atkins's part cannot be overlooked – if her mother was the convict woman living with him at the time of his wife's arrival, the younger Mary's continuing presence may simply have proved one embarrassment too many for the 'proud' Mrs Atkins.

For Baudin's officers the commandant's motive for bringing Mary on board is transparent. Midshipman Breton, officer of the watch on the Naturaliste, *notes Captain Hamelin's return to his ship at 8 p.m. on 16 November. Having spent the day ashore with the commandant, the captain returns with an English woman wearing a long, hooded cloak. She passes the night in his cabin, and next day remains hidden in the gunroom to avoid being seen by the English passengers, Surgeon Thomson and his wife. Members of the crew recognise her as the* 'fille de joie' *with whom the commandant has been living in Sydney. During the night Captain Hamelin escorts her across to the* Géographe.

Lieutenant Henri de Freycinet, senior deck officer on the flagship, is more explicit in his comments: 'M.B. le Cen Comdt embarque avec lui une fille publique pour son usage particulier' *('The Citizen Commandant embarks a prostitute for his personal service').*[40] *As the French officers have learned from their British counterparts, female convicts are to be classed as whores.*

Departure

At Port Jackson, Tuesday, 16 November 1802. Commandant Baudin to Governor King:

> *Sir, In leaving this colony, I bequeth to the French nation the duty of offering you the thanks which are due to you as Governor for all you have done, as well for ourselves as for the success of the expedition; but it is for*

*me to assure you how valuable your friendship has been and will ever be
to me, if you will allow me to put you in mind of it whenever an opportunity offers itself.*

*The sincerity and honorableness of all your dealings with me leave no
doubt in my mind that you will give me the permission I ask for, the more
so as the opportunities of my meeting you after leaving this port will be
exceedingly rare. It will therefore be a satisfaction for me to correspond
with you from whatever country events may bring me to. It is, as you know,
the only means which men who love and esteem one another can make
use of, and it will be the one we shall reciprocally avail ourselves of, if ...
I have been able by my conduct to inspire you with the feelings which yours
has inspired me with.*

I have the honour, &c. N. Baudin.

[Historical Records of New South Wales, *vol. IV, 1800-1802, p. 1006*]

The French prepared to weigh anchor at daybreak on 17 November.
Baudin had arranged with Hamelin that the *Naturaliste* would accompany the *Géographe* and *Casuarina* as far as King Island, but would
then sail directly for Isle de France with the expedition's precious
natural-history collections. The other vessels would follow Fleurieu's
original instructions (now a year behind schedule), retracing their route
south and west round the continent to examine the coasts of Arnhem
Land, the Gulf of Carpentaria and southern New Guinea before finally
heading homewards.

A general search of the flagship the night before departure discovered twelve stowaways hidden below deck; a search the preceding
day had found ten others, nine men and a woman; all were put ashore.
At first light, with the tide in his favour, Baudin ordered the anchors
raised, but the bower was so firmly embedded in mud that the heavy
tackle used to raise it snapped. The mishap delayed their departure
for another day.

In the meantime, yet another search uncovered three more convict
stowaways hiding beneath the longboat's tarpaulin. Baudin sent his
dinghy ashore with the three men under guard – 'it did not occur to
me that any attempt would be made to carry them off' – and was infu-
riated when the harbourmaster's barge intercepted the dinghy and took
the prisoners off by force. Signalling the *Casuarina* to come alongside, he boarded her and went into Sydney, to demand an explanation from the governor:

he openly condemned the behaviour on this occasion; and as proof of his disapproval he granted a free pardon for this time to all those who had deserted, even though their departure and names were known before they were taken out of my boat. I think that when he returns, the harbour master will receive a lesson that will teach him to be more prudent in future.[41]

By eight next morning, the 18th, the three ships had passed through the heads and set course for Bass Strait. They were barely hull-down on the horizon when a rumour reached the governor that the French contemplated a settlement in the strait or in Storm Bay, Van Diemen's Land. Its likely source was Captain Kemp, still smarting from his humiliation by Baudin's officers two months before.

King could not ignore the gossip, whatever its origin, and wrote at once to Colonel Paterson:

Understanding that he is in possession of some information respecting the intention of the French nation settling on Van Diemen's Land, [His Excellency] requests the Lieut.-Governor to furnish him with what information he possesses … in order that [he] may take the necessary steps.[42]

The request surprised Paterson: 'The conversation was so general among the French officers respecting their making a settlement in the Straits of d'Entrecasteaux,' he replied, 'that [he] could not suppose it was unknown to Governor King.' One of the officers (probably Péron) had pointed out the site for the proposed settlement at Frederick Hendrick Bay, but Paterson very much doubted whether any such action was intended on the present voyage.

King was not prepared to take the risk. With HMS *Porpoise* away at Tahiti, the only naval vessel available was the 29-ton armed schooner *Cumberland*. He hurriedly cobbled together a scratch crew, added three marines, and gave the command to Midshipman Robbins, master's mate of the *Buffalo*, promoted acting lieutenant for the mission. The day Robbins sailed, 23 November, King wrote to Lord Hobart, HM Secretary of State:

A few hours after the French ships were out of sight I was informed that some of the French officers during their stay here had informed L't Col. Paterson and others that it was the intention of the French to make a settlement in what is called by us 'Storm Bay Passage' and by the French 'Le Canal d'Entrecasteaux', on the east side of Van Diemen's Land … I have lost no time in expediting the *Cumberland*, armed Colonial Schooner. She sails this day, and from the arrangements I have made His Majesty's claim to that part of this territory cannot be disputed …

The officer I have entrusted with this expedition is directed to proceed immediately to Van Diemen's Land, [with instructions] I have given him to communicate to Mons. Baudin if he falls in with him, as I know his intention is to go immediately through Basses Straits, and whatever may be in contemplation it cannot be performed by him. How far he may have recommended it to the French Government I do not know. It seems by Col'l Paterson's information that they do intend it. It is my intention, as soon as the *Porpoise* arrives, to dispatch her with a small establishment to the most eligible place at Storm Bay Passage, and one at Port Phillip or King's Island. Your Lordship's instruction on these points I shall be glad to receive as soon as possible.

Philip Gidley King.[43]

Thus the immediate consequence of the French voyage of discovery proved to be the British settlement of Tasmania. In September 1803, Lieutenant John Bowen landed at Risdon Cove with the first settlers, forty-nine in all, among them eight soldiers and twenty-four convicts.

EPILOGUE, 1813

Downing Street, London, 19 August 1813. HM Secretary of State, Earl Bathurst, writes to Governor Macquarie of New South Wales. The dispatch is marked 'Separate and Secret':

Sir, I cannot close my Dispatches … without communicating to you a Paper of information respecting a Plan said to be entertained by the Enemy of attacking the Settlements under your Government. The Person who has given this information is a Dane named Jorgensen, who has been much in the South Seas, and must be known at Port Jackson, having served as Master's Mate in the Lady Nelson under L't Grant, when that Vessel was employed in surveying part of the Coast of New Holland. It is believed that Jorgensen then went under the name of Johnson or Jansen.

The doubtful Character of this Individual, the great improbability of the plan itself, and the still greater improbability of such a Person being minutely acquainted with its Details, supposing it to be in Agitation, have led H.M.'s Government to refuse any Credit to the information. They have nevertheless thought it proper to communicate it to you, in order that you may not be ignorant of any thing which may possibly affect the Welfare and Security of the Colony under your Charge.

I have, & c., Bathurst
[Historical Records of Australia, series I, vol. 7, p. 72]

Information respecting a Plan for Attacking the Colony

Previous to my making any mention of the Instructions given to the Commanders of several French Ships of War, now destined for an Expedition to the South Seas, I shall say a few Words on the Causes which have induced the French government to turn their attention to a distant Quarter of the Globe.

* * * * *

Bonaparte, ever attentive to all that can ... prove injurious to British commerce, sent two French Brigs of War, the Geograph and the Naturalist [*sic*] to the South Seas under a pretence of making Discoveries, but in reality to espy the Situation of the English Colonies in New South Wales. Captain Baudin commanded the Expedition; he was no Seaman, but excellently skilled for the Duty he was sent upon: he was a good natural historian, and well qualified to judge of the real situation, political and Commercial, of any country. He proceeded straight to Port Jackson, where he met with very liberal treatment from the Governor who had orders to afford him every assistance in his Discoveries.

Commodore Baudin had expected to see nothing but a few miserable huts, and a people destitute of all the Conveniences and pleasures of life, laboring under all the disadvantages of Slavery, and a long distance from the more civilized parts of the World. He was deceived, nothing could exceed his Astonishment, when instead of finding huts he observed Palaces and good buildings everywhere. The farm houses were well and strongly built, and the Banks on the River Hawkesbury presented for many miles the Sight of fertility and cultivation. He observed every thing with the curious Eye of an Observer ... The colony had not then been established for any greater length of time than twenty years, and yet it presented a picture of increasing wealth and prosperity. These circumstances, with the excellent situation of New South Wales for trade and commerce, could not escape the penetrating Eye of Captain Baudin.

* * * * *

The Statement, which was afterwards laid before Bonaparte relative to the English Colonies in the South Seas, contained an exact and true account of all the above circumstances, whilst Baudin's discoveries and remarks on natural history were slightly noticed. Bonaparte heard with astonishment and was vexed to learn that, whilst he made every attempt to destroy British Commerce in Europe, new resources were open to the trade of Great Britain, and colonies established, which might ... become an Empire powerful enough to exclude all other nations from the fisheries, trade and commerce of the vast and extensive Seas beyond Cape Horn.

Bonaparte contemplated an Expedition against the South Seas, but at that time so many obstacles presented themselves to his Schemes on this head, that nothing could be done with any reasonable prospect of Success. The case is now much altered; the War with America, the Insurrection in South America, and other causes favorable to his views, point out this as the most favorable moment for executing his plans ... There are now four French frigates fitting out, and almost ready for Sea ... mostly commanded by those Officers who attended Captain Baudin on his Expedition to the South Seas. It is determined that they shall sail from their port about the Month of November, the Winter Season being by far the most favorable for making an attempt unnoticed. Each frigate will take on board 250 soldiers, and in all 25 horse. Thence they are to proceed with all speed towards the Falkland Islands ... there to await the arrival of an American frigate and a Storeship, which will leave America about the same time to join the French. The French frigates are accompanied by two Americans, named Kelly and Coleman, who have for years been engaged in South Sea Whalers, and who know the South Seas and the Coasts very well. Being joined by the Americans the Squadron will proceed Round the Cape of Good Hope ... and make for Basses' Straits ... the Expedition is immediately to proceed for Port Jackson and attack that place; but as there is a large battery on the North Shore of Port Jackson, and a battery on the point going into Sydney Cove, it is deemed adviseable for the Squadron to proceed to Broken Bay, there to anchor, and proceed with the Troops, horse and Artillery, up the River Hawkesbury, to effect a landing, so as to cut off the Settlers from Sidney [*sic*], and to obtain possession of the Wheat and Grain; Parramatta of course must fall. 1500 Stand of Arms are to be delivered into the hands of such First [Irish?] Convicts who are willing to join the French. It is also proposed that, if circumstances will permit, a detachment of 250 men should be landed in Botany Bay, to make a diversion ... but as there are many marshes about that place much care is to be taken to run no risk, and rather than do that, all troops are to be landed up the River Hawkesbury ... Two or

Three frigates are to proceed without delay to South America, where they are to endeavour to persuade the Insurgents to declare for Joseph [Napoleon's brother, King of Spain], but if they cannot succeed in that, to deliver them 6000 Stands of Arms, so at all events to enable them to revert King Ferdinand's power. All the Whalers in the South Seas are to be destroyed or burnt, unless they should prove of Service to the French.[44]

Governor Macquarie replies on 30 April 1814:

It has been generally supposed that the Geographe *and* Naturaliste, *French Ships of War under Capt. Baudin … came solely for the purpose of ascertaining how far it might prove expedient for the then Ruler of France to establish a Colony on some part of New Holland, to counteract the views of the British Government in this Country; and I have no doubt Bonaparte would long since have prosecuted his views in this respect, had his more important Engagements in Europe admitted of his sending out a sufficient Force for the Conquest of this Colony …*[45]

Notes

1. *Historical Records of New South Wales (HRNSW),* vol. IV, p. 945.
2. Grant had made a running survey from Cape Banks to Western Port in December 1800.
3. Flinders, 1814, vol. I., p. 228.
4. *HRNSW*, op. cit., pp. 776-779.
5. Scott, 1914, pp. 265-266.
6. Ingleton, 1986, pp. 176-177.
7. The news was premature, and related to a provisional armistice negotiated between the powers the previous October.
8. *HRNSW*, op. cit., p. 949.
9. Péron, 1809, p. 262.
10. Milius, 1987, p. 43. (Passage translated by J. Treloar.)
11. Information provided by Mr R.T. Sexton, 1998.
12. The only record of this excursion is Jorgenson's account in his unreliable memoirs, *A Shred of Autobiography,* published in Hobart in two parts, 1835 and 1838. Baudin's letter to Jussieu provides corroborative evidence that he journeyed inland, but gives no details. The most likely time would seem to be early in July 1802, between Hamelin's return to Port Jackson on 26 June and Flinders's reception for the French officers on 14 July. King provided boats and guides for exploring parties from both expeditions.

13. R. Brown, in preparation. (Transcribed from the original in the Natural History Museum, London.)
14. Good, 1981, p. 80.
15. Péron, op. cit., p. 297.
16. Flinders, op. cit., pp. 236-237.
17. Clark, 1962, vol. I, p. 162.
18. Hélouis transcripts.
19. *HRNSW*, op. cit., p. 978.
20. Ibid., p. 976.
21. Ibid., p. 977.
22. Ibid., p. 980.
23. Ibid., pp. 983-984.
24. Scott, op. cit., p. 442.
25. Milius, op. cit., p. 44.
26. General Humbert commanded the French force that landed at Killala Bay, County Mayo, in August 1798 to support the Irish rebels. His manifesto called on Irishmen to join the common cause: ' ...behold Frenchmen arrived amongst you! ... The contest between you and your oppressors cannot be long. Union! liberty! the Irish republic! – such is our shout, let us march – our hearts are devoted to you ...' He surrendered to the British a fortnight later.
27. Péron, op. cit., p. 289.
28. Ibid., p. 307.
29. Ibid., p. 309.
30. Jane Austen, *Northanger Abbey*.
31. Scott, op. cit., p. 438.
32. See Rusden, 1874, pp. 86-89.
33. If Hamelin was indeed one of the officers present, he would have spoken with feeling – he sailed on the ill-fated Irish expedition of 1796, under General Hoche.
34. See Horner, 1987, p. 303.
35. Scott, op. cit., p. 261.
36. Surgeon James Thomson had received permission to return to Europe with his wife and servant on the *Naturaliste*.
37. Giving evidence at the trial, Mary Beckwith senior stated she was the girl's stepmother.
38. Catherine Haggarty was granted an absolute pardon by Governor Hunter in February 1800, and returned to England on board HMS *Reliance* (Flinders's ship) – surely as a result of Atkins's influence. (See Atkinson, 1999.)

39. John Grant. Letters to his mother and sisters. MS, National Library of Australia.

40. Hélouis transcripts.

41. Baudin, 1974, p. 425.

42. *HRNSW*, op. cit., p. 1006.

43. Ibid., p. 1008.

44. *Historical Records of Australia*, series I, vol. 7, pp. 73-77.

45. Ibid, p. 241.

CHAPTER EIGHT

⚓

Flinders – Triumph and Tragedy,
July 1802 to December 1803

The east coast

Port Jackson, Thursday, 22 July 1802.

> *Lieutenant John Murray, commander of the brig* Lady Nelson, *having received orders to put himself under my command, I gave him a small code of signals and directed him, in case of separation, to repair to Hervey's Bay; which he was to enter by a passage said to have been found by the south-sea whalers, between Sandy Cape and Break-sea Spit. In the morning of July 22 we sailed out of Port Jackson together; and the breeze being fair and fresh, ran rapidly to the northward, keeping at a little distance from the coast. [Flinders, 1814, vol. II, pp. 1-2]*

The two ships passed Port Stephens shortly before midnight. In the fresh breeze the brig was left astern, and the *Investigator* lay to for an hour next morning until she caught up. The following night they were again separated, and Flinders did not see his consort for a week, when she rejoined him at anchor in Hervey's Bay, some 700 miles to the north of Port Jackson. The delay convinced him of 'the *Lady Nelson* being an indifferent vessel', and also of Lieutenant Murray's inexperience as a surveyor, 'not much accustomed to make free with the land'.

They remained in the bay another day, while three boat parties went ashore. The brig anchored offshore, with orders to cover the landing parties – 'the natives here being very numerous'. Flinders, accompanied by his Aboriginal friend and interpreter Bongaree and several others, made for Sandy Cape at the extremity of the bay; Murray and his people were to cut wood for fuel; while Robert Brown with a smaller group set off to botanise on Fraser Island.

Flinders's party had not progressed far when they were confronted by a group of natives, naked and angrily waving branches of trees at the intruders. Bongaree stripped off his clothes and laid down his spear, to show that they came in peace; finding they did not understand him he switched 'in the simplicity of his heart' to broken English, hoping to succeed better. At length the 'Indians' allowed him to approach, and by degrees the rest of Flinders's party joined them. Hatchets and presents were handed around, and more natives came up. In good humour they returned together to the boats, and feasted on the blubber of two porpoises brought ashore purposely for them.

As he continued the survey Flinders found that his observations, made with the benefit of the two Earnshaw chronometers, differed increasingly from the longitudes noted on Cook's charts:

> From Port Jackson to Sandy Cape, Captain Cook's positions had been found to differ from mine, not more than from 10′ east to 7′ west; which must be considered a great degree of accuracy, considering the expeditious manner in which he sailed along the coast, and that there were no time keepers on board the *Endeavour*; but from Sandy Cape northward, where the direction of the coast has a good deal of westing in it, greater differences began to show themselves.[1]

At Hervey's Bay, Cook's error for longitude was about 16 miles, but it had more than doubled (to 35 miles, or half a degree) when they reached Cape York – a clear demonstration of the chronometer's value in solving the problem of longitude.

Four days were spent at Port Curtis (south of the present Gladstone). Here the *Lady Nelson* grounded on a shoal and lost her main sliding keel. The botanists took advantage of the delay to make daily excursions, while the crew were allowed ashore 'to divert themselves'. As they prepared to sail at daylight on 9 August, the *Investigator*'s anchor came up with one arm broken off, the result of a flaw extending two-thirds through the iron – the manufacturer's negligence, Flinders noted angrily, 'might [well] have caused the loss of the ship'. Later that day the two vessels entered Keppel Bay, named by Cook in 1770, and remained for nine days.

The terrain is arduous – shoals, muddy islands, dense mangrove swamps. Sluggish, shallow creeks reach far back into the land, ending in blind culs-de-sac. Flinders probes the various arms of the bay, exploring by boat and on foot, attempting to lay down an accurate plan: 'there are few places where it was not necessary to wade some distance in soft mud', he writes, 'and afterwards to cut through a barrier of mangroves, before reaching the solid land'.

Signs of the 'Indians' are everywhere. A large group falls in with a party of seamen near the ship; they are stout, muscular men who seem to understand bartering better than most natives they have met. At first 'they menaced our people with spears, [but] finding them inclined to be friendly laid aside their arms, and accompanied the sailors to the ship in a good-natured manner'.

At nightfall a master's mate and a seaman with him are reported missing. A signal gun is fired at intervals throughout the night, but without result. Search parties set out at first light, and Flinders himself takes an armed boat to Cape Keppel to look for the missing men. Fears grow for their safety, until about noon the men working on the ship see some twenty natives walking along the beach, their missing shipmates among them. A boat is launched at once 'with the drum, fife and fiddle', and with presents for the rescuers.

The lost men are unharmed but exhausted, and covered with mud and mosquito bites. They had strayed from the main party and at sunset, when they should have been at the beach, were entangled in a slimy mangrove swamp several miles distant. Unable to find their way back in the darkness, they had spent the night persecuted by swarms of mosquitoes and sandflies, and went without sleep. Making their way out of the swamp at dawn, they soon found themselves at the natives' camp; here their kindly hosts broiled two ducks for them over a fire, and 'after the wanderers had satisfied their hunger ... conducted [them] back to the ship'.

The sailors from the boat embrace their comrades, hand out presents to their rescuers, and call on the musicians to celebrate the occasion with a suitable tune: 'Upon the Drum beating they began to walk off, the Fife they did not pay much attention to, the Fiddle none at all.' In a short while the natives have disappeared into the woods, and no more is seen of them.

The two ships sailed from Keppel Bay on 17 August, and steered northwards. The numerous shoals in the bay had rendered 'the services of the *Lady Nelson* [largely] useless to the examination', but Flinders still hoped she would provide better assistance in the future, as they ap-

proached the treacherous waters of the Great Barrier Reef. Soon, however, the vessels again parted company – 'their [*sic*] being a stiff Breeze
& she [the brig] a dud sailing vessel', wrote Seaman Smith in his diary.
She caught up with the *Investigator* on 21 August, and they anchored
in a deep inlet that Flinders named Port Bowen (now Port Clinton).
Shore parties were landed to search for fresh water, collect botanical
samples for the naturalists and fish for the crew, and cut timber to replace the brig's main sliding keel, carried away in Port Curtis. The voyage resumed on 24 August, the *Lady Nelson* as usual falling to leeward.

Lost anchors, broken keels

Shoalwater Bay, 26 August to 4 September 1802.

> We hove our Anchor every Tide stil [sic] Working further in on the 26th we
> lost our Cutter by being Swamp'd astern of the ship, one Man being in the
> Boat that cou'd not swim, but the Nelson's Boat being sent away expediti-
> ously the Man was saved but the Boat Inevitably lost. On the 28th our Boat-
> swain was confined & order'd to prepare for a Court Marshal [sic] on the
> first Opportunity, for Drunkenness & our Boats Imploy'd on different Duties
> on shore, & those on Board are working constant Tides Work with the ship
> in compy. with the Lady Nelson, the Time slippt away in this Imploy almost
> Imperceivable. [Seaman Samuel Smith's journal, MS, Mitchell Library]

Each day brought fresh hazards. At the entrance to Cook's Shoalwater
Bay the rapid tides in Strong-Tide Passage – 6 miles long and from
1 to 2 miles broad, with half the width taken up by rocks and shoals –
left Flinders with little choice of a course. The cutter at *Investigator*'s
stern was swamped as she was about to be hoisted up, and was swept
away by a 4-knot tide. The man in her was thrown into the water, but
was saved by the brig's boat.

The ships remained in the bay for nine days. Flinders landed each
day to take his bearings, correcting Cook's positions with the more
accurate longitudes made possible by his chronometers. The botanists'
plant collections grew steadily, some 500 species being gathered on
the east coast alone. Here too they had a peaceful meeting with the
land's inhabitants:

> the naturalist and other gentlemen had gone over in the launch to the
> west side of the bay, where they had an interview with sixteen natives;

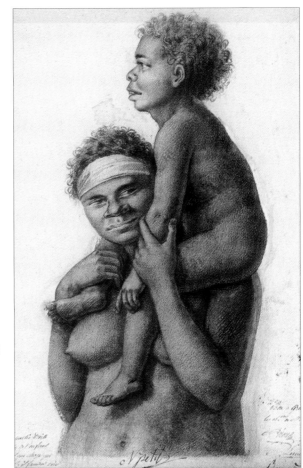

Right: *Aboriginal woman and child, New Holland, by N.-M. Petit. [Collection Lesueur, Muséum d'Histoire Naturelle, Le Havre]*

Below: *'View of Port Jackson taken from South Head', by William Westall, engraved by John Pye, 1814. [In* Views of Australian Scenery; *by permission of the National Library of Australia]*

Top: *Government House,*
Sydney, by William Westall,
circa 1802. [By permission of
the National Library of
Australia]

Above: *'Hawkesbury River,*
view no. 13', by Westall, 1802.
[By permission of the National
Library of Australia]

their appearance was described as being much inferior to the inhabitants of Keppel and Hervey's Bays, but they were peaceable, and seemed
to be very hungry ... The number of bones lying about their fire places
bespoke turtle to be their principal food; and with the addition of shell
fish, and perhaps fern roots, it is probably their sole support.[2]

*At every port they enter, Flinders's first concern on landing is to exam-
ine the jetsam at the water's edge. The memory of La Pérouse is with
him always. Many believe the ill-fated navigator had been wrecked off
New Caledonia or some neighbouring island; if so, perhaps the remnants of his ships have been driven here by the trade winds. There can
be no rational hope of finding survivors after so many years, yet certain knowledge of their fate would at least do away with the pain of
suspense – and there might possibly still be time to retrieve documents
relating their discoveries.*

*The day before he sails, Flinders anchors in the west bight. Too fa-
tigued to face a laborious walk on the mainland, he sends Brown and
his colleagues ashore in the launch; 'as an easier task' he lands on
nearby Aken's Island to take some bearings. First he sifts through a mass
of rubbish on the south-east side: a quantity of pumice, several varieties of coral and numerous shells, skeletons of fish and sea snakes, and
a coconut shell – there is nothing to indicate a shipwreck.*

*Later a seine net is hauled upon the beaches at the island's south end,
and brings ashore a large quantity of fish – among them a swordfish
nearly 5 feet long. Its sword alone is 1½ feet long, and fringed with strong,
sharp teeth. The sailors keep their distance as the fish thrashes from side
to side in such fury that none can subdue it.*

They entered Thirsty Sound, so named by Cook, on 6 September, and
Broad Sound two days later. The latter was the most extensive complex of bays, inlets and small islands on this part of the coast, and
Flinders remained for three weeks exploring and charting its shores.
He was intrigued by the vast areas of mud left dry at low tide on both
sides of the channel in which the ships lay at anchor. The difference
between high and low tide was 'no less than thirty-two feet, and it
wanted a day of being full moon'. The spring tides, he thought, would
reach 2 or 3 feet higher.

Both vessels moved further into the sound, and the observatory tents
were set up onshore, protected by a marine guard. Flinders took the

Lady Nelson and a whaleboat to explore the head of the sound with Robert Brown, leaving Lieutenant Fowler to prepare the *Investigator* for sea; Flinders's brother Samuel remained at the observatory, charged with checking the accuracy of the timekeepers and taking the usual astronomical observations.

The *Lady Nelson* grounded 11 miles from the ship. The new main keel, swollen by the wet, could not be raised, and when the brig settled on the bottom the keel broke off. At low water the men found the best-bower cable parted, and the anchor so far buried in quicksand that it too could not be raised.

More bad news awaits Flinders at the encampment. Samuel has allowed the timekeepers to run down, and rerating them will take another week. 'Being unwilling to remain so long inactive', he decides to sail to the sound's inner end in the Investigator, *instructing his brother to re-establish the rates while he is away. Murray, meanwhile, is to repair the damage to the brig's hull and be ready to sail in a week's time for Torres Strait.*

The next few days are spent examining the inner recesses of Broad Sound. Flinders and Brown make several excursions ashore, the bota-nist observing at one landing 'a tree of moderate size ... with ripe cap-sules', which he later names Flindersia australis *in honour of his cap-tain. Returning to the tents on 24 September, Flinders discovers to his 'surprise and regret' (and surely his incredulity) that the timekeepers have again been let down; only one day's rates have since been obtained. It appears that Second Lieutenant Flinders was so intent on his astrono-mical observations that he forgot to wind them up on the 22nd at noon.*

Flinders finds this latest difficulty highly embarrassing. To sail for Torres Strait and the Gulf of Carpentaria without precise rates will 'crip-ple the accuracy of all our longitudes'. On the other hand, it is impera-tive to reach the north coast before the north-west monsoon sets in, and the time already lost leaves no room for further delay. He decides to depart as fast as possible, 'with the best rates we could scrape together'.

Quitting Broad Sound on 27 September, the ships steered for the North-umberland Islands, lying at the southern end of the Great Barrier Reef. For two weeks they sailed northwards inside the line of reefs, searching in vain for a safe passage to the open sea. The unceasing roar of high breakers on the far side belied the tranquil waters within. Landing on the reef one afternoon, Flinders and the botanists found a new creation presented to their view:

> We had wheat sheaves, mushrooms, stags horns, cabbage leaves, and a variety of other forms, glowing under water with vivid tints of every shade betwixt green, purple, brown and white, equalling in beauty and excelling in grandeur the most favourite *parterre* of the florist. These were different species of coral and fungus … and each had its peculiar form and shade of colouring; but whilst contemplating the richness of the scene, we could not long forget with what destruction it was pregnant.[3]

His hopes of finding an easy passage through the reefs soon vanished. 'The prospect at noon today,' he wrote in his journal on 8 October, 'is still worse than ever; for except to the westward, whence we came, the reefs encircle us all round'; and the following day: 'There was no appearance of any passage this morning.' The 11th brought near-disaster; attempting a narrow opening between two sections of reef, through which the tide gushed at a fearful rate, both ships lost anchors – the *Investigator* her stream anchor and part of its cable, and the *Lady Nelson* a kedge anchor and one arm of the bower. The mishap decided Flinders to send the brig back to Port Jackson:

> The *Lady Nelson* sailed so ill, and had become so leewardly since the loss of the main, and part of the after keel, that she not only caused us delay, but ran great risk of being lost; and instead of saving the crew of the *Investigator*, in case of accident … it was too probable we might be called upon to render her that assistance.[4]

The two ships parted company on 18 October, the *Investigator* displaying her colours to bid the brig farewell. Flinders watched *Lady Nelson*'s sails dip below the horizon with mixed feelings. She had become 'a burthen rather than an assistant', while Murray had shown himself 'not much acquainted with the kind of service on which we were engaged'. Yet the younger man's zeal

> to make himself and his vessel of use to the voyage made me sorry to deprive him of the advantage of continuing with us; and increased my regret at the necessity of parting from our little consort.[5]

Murray carries with him Flinders's dispatches to the Admiralty and to Governor King. With the latter Matthew has enclosed a letter to Ann, with a note asking the governor to forward it to England by the first ship:

My dearest love,

Up to this day we are all well, and the accomplishment of the objects of the voyage is advancing prosperously.

Amidst my various and constant occupations, thou art not one day forgotten. Be happy my beloved, rest assured of my faith, and trust that I will return safely to sooth [sic] thy distress, and repay thee for all thy anxieties concerning me.

Beg thy good father and mother to accept my affectionate and respectful regards, as well as my friend Belle, and believe me to be thy own,

Matthw. Flinders[6]

Threading the needle – Torres Strait

On board Investigator, *at sea, Wednesday, 20 October 1802.*

The reefs in sight were small, and could not afford shelter against the sea which was breaking high upon them; but these breakers excited a hope that we might, even then, be near an opening in the barrier; and although caution inclined to steering back towards the land, this prospect of an outlet determined me to proceed, at least until four o'clock, at the chance of finding either larger reefs for shelter, or a clear sea. We were successful. At four, the depth was 43 fathoms, and no reefs in sight; and at six, a heavy swell from the eastward and a depth of 66 fathoms were strong assurances that we had at length gained the open sea. [Flinders, 1814, vol. II, p. 100]

The breakthrough came just two days after Murray's departure. The wind changed on the 19th and blew fresh from the eastward all night, 'raising a short swell which tried the ship more than any thing we had encountered from the time of leaving Port Jackson'. *Investigator*'s former leakiness returned, and she took in up to 5 inches of water an hour. Next morning, however, they came to a break in the great reef some 3 miles wide; bearing through it they found another. By midafternoon Flinders dared to hope he had found the long-sought passage,[7] and by nightfall he was confident the *Investigator* was safely through. In the *Voyage* he allowed a rare hint of hubris to creep into the text:

The commander who proposes to make the experiment must not, however, be one who throws his ship's head round in a hurry, so soon as breakers are announced from aloft; if he do not feel his nerves strong enough to thread the needle ... amongst the reefs, whilst he directs the steerage from the mast head, I would strongly recommend him not to approach this part of New South Wales.[8]

Soundings were taken at each change of watch during the night: no bottom with 75 fathoms at 8 p.m., none with 115 fathoms at midnight, and none with 100 fathoms at 4 a.m. At daylight no reefs were visible from the masthead, and the ship made all sail possible to the north-east. Two days later they entered the trades, the ship's course set for Torres Strait. Flinders knew he must pass the strait before the onset of the north-west monsoon in November.

The run northwards in more open water brought the seamen some relief from the stresses of the past weeks. Even so, when off watch at night, their few hours' rest was all too often broken into by the working of the ship, as orders called them on deck to go to their stations, to tack, to shorten sail. At different times, wrote Seaman Smith,

[we] was Oblidg'd to steer different courses upon acct of the different shoals we met with at sea, at Night our Captn was very cautious of running on, sometimes hove too, at other times Tackt & stood off & on.[9]

At noon on 28 October, *Investigator* was off Pandora's Entrance to Torres Strait, Flinders preferring to use this approach to the more northerly one discovered by Bligh in 1792, during the *Providence* voyage. Next day, towards evening, he dropped anchor in the lee of the Murray Islands.

No sooner has the ship anchored than three canoes leave the shore, manned by between forty and fifty 'Indians'. Recalling the islanders' attacks on the Providence *ten years before, Flinders keeps the guns ready and his marines under arms; officers are stationed to watch each canoe so long as it remains near the ship. The natives, however, seem intent on barter, holding up coconuts, plantains, joints of bamboo filled with water, and bows and arrows, and shouting* tooree! tooree! *and* mammoosee!

The islanders are dark chocolate in colour, active muscular men about the middle size. Like the natives of New South Wales they go quite naked, but many of them wear ornaments – of plaited hair, bark fibres, or shells – about the waist, neck or ankles. Hoping to secure their friendship, and unable to distinguish any chief among them, Flinders selects the oldest man and presents him with a handsaw, a hammer and nails, and other trifles; but the old man becomes frightened on finding himself singled out in this fashion.

Bartering begins. A hatchet or some other iron piece (tooree) being held up, the natives offer in exchange a bunch of plantain, a bow or quiver of arrows, or some equivalent item. Upon its acceptance in sign language, the man leaps into the sea with his barter and swims to the ship, handing it to the sailor who climbs down the side with his exchange. At first any item of iron is accepted, but later, if a nail is held up on its own, the 'Indian' shakes his head, 'striking the edge of his right hand upon the left arm in the attitude of chopping; and he [is] well enough understood'.

Seven canoes arrive next morning; several of them come in under the stern and close to twenty natives clamber on board, bringing with them pearl-oyster shells and necklaces of cowries to exchange for the precious toree. *The ship is preparing to sail and the seamen make signs to the visitors to return to their canoes. At first they seem 'unwilling to comprehend; but on the seamen going aloft to loose the sails, they hastily [descend] the ship's sides, and shove off'. Before the anchor is up they have paddled back to the shore.*

Remembering Bligh's difficulties in the northern part of the strait, Flinders kept as far to the south, towards Cape York, as the direction of the reefs allowed. Late in the afternoon, he came upon a small, thickly wooded island, about a mile in circumference, which promised shelter from the south-east winds and offered a convenient anchorage overnight; he named it Halfway Island. Next day, 31 October, the *Investigator* passed Mount Adolphus Island, off the tip of Cape York, then steered north-west and anchored in the lee of Wednesday Island (one of the Prince of Wales group named by Cook). Here she remained all the following day, the wind so strong that it was prudent neither to quit the anchorage nor to land.

On 2 November, Flinders sailed west again, through the present Prince of Wales Channel, landing briefly with the botanists on a small island he named after Peter Good the gardener. Booby Island was sighted on the 3rd, and Flinders relaxed, considering 'all the difficulties of Torres Strait to be surmounted, since we had got a fair entry into the Gulph of Carpentaria' before the onset of the monsoon. With a route 'almost wholly to seek', he had cleared the strait in six days – with charts and good weather it could, he thought, probably be sailed in three. The discovery offered a new route from Port Jackson to India, which might cut as much as five or six weeks' sailing time from the usual passage by the north coast of New Guinea or the more eastern islands.

His satisfaction at the achievement – 'perhaps the most advantageous thing we have done for navigation', he wrote to Banks in 1803 – was soured by the carpenter's report that the ship's leak was worsening fast. It now measured more than 10 inches an hour, even in good weather, and required almost continuous use of the pumps. The cause must lie below the waterline, and would not easily be remedied.

The Gulf of Carpentaria I – a rotten ship

To Matthew Flinders Esq., Commander of HM *Sloop* Investigator. *Friday, 26 November 1802.*

> *Sir, … Mr. Aken has known several ships of the same kind, and built at the same place as the* Investigator, *and has always found, that when they began to rot they went on very fast. From the state to which the ship now seems to be advanced, it is our joint opinion, that in twelve months there will scarcely be a sound timber in her; but that if she remains in fine weather and happens no accident, she may run six months longer without much risk. [Extract from a report by John Aken, master, and Russel Mart, carpenter, in Flinders, 1814, vol. II, p. 142]*

With the perils of the strait behind him, Flinders 'steered in for the coast of Carpentaria on the east side of the Gulph' on 4 November. For ten days the *Investigator* crept south down the low-lying, almost feature-less coast. Shoals ran far out to sea, forcing the ship into deeper water 5 or 6 miles from shore as the leadsman sang out the soundings. From an old Dutch chart of the gulf, dating from 1663, Flinders checked his observations against the Dutchmen's, identifying their landmarks where possible, and elsewhere mapping any deviation that broke the monotony of the coast.

The land was parched – six months or more with little or no rain, awaiting the onset of the rainy season. Flinders could not know how this changed the landscape. Where the Dutch chart showed river mouths, he saw only stagnant coastal lagoons, and judged his predecessors were wrong; not for the first time, he failed to notice the major river systems of the region. The waters became still shallower, and the ship steered even further out from the shore, to a distance of from 6 to 9 miles. At length, after ten days, the direction of the coast changed westwards, leading Flinders to comment:

this, with the increasing shallowness of the water, made me apprehend that the Gulph would be found to terminate nearly as represented in the old charts, and disappoint the hopes formed of a strait or passage leading out at some other part of Terra Australis.[10]

On 17 November a small hill was seen ahead; the highest land yet sighted in the gulf, it scarcely exceeded the height of the ship's masthead. Flinders went ashore to take observations, and found to his surprise he was on an island (later named Sweers Island, after a member of the Batavia Council in Tasman's time). Five miles long, it was separated from the mainland by a wide channel that offered a sheltered anchorage for the ship. Anxious to bring the leaks under control, Flinders took the opportunity to careen her and recaulk the starboard seams, which he supposed were responsible. While the caulking and other needed repairs got under way, and the scientists went ashore exploring and botanising as usual, Flinders landed with his boat's crew on another island nearby.

They come across three Indians dragging rafts towards some rocks at the water's edge, where three other natives are sitting. Not prepared to abandon their rafts, the men watch uneasily as the white men approach; they lay down their spears when signs are made for them to do so. Two of them seem advanced in years, and from their appearance are probably brothers. Apart from two Tahitian chiefs they are the tallest Indians Flinders has met – 3 to 4 inches taller than his coxswain, who is 5 foot 11 inches; 'like most of the Australians their legs do not bear the European proportion to the size of their heads and bodies'. Each has lost two front teeth from the upper jaw, and their hair is short, but not curly. To Flinders's surprise the older men appear to be circumcised, but he cannot make an observation on the youngest, who remains seated.

The older men begin walking towards the boat, Flinders accompanying them hand in hand. They stop halfway, then turn back and make towards their companions at the rocks. Flinders now judges these to be women, and the men's proposal to go to the boat 'a feint to get us further from them'. It seems, though, that 'the women were [not] so much afraid of us, as the men appeared to be on their account', for they remain seated quietly picking oysters. Not wishing to disturb them, he and his men continue their exploration of the island.

On his return to the ship, Flinders was told a spring of clear fresh water had been discovered nearby, while the fishing parties had enjoyed success with the seine. This good news was eclipsed by the carpenter's alarming report of 'rotten places found in different parts of the ship' – in the planks, timbers, beams, bends, and treenails. It was far worse than Flinders had expected, and he directed the master and the carpenter to make a thorough examination 'into all such essential parts', and submit their findings as soon as possible.

The report from John Aken and Russel Mart was devastating:

> ... we have taken with us the oldest carpenter's mate of the *Investigator*, and made as thorough an examination into the state of the ship as circumstances will permit, and which we find to be as under – out of ten top timbers on the larboard side near the fore channel, four are sound, one partly rotten, and five entirely rotten ... we have seen but one timber on the larboard quarter, which is entirely rotten ... on the starboard bow close to the stem, we have seen three timbers which are all rotten ... the stem appears to be good, but the stemson is mostly decayed ... under the starboard fore chains we find one of the chain-plate bolts started, in consequence of the timber and inside plank being rotten ... the ends of the beams we find to be universally in a decaying state ... the treenails are in general rotten.[11]

So it went. The two warrant officers judged that 'in a strong gale, with much sea running, the ship would hardly escape foundering', and that if she were to get onshore 'under unfavourable circumstances' she would at once go to pieces. If the weather remained fine, and she met with no accident, she might run for another six months.

How was it possible, Flinders asked himself, that some indication of the *Investigator*'s decayed state 'should not have been found before, when [she] was in dock at Sheerness, or when she was caulked at the Cape of Good Hope, or at least at Port Jackson, when the barricade was removed'? In hindsight, there can be little doubt that the dockyard contractors at Sheerness – notorious for their corruption – were at least partly to blame. Cook and Bligh were aware of the risks and kept a close eye on their work. Flinders, a younger man, lacked their experience and was without a master to deputise for him; also he was newly wed and had his wife on board – which, as Banks observed, tended to act as a distraction from duty.

Recriminations are pointless. Aken and Mart are experienced officers, practical seamen not prone to exaggeration. Flinders ponders the gravity of their findings, and the implications for the future of the voyage. From the dreadful state of the ship, a return to Sydney seems almost immediately necessary:

> *as well to secure the journals and charts of the examinations already made, as to preserve the lives of the ship's company; and my hopes of ascertaining completely the exterior form of this immense, and in many points interesting, country, if not destroyed, would at least be deferred to an uncertain period.*

From the start of the survey, Flinders's object has been 'to make so accurate an investigation of the shores of Terra Australis that no future voyage to this country should be necessary', and with this always in view he has endeavoured

> *to follow the land so closely, that the washing of the surf upon it should be visible, and no opening, nor any thing of interest escape notice. Such a degree of proximity is what navigators have usually thought neither necessary nor safe to pursue, nor was it always persevered in by us; sometimes because the direction of the wind or shallowness of the water made it impracticable, and at other times because the loss of the ship would have been the probable consequence of approaching so near to a lee shore.*

Such has been his plan; 'and with the blessing of God, nothing of importance would have been left for future discoverers'. But with the voyage still not half completed, it seems that Fate has deprived Flinders of his dream:

> *with a ship incapable of encountering bad weather – which could not be repaired if sustaining injury from any of the shoals or rocks upon the coast – which, if constant fine weather could be ensured and all accidents avoided, could not run more than six months; – with such a ship I knew not how to accomplish the task.*[12]

Flinders kept his agonising to himself – the responsibility for the safety of the ship and all her company was his alone. Given the *Investigator*'s rotten state he was effectively trapped in the gulf for the next three to four months by the approaching monsoon. 'When the fair wind should come', in March, he would have the option of proceeding 'by the west to Port Jackson, if the ship should prove capable of a winter's passage along the South Coast', or, failing that, running for the nearest port in the East Indies. Until then he would salvage what he could from this disaster: he resolved to proceed with the survey and 'finish, if possible, the examination of the Gulph of Carpentaria'.

The gulf posed a sufficient challenge in itself. Though a succession of Dutch navigators (from Willem Jansz in 1606 to Hendrick Swaar-Decron and Cornelius Martin van Delft in 1705) had explored and charted the eastern side, the outline of the southern and western coasts was suspect: 'no one could say, with any confidence, upon what authority its form had been given in the charts'. At the least, thought Flinders, he would complete its charting, and in the process demonstrate, as he now believed, that it was not the entrance to some vast strait or river leading to the interior of the continent.

The Gulf of Carpentaria II – prisoners of the monsoon

Aboard Investigator *in the gulf, Wednesday, 8 December 1802.*

> *We weigh'd Anchor & came too at night, which is our Dayly custom owing to the Dangerous & unnavagable Coast, on the 8th A Black squall appear'd astern of us, it being then A perfect calm it came on very fast, the hands was call'd & Double reef'd the Topsails & took in Topgallant Sails & haul'd up the Mainsail, a Large Water spout appear'd, but soon broke & the squall was of little consequence. [Seaman Samuel Smith's journal, MS, Mitchell Library]*

By the end of November, wooding and watering of the ship was completed, the carpenter and his mates had carried out running repairs to her planks and timbers 'to the best advantage', and the gunner had dried his powder in the sun. A sailor with time to spare carved the ship's name on the trunk of a tree near the shoreline.[13] On 1 December they got under way and steered north-west along the coast, sighting at noon 'that great projection of the main, represented in the old chart under the name of *Cape Van Diemen*'. As the ship came closer it was seen to be an island, one of several lying off the coast. Unable to double the cape because of a shift in the wind, Flinders anchored in the lee of the island and prepared to go ashore.

Turtle tracks were seen on the beach as they rounded the island's north-east point,

> and afforded us the pleasurable anticipation of some fresh food. We had explored tropical coasts for several months, without reaping any of the

advantages usually attending it … we now hoped to have found a place where the Indians had not forestalled us, and to indemnify ourselves for so many disappointments.

His boat's crew row Flinders towards the beach, sweat streaming down their bodies in the sultry air. They glance hungrily at the turtles swimming past, looking to supplement their wretched rations of salted horse and pork and weevil-ridden ship's biscuit. Beaching, they run to capture some of the animals resting above high-water mark, 'but found them dead, and rotten'; they must have fallen on their backs climbing the steep bank and, unable to right themselves, perished miserably.

Flinders takes his observations from a small hillock near the centre of the island. The impatient sailors, not waiting for the swimming animals to come ashore, wade out and attack them in the water. When the captain returns he finds three large turtles dead on the beach, and his harpoon broken in the process. Others of the crew have busied themselves

> *scratching out some holes [in the sand]; from one they filled a hat with turtles eggs, and from another took a swarm of young ones, not broader than a crown piece, which I found crawling in every part of the boat … I hastened to the ship, and sent Lieut. Fowler with a party of men, to remain all night and turn them.[14]*

Fowler's party continues the hunt by moonlight. Their success is so great that the two boats sent to bring them off at daybreak return filled, and the launch has to be hoisted off as well. It takes most of the day to bring on board as much as the decks and holds can contain. For Seaman Smith and his shipmates it was a memorable event, Fowler and his men, 'having turn'd 60 in the course of the Night … they was made into soup for the ships company, which was an excelent refreshment'.[15] Flinders inscribes the name 'Bountiful Island' on his chart.

In rainy weather, with much thunder and lightning, *Investigator* sailed on 5 December. The mainland now trended west-north-west; it was low, woody country, fronted by a sandy beach, of 'tedious uniformity'. The sea remained shallow, the soundings varying from 3 to 6 fathoms – 'the distances from the land being in miles, as nearly as might be what the depth was in fathoms'. Day followed day, the sun a bronze hammer in the sky beating down on land, sea and ship. The monsoon

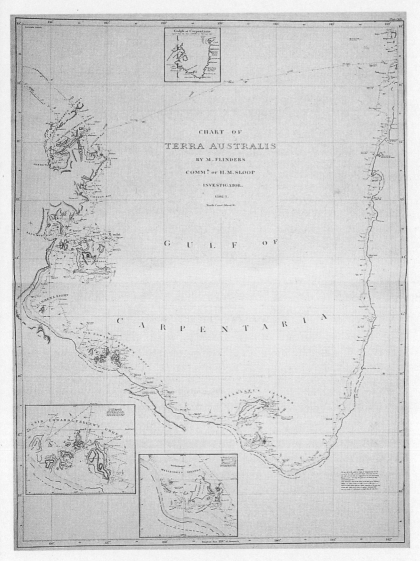

Gulf of Carpentaria, 1802-03. Chart by Flinders.
[Courtesy of the Royal Geographical Society of South Australia]

brooded beyond the horizon, black skies rent by lightning, thunder rolling behind the low hills.

On the evening of the 11th, 3 miles from shore, the ship was under light sail in $3\frac{1}{2}$ fathoms when the water shoaled without warning;

> before the helm was put down [she] touched upon a rock, and hung abaft. By keeping the sails full she went off into 3 fathoms, but in five minutes hung upon another rock; and the water being more shallow further on, the head sails were now laid aback. On swinging off, I filled to stretch out by the way we had come; and after another slight touch of the keel we got into deep water, and anchored in 4 fathoms, on a bottom of blue mud.[16]

All on board recognised how narrowly they had escaped destruction. In the captain's view:

> The bad state of the ship could have made our situation amongst these rocks very alarming, had we not cleared them so quickly; but the water was very smooth at the time, and it could not be perceived that any injury had been sustained.[17]

Samuel Smith, with a seaman's fatalism, wrote prosaically: 'no sooner was we of[f] one Shoal we was on Another, but by good Manadgement & a Steady Breeze we got over it'.

As Flinders prepared to take altitudes of the star Rigel to ascertain the longitude of the shoals, he found that the time keepers had stopped, 'my assistant having forgotten to wind them up at noon'. Again the fault was Samuel's, but after the previous episodes the commander should have brought in some system to prevent a recurrence. Before the voyage, Alexander Dalrymple at the Admiralty had suggested that one clock should be wound at noon and the other at 6 p.m., but this sensible idea was not adopted.

In the almost unbearable heat and humidity, conditions on board worsened by the day. Three turtles were thrown overboard, 'their Flesh allow'd by the Doctor to be unfit for use'; Surgeon Bell was also worried by the condition of the men in the oppressive climate. Flinders was unwell, and walked with a limp. Fruit and vegetables, the best antiscorbutics, had been missing from their diet for nearly five months, since leaving Port Jackson. Trapped in the gulf, there was no likelihood of improvement.

Cape Vanderlin, a large promontory on the Dutch chart, was reached on 14 December. This cape too was found to be an island – Flinders presumed that Tasman had 'made a distant and cursory examination, and brought conjecture to aid him in the construction of a chart'. He retained the name for the island, but later called the small archipelago to which it belonged Sir Edward Pellew's Group 'in compliment to a distinguished officer of the British Navy, whose earnest endeavours to relieve me from oppression in a subsequent part of the voyage demand my gratitude'. Twelve days were spent exploring the group. The monsoon now set in, with frequent heavy squalls accompanied by rain, thunder and lightning. On board ship the temperature averaged 85°F; onshore it was hotter still, but the routine work of the expedition continued without a break.

The Gulf of Carpentaria III – the fatal shore

On board HM Sloop Investigator, *Christmas Day 1802.*

> *'25th being Christmas Day, we sighted boath Anchors & again moor'd ship; & had a thorough clean of Decks – in the Afternoon liberty was given for any one to go on shore that thought proper & likewise a Fishing in this manner Christmas day was spent; being still in the Gulf of Carpentaria, on the 27th we unmoor'd Ship & in the Evening came to An Anchor & here spent the remainder of this Month. [Seaman Samuel Smith's journal, MS, Mitchell Library]*

If the crew had looked forward to the usual Christmas festivities, they were sorely disappointed. At dawn they were mustered on deck to holystone and swab as on any other day, and at noon were served only sherbet (lime juice and sugar), but no grog. Those who took liberty on the desolate shore found little to celebrate, although some caught a few fish.

Flinders sailed from the Pellew Group on 27 December and steered west and north. During their stay the scientists had landed to botanise and explore, Westall to sketch, and Samuel Flinders to take bearings from a small island in mid-channel. Flinders landed on several islands in the group, to take observations from various stations. He also went on a four-day excursion with Westall to explore the outer islands, and if possible make contact with the local natives. Though a few had been seen at a distance, they had avoided any contact with the land parties.

The reason became clear when he and Westall found numerous signs of some 'foreign people' being frequent visitors to the islands:

Besides pieces of earthen jars and trees cut with axes, we found remnants of bamboo lattice work, palm leaves sewed with cotton thread into the form of such hats as are worn by the Chinese, and the remains of blue cotton trowsers ... A wooden anchor of one fluke, and three boat rudders of violet wood were also found.[18]

Most puzzling were a row of low stone structures, divided into compartments, in which were the remains of charcoal fires. Evidently the foreigners were Asiatics, 'but of what particular nation, or what their business here, could not be ascertained'. Flinders suspected they were Chinese.

The state of the ship, the shoaling waters, and the monsoon frustrated all his efforts to closely survey the mainland in this part of the gulf. His charts carry the notation: 'These parts of the coast seen indistinctly from the mast head.' As a result he again missed several important river systems, including the McArthur and Roper rivers.

The year 1803 opened inauspiciously and grew progressively worse. Three days into the year the first death occurred since leaving Port Jackson. The *Investigator* had safely ridden out a night of strong squalls of wind, driving rain, thunder and lightning. Conditions improved in the morning, and Flinders stood in to the edge of the shoal to continue charting. Two men were sent in the whaleboat to sound ahead of the ship, but were recalled when a fresh wind set in from the north-east.

As the whaleboat veered astern she was swamped alongside and broke adrift, and the two men in her were thrown into the sea. A second boat was lowered to save them, while Flinders ran the ship to leeward and came to anchor. The boat was soon picked up, though much damaged, and one of her crew was hauled aboard. The other man, William Murray, captain of the foretop, could not swim; to save himself he grabbed the ship's hawser as she passed, but was dragged through the waves at such a rate that he could not keep his head above water. His body was not found.

Murray was a good seaman, rated able, and popular in the fo'c'sle. The circumstances of his death reminded his shipmates yet again of Mr Pine's prophecies on the eve of their leaving England. A morose unease settled over the lowed deck.

Groote Eylandt, Tasman's Great Island, was reached on 4 January. Flinders spent the rest of the month on this part of the coast, charting the waters around the island, several smaller islands close by, and Blue

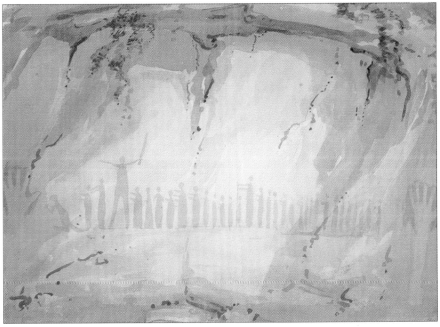

Top: 'Murray Isles', by
William Westall, circa 1802.
[By permission of the National
Library of Australia]

Above: 'Chasm Island,
native cave painting', by
Westall, circa 1803. The island
is off the northern end of
Groote Eylandt, in the Gulf of
Carpentaria. [By permission of
the National Library of
Australia]

Top: 'View of Malay Road
from Pobassoo's Island', by
William Westall, engraved by
Samuel Middiman. [Rex Nan
Kivell Collection; by permission
of the National Library of
Australia]

Above: 'The English
Company's Islands, Malay
proa', by Westall, 1803. [By
permission of the National
Library of Australia]

Mud Bay on the mainland. For the first time since November the botanists were able to go ashore on the main. Robert Brown was greatly cheered by his harvest of new plants on the borders of a small freshwater lake:

> The no. of plants in this neighbourhood must be very great as I was able in the few hours I was on shore to count upwards of 200 about 26 of these we had not seen before …[19]

More traces of the same foreign people were found along the beach. Nearby were three small huts, entirely covered with grass; they were empty, and nothing was uncovered beneath. No Indians were encountered, but Brown was intrigued by the discovery a mile or so beyond the beach of what seemed to be a native burial site, unlike anything seen so far. His party found several human skeletons standing upright in the hollow trunks of trees; the trunks were in sets of three and about 14 feet tall. All of one set had fallen and were much decayed, but the one he examined was still in good condition:

> It was painted externally with transverse bands & stripes of ochre red Each of the trees contain a human skeleton. That in the painted tree was also painted generally red but some of the bones with singular white stripes towards one of their extremities.[20]

Thursday, 14 January dawns fine and calm. The ship anchors near a high cliffy islet a mile and a half from the northern end of Groote Eylandt, and Flinders lands with the botanical gentlemen, 'intending to take bearings from the uppermost cliffs'. Deep chasms intersecting the upper reaches make it impossible to reach the top in the time available. While the botanists investigate various trees and bushes growing in the recesses of these rifts, Flinders explores several deep holes or caverns undermining the cliffs. To his astonishment, many of the walls are covered with rude drawings of porpoises, turtles, kangaroos, and a human hand, made with charcoal and ochre upon the whitish ground of the rock. Here indeed is something to excite Sir Joseph's curiosity.

Westall is sent to Chasm Island to make true copies of the Aboriginal art. He finds further drawings not seen earlier – one representing a kangaroo hunt, and another the pegging of a turtle. In the former, a file of thirty-two ill-drawn men follow the crude figure of a kangaroo, the whole apparently drawn with charcoal. One of the figures in the

*file is twice the height of his companions, and holds in his hand a
weapon 'resembling the whaddie, or wooden sword of the natives of Port
Jackson'. His dominance in the band suggests to Flinders that these natives possess a primitive concept of chieftainship:*

> They could not, as with us, indicate superiority by clothing or ornament,
> since they wear none of any kind; and therefore, with the addition of a
> weapon, similar to the ancients, they seem to have made superiority of
> person the principal emblem of superior power, of which, indeed, power
> is usually a consequence in the very early stages of society.[21]

Returning to the mainland, the *Investigator* dropped anchor in the lee
of an island in Blue Mud Bay. The usual wooding and watering parties
were sent ashore; the botanists also landed, while Flinders walked to
the west end of the island to take his bearings. Every man was armed,
'for marks of feet had been perceived, so newly imprinted, that we
expected to meet with Indians'.

Westall was sketching on the island's far side when he saw six natives land from a canoe and disappear into the woods. He hurried back
to the beach to report their landing to Flinders, who set off 'to obtain
a friendly interview with these unfortunate people'. After several vain
attempts to make contact, Flinders gave up and returned to the ship
with the botanists, leaving the mate, Mr Whitewood, and the working
parties to complete their tasks.

*Musket shots ring out from the shore, followed shortly by a signal for
a boat. Flinders at once dispatches the master with two boatloads of men
– one to bring off any of the crew who may be injured, the other to seize
the natives' canoe if it was they who launched an attack. Any dead or
injured Indians are also to be brought back. Suspecting that his men
were most likely the aggressors, Flinders instructs Mr Aken, if he should
encounter the natives, to be friendly 'and by no means to pursue them
into the wood'.*

*The first boat returns at five, carrying Mr Whitewood, who has several
spear wounds, and Benjamin Morgan, a marine, quite senseless from
sunstroke. Whitewood's wounds luckily are not serious, but Morgan dies
without regaining consciousness. Mr Aken's boat returns towards 10 p.m.,
and Flinders pieces the story together from the men's reports.*

*It seems the Indians were seen at the top of a hill, and were approached
by the mate and Mr Allen the miner. Some of them laid down their*

weapons as a sign of peace, and Whitewood signalled to the nearest native to do the same. The man held out his spear with its point outwards, and when Whitewood went to lay hold of it pushed it into his side. The mate snapped his musket at his attacker but it did not fire. He and Allen then fled for the safety of the beach, spears flying past them. The mate, with the spear still in his side, was hit several times. Finding himself closely pressed, he turned and fired again, this time successfully, wounding one man. By now the rest of the party had come up, more shots were fired, and the Indians fled.

Later, as Aken's party is preparing to return to the ship, three natives are seen paddling their canoe away from the island. The men open fire with ball and buckshot, and see one of the Indians fall; the others leap overboard and dive away. A seaman, claiming credit for the shot, swims to the drifting canoe and finds the man dead in the bottom, with a straw hat on his head which he recognises as his own. Waving this in triumph he upsets the frail craft, and the body sinks. He tows the canoe to the shore, whence it is brought back to the ship.

Flinders was greatly distressed by the unfortunate incident, and displeased with Aken for acting so contrary to his orders. The damage was done, however, and next morning Flinders sent a boat to search for the dead man's body, 'the painter being desirous of it to make a drawing, and the naturalist and surgeon for anatomical purposes'. The corpse was found onshore at high-water mark, and carried on board. 'He was dissected & his head put in Spirits,' wrote Peter Good, '& the body of Benjamin Morgan committed to the deep with the usual ceremony – about $\frac{1}{2}$ past one p.m. weighed & stood westward – Anchored in evening.'

Malay Road

Off Cape Wilberforce, Thursday, 17 February 1803.

Thus was the examination of the Gulph of Carpentaria finished, after employing one hundred and five days in coasting along its shores and exploring its bays and islands. It will be remarked that the form of it, given in the old charts, is not very erroneous, which proves it to have been the result of a real examination ; but as no particulars were known of the discovery of the south and western parts ... though opinion ascribed it with reason to Tasman, so the chart was considered as little better than a representation of fairy land. Henceforward the Gulph of Carpentaria will take its station amongst the conspicuous parts of the globe. [M. Flinders, Voyage to Terra Australis ..., *Vol. II, 1814; p.228]*

Coasting north from Blue Mud Bay, the *Investigator* entered a wide, deep inlet with the appearance of a river estuary (later named Caledon Bay) on 2 February. They spent a week at the anchorage, two tents being erected ashore, with a marine guard, to house the timekeepers and provide a land base for the botanists and artists. A group of natives came to meet them 'without showing that timidity so usual with the Australians ... and expressed much joy at seeing Bongaree, our good-natured Indian from Port Jackson'.

With confidence apparently established, Flinders made for a nearby sandhill to take bearings; 'but whilst making the circuit of a salt swamp ... the natives were heard running in the wood, and calling to each other'. Soon after a musket was fired, and he returned to the tents 'with all expedition'.

The botanists, artists and their servants have made an excursion into the wood, each walking arm in arm with an Indian. Without warning, one of the natives snatches a hatchet out of a servant's hand and makes off with it through the trees. Seeing no pursuit, nor much notice taken, the others become still friendlier. Two of them pay particular attention to Mr Brown's servant – so much so, that one holds him by the arm while his companion snatches the musket from his shoulder. They all run off together, Bauer firing his musket after the thief, but with no result other than to make him run faster. Judging it imprudent to pursue them, the botanists return to the tents. Later two natives bring back the musket, its stock broken and ramrod gone, and are rewarded with a hatchet.

The weather continues extremely hot and humid. Two uniformed marines, chasing a thief, who has stolen an axe, through the wood without their hats, collapse from sunstroke and are taken to the ship in a state of delirium; happily they recover. To curb the continual thieving, Flinders decides to follow Captain Cook's example and takes two hostages. The elder is made to understand that his companion will be held until the stolen axe is returned, and is then released. The captive, a youth of about fourteen named Woga, is bound and taken by boat to a spot

frequented by the group. Shortly two older men come forward, bring-
ing a young girl whom they offer to Bongaree with expressive signs –
no doubt hoping to entice him ashore and seize him in retaliation. They
have not brought the axe, and Woga is taken back to the ship with 'a
great deal of crying, intreating, threatening, and struggling on his part'.

Once aboard the ship, the boy's demeanour changed completely: 'he
ate heartily, laughed, sometimes cried, and noticed everything; and especially at the sheep, hogs, and cats'. Flinders repeated the experiment
next day, leaving Woga tied to a tree, eating rice and fish, and with
Bongaree for company. The natives kept away, and with no sign of
the axe being restored Flinders freed the boy before nightfall. Clearly
his use of Cook's tactic had failed; moreover, with Baudin's expected
arrival in the gulf, he did not wish to jeopardise the Frenchmen's safety
by his actions.

Flinders sailed from Caledon Bay on 10 February, the ship's water
barrels refilled and her hold stacked with firewood, and steered north
for Cape Arnhem, the easternmost extremity of the west coast. A week
later, the 17th, he rounded Cape Wilberforce at the north-eastern tip
of Arnhem Land and set his course westwards. His survey of the gulf's
lonely waters had taken three and a half months.

Beyond the cape, in a roadstead between the mainland and some
offshore islands, they were astonished to find six vessels 'covered over
like hulks, as if laid up for the bad season'. Flinders sent Samuel across
in an armed boat to discover who they were, while he dropped anchor within musket shot, and with all hands at quarters. Alert for any
sign of treachery, he kept his glass trained on the whaleboat and on
the vessel she lay alongside. All passed quietly, and Samuel returned
with the news that they were Malay proas from Macassar, fishing for
trepang (*bêche-de-mer* or sea cucumber). The six Malay captains soon
followed and were welcomed aboard the *Investigator*, Flinders's Malay
cook acting as interpreter. The visitors, being Muslims, were horrified
at the sight of the hogs penned in the launch, but surprisingly had no
objection when offered port wine; they even requested a bottle to take
back to their ships.

Pobassoo, their elderly chief, explains that his vessels are part of a fleet
of sixty proas upon the coast. He left Macassar two months before, at
the onset of the monsoon, and plans to remain another month. This is

his seventh voyage in the past twenty years to collect trepang, which brings a great price among the Chinese. He has never before met with any ship here, and cautions Flinders to beware of the natives, who cannot be trusted. Pobassoo has been speared in the knee during a skirmish with them.

A further five proas sail in next day and anchor close by. They too send canoes alongside to barter, and Flinders's concern grows at the number of Malays around the ship. There are far more than he cares to allow on board, for each man wears a dagger, or kris, at his side. He keeps his men under arms, and exercises the guns at the request of the Malay chiefs. At dusk they retire quietly to their vessels, but the guns are kept ready and half the ship's company are at quarters throughout the night. The proas sail at daylight, directing their course south-east for the gulf. Flinders farewells Pobassoo with an English jack to fly at his masthead, and gives him a letter for Captain Baudin, should he happen to meet him on his voyage.

Time was now fast running out. Almost three months of the six the master and carpenter had judged the ship might run without much risk had expired. Determined to continue to the last possible moment, Flinders steered west along Arnhem Land, charting as he went. The weather was often rainy, with strong squalls that made it unsafe to take the ship close in. The routine followed a set pattern; each day, wrote Seaman Smith,

> we got under Weigh, & on Aproaching an Island boats was Imediately sent on shore with the Captn. & the Bottanist; if found worth perticulour Investigation the Anchor is let go, & remain until the Business is Done …[22]

On Sunday, 6 March, off the southern end of the Wessel Islands, Flinders surrendered to the inevitable. The time remaining 'being not much longer than necessary for us to reach Port Jackson, I judged it imprudent to continue the investigation longer'. He gave the order to set course for Timor:

> In addition to the rottenness of the ship, the state of my own health and that of the ship's company were urgent to terminate the examination here; for nearly all had become debilitated from the heat and moisture of the climate – from being a good deal fatigued – and from the want of nourishing food. I was myself disabled by scorbutic sores from going to the

masthead, or making any more expeditions in boats; and as the whole surveying department rested upon me, our further stay was without one of its principal objects …

The accomplishment of the survey was, in fact, an object so near to my heart, that could I have foreseen the train of ills that were to follow the decay of the *Investigator* and prevent the survey being resumed – and had my existence depended upon the expression of a wish, I do not know that it would have received utterance; but Infinite Wisdom has, in infinite mercy, reserved the knowledge of futurity to itself.[23]

Kupang to Port Jackson – race against death

On board Investigator, *at sea, Saturday, 26 March 1803. Surgeon Bell presents his report to Captain Flinders.*

> *For the last eight months, we have had no refreshments but what chance threw in our way, and fruit and vegetables, the best antiscorbutics, formed no part of what was procured. During this period, the ships company have been exposed to almost incessant fatigue in an oppressively hot climate, as also to an exceedingly deleterious atmosphere since Dec. 16 when the weather became dark and cloudy, with thunder, lightning, and rain. The ill effects of this alteration in the weather were perceptible in a short time amongst the ships company; a violent diarrhoea being produced, attended frequently with symptoms of fever, which, had it not been for timely remedies and the great attention paid to cleanliness, would soon have generated the worst of dysenteries. [Mack, 1966, p. 148]*

The voyage across the Arafura Sea to Timor took twenty-five days. Though well aware from his visit on the *Providence* to the Dutch outpost in 1792 how limited were the port's facilities, Flinders had little choice in the light of Bell's report. Twenty-two men had obvious symptoms of scurvy – lassitude, livid sores, and spongy gums – although no more than four or five were unfit for duty. Flinders's 'incorrigible ulcers' prevented him wearing boots or climbing to the masthead. The surgeon's advice was clear enough:

> a body, though in health, may at the same time be losing strength; and consequently be likely to fall under any violent and long-continued exertions. If you should dread such an event, it would be well, if possible,

to provide against it by refreshing the ships company, and procuring those articles of provision [rice, sugar, molasses, and peas] of which the ship is deficient, or substitute others in their room.[24]

There was also the chance that he might find there a vessel bound for Europe. If so, he proposed to send Lieutenant Fowler to the Admiralty 'with an account of our proceedings, and a request that he might return as speedily as possible, with a vessel fit to accomplish all the objects of the voyage'.

The *Investigator* anchored in Kupang roads on 31 March. The little town had been occupied for a short time by British troops during the late war, but had been evacuated after a native uprising in which many of the garrison were massacred. Despite this unhappy recent history. Governor Giesler received Flinders courteously, 'with that polite and respectful attention which the representative of one friendly nation owes to that of another'.

From a Dutch brig and an American ship just arrived from Europe, Flinders learnt that the Peace of Amiens still held, and a lasting settlement between the powers seemed possible. His hopes of sending Fowler back to London were dashed, however – a ship had sailed for Cape Town only the week before. All he could do was dispatch a packet of letters with the brig's captain, who was about to leave for Batavia and promised to pass them on to a home-bound vessel.

Flinders's concern now is to reprovision the ship and sail without delay for Port Jackson, before the winter gales strike the south coast. The carpenter has found her timbers not perceptibly worse, 'so that I was led to hope … that the ship might go through this service without much more than common risk, provided we remained in fine-weather climates'.

Supplies of rice, arrack, sugar, palm syrup, fresh meat, fruit and vegetables are procured and stored aboard. Tea, sugar candy, and other articles for the messes are purchased from the Chinese-Malay stalls in the town, while poultry is obtained from bumboats alongside. The change in diet serves to banish the symptoms of scurvy almost overnight.

From the appearance of the few Europeans in the colony the climate is harmful: 'for even in comparison with us, who had suffered considerably, they were sickly looking people'. Recalling that Baudin had lost several men from dysentery during his stay in 1801, Flinders visits the cemetery and finds the memorial erected by the French captain to his friend Anselm Riedlé and the Bounty*'s gardener David Nelson; already it is beginning to decay.*

The day before departure some of the crew are allowed ashore after they have finished work. Two men – Flinders's Malay cook and a youth from Port Jackson – fail to return in the evening. The town is searched for them, but in vain. They are still missing when the ship sails next morning, 8 April – the anniversary of his meeting with Captain Baudin off the south coast. Flinders stands off and on in the roads while Fowler makes one last search onshore, again without success.

In the crowded, fetid fo'c'sle Seaman Smith, aged thirty-two, watched his shipmates sicken and die as the ship ran for the safety of Port Jackson:

> On the 9th [April] we stood out, our live Stock consisting of Caraboos [Timor buffalo], Sheep, Goats &c. – the Ships Company in good health but in a few days they began to be in a bad State, the chief part having the Flux whether from the Fruit, or the Water, I cannot Ascertain, but our Ships Compy was reduced so by Sickness & Deaths that we cou'd scarce find hands to work the Ship …
>
> On the 1st May we having so many Sick the remainder was put into 3 Watches – 2nd we Cross'd the Trophical line of Cancer & continued our course, with squally Weather at Intervalls – at other times calms & little Wind, on the 14th of May at 6 o'C – we espied the Land upon our Larb'd Beam, it proov'd to be the South West cape of New Holland; the ship being in such a rotten State we was happy to make the Land, on the 17th we came to an Anchor near Goose Island [Recherche Archipelago] – the same day Died Chas Douglass, Boatswain; our Boats Imploy'd, in Killing Seals, in Order to get Oil which we are in great Want, 19th Mr Douglas was taken on shore, & Interr'd; like-wise a Copper plate Ingraved, with his Name & time of his Death –
>
> 21st in the morning Died Willm Hillier, the same Morng we attempted to get under Welgh, & after heaving some time on our small Bower, we found the Ship drifting; we then let go the Best Bower, it blowing a strong Breeze, again we tried to Weigh but fail'd we run out a Kedge & hove upon our Bower but found her drifting, let go the Stream & smal Bower but to no purpose, finding we was drifting bodily on shore, where there was a terrible Surf – we was oblidg'd to Cut & run; leave Anchors & Cables Buoys; & Buoy Ropes, made Sail & soon got clear, only for this method, the ship wou'd without doubt been Lost –

every Day, more distress came upon us in loss of Men carried off by the Flux; we made all the haste we possibly could to Arrive into Port & arriv'd in Port Jackson on the 9th of June – our Sick was Imediately sent on Shore to the Hospital, which Died Dayly, very few recover'd.[25]

Lying in his sick berth, the gardener Peter Good makes the last entry in his journal on 17 May:

At day break past Termination Island & about 8 A.M. past Bay No 1 keeping to the South of all the Islands – past noon departed this life Charles Douglas Boatswain of a Dysentery with which he had laboured since the middle of Aprile – Self and several of the Crew labouring under the Same disorder In the evening Anchored to Leward of Salt Island or Bay II near where we Anchored in the middle of January 1802.[26]

He dies soon after their arrival at Port Jackson. Flinders writes in his journal: 'Peter Good, botanical gardener, a zealous worthy man ... regretted by all.'

'Not worth repairing in any country'

Sydney Cove, Thursday, 9 June 1803.

On Thursday arrived His Majesty's Ship Investigator, Captain Matthew Flinders; she sailed from hence in July last, to continue the survey of the coasts of New Holland ...

The Officers and Ship's Company have generally been very healthy, until a short time before their arrival, when getting into cold weather, after being so many months in the Torrid Zone, they were generally attacked with a Dysentery, which we are sorry to say carried off Mr. Charles Douglas, Boatswain, a very good Officer; Serjeant James Greenhalgh of the Royal Marine Forces, a very valuable Non-Commissioned Officer; W. Hillier and John Draper, Quartermasters; and C. Smith, a seaman; the loss of whom is much lamented by Capt. Flinders. Twelve sick seamen were landed on her arrival, of whose recovery there is every hope ...

We are sorry to add, that the future advantages expected from Capt. Flinders' Perseverence and Activity in his pursuits, are likely to suffer a delay, owing to the state of the Investigator's hull, which will be surveyed as soon as possible.

His Excellency having given Captain Flinders Permission to take 11 Seamen Prisoners on a Provisional Emancipation, we are happy to state, *from Capt. Flinders' authority, that their conduct has given him and his Officers great satisfaction; especially that of Francis Smith, who received a Free Pardon on the ship's anchoring in the Cove. [Sydney Gazette, vol. I, no. 15, Sunday, 12 June 1803]*

The *Investigator* returned to Sydney on 9 June 1803; she had completed the first close circumnavigation[27] of Australia in a voyage lasting ten months and eighteen days. Flinders's immediate concern was for his sickly crew. Calling on Governor King, he arranged for the worst cases to be removed at once to the colonial hospital, where 'they received that kind attention and care which their situation demanded'. Unfortunately this came too late for some of them; debilitated by dysentery, fever and scurvy, four men died within a few days. The first to succumb was the cheerful young gardener, Peter Good, who died before he could be taken ashore. The officers 'attended in procession to the place of burial, where after the funeral ceremonies … a party of marines fired three vollies over the grave'.

Flinders next submitted to the governor the report on the ship's condition prepared by the master and carpenter in the Gulf of Carpentaria. King appointed a local board to make a thorough survey of her planking and timbers. Their report confirmed everything Aken and Mart had found, and concluded:

> The above being the state of the *Investigator* thus far, we think it altogether unnecessary to make further examination; being unanimously of opinion that she is not worth repairing in any country, and that it is impossible in this country to put her in a state fit for going to sea.[28]

Flinders is ill, 'debilitated in health and I fear in constitution', with scorbutic sores on his legs and feet, and much distressed by the hindrance to the voyage and the melancholy state of the crew. His pleasure at finding many letters from home awaiting him at Sydney is shattered by his stepmother's news:

> *… the joy which some letters occasioned is dreadfully embittered by what you, my good and kind mother, had occasion to communicate. The death of so kind a father, who was so excellent a man, is a heavy blow, and strikes deep into my heart. The duty I owed him, and which I had now a prospect of paying with the warmest affection and gratitude, had made me look forward to the time of our return with increased ardour … One of*

*my fondest hopes is now destroyed. O my dearest, kindest father, how much
I loved and reverenced you, you cannot now know!*[29]

*Matthew's pain at his father's death is relieved in part by six loving
letters from Ann. He writes to her playfully:*

*If I could laugh at the effusion of thy tenderness, it would be to see the
idolatrous language thou frequently usest to me. Thou makest an idol and
then worshippest it, and, like some of the inhabitants of the East, thou also
bestowest a little castigation occasionally, just to let the ugly deity know
the value of thy devotion. Mindest thou not, my dearest love, that I shall
be spoiled by thy endearing flatteries?*

In another letter to her he comments revealingly on his companions:

*Mr Fowler is tolerably well (and is a good-natured fellow and suits me
very well), my brother is also well, is becoming steady, and is more friendly
and affectionate with me since his knowledge of our mutual loss. Mr Brown
is recovered from ill health and lameness (we are not altogether cordial,
but our mutual anxiety to forward far the complete success of the voyage
is a bond of union; he is a man of abilities and knowledge, but wants
feeling kindness). Mr Bauer, your favourite is still polite and gentle (and
is so to a considerable depth, but I fear there is a dreadful disposition at
the bottom). Mr Westall wants prudence, but is good-natured; the last two
are well and have always remained upon good terms with me. Mr Bell is
misanthropic and pleases nobody (he may possibly leave us). Elder continues to be faithful and attentive as before. I like him and apparently he
likes me … Trim, like his master, is becoming grey. He is at present fat
and frisky, and takes meat from our forks with his former dexerity [sic];
he is commonly my bed fellow. The master we have in poor Thistle's place,
is an easy good-natured man.*[30]

King was as keen as Flinders to continue the survey by any available
means. He assured his young friend 'how much I feel it a duty and a
pleasure to render you every assistance to forward the useful and
beneficial service you are employed on'. With the *Investigator* ruled
out, the governor had several other vessels at his disposal – the *Lady
Nelson*, the colonial schooner *Francis*, and HM ships *Buffalo* and *Porpoise*. The two men agreed on HM Armed Vessel *Porpoise* as best suited
to the task.

After his experience with *Investigator*, Flinders thought it prudent to request that the *Porpoise* too be examined, to ascertain 'whether she is now or can be in a short time made sufficiently strong and sound to take the risk of any weather for two-and-a-half years to come' – the time he judged necessary to complete the survey and return to England. To his and King's dismay the board's findings were almost as damning as those on his former ship. They found her to be 'very weekly [*sic*] and slight built', part of her timbers needed replacement, and it would take some twelve months to complete her refit. The governor, aware of Flinders's impatience to continue, suggested he should immediately return to England in the *Porpoise*; once in London he could present his charts to the Admiralty and obtain another ship 'to complete the service you have so beneficially commenced'.

Flinders accepted the offer, 'notwithstanding the reluctance I felt at returning to England without having accomplished the objects for which the *Investigator* was fitted out', and chose to return as a passenger on the voyage. Lieutenant Fowler was appointed the ship's commander, with orders to sail through Torres Strait 'by the route captain Flinders may indicate'. King wished to give the navigator the opportunity to establish beyond doubt that the strait could become 'a safe general passage for ships from the Pacific into the Indian Ocean'.

Shipwreck

At Port Jackson, August 1803.

> *In the beginning of August, the* Porpoise *was nearly ready to sail; and two ships then lying in Sydney Cove, bound to Batavia, desired leave to accompany us through the Strait. These were the Hon. East-India-Company's extra-ship* Bridgewater, *of about 750 tons, commanded by E. H. Palmer Esq., and the ship* Cato *of London, of about 450 tons, commanded by Mr. John Park. The company of these ships gave me pleasure; for if we should be able to make a safe and expeditious passage through the strait with them, of which I had but little doubt, it would be a manifest proof of the advantage of the route discovered in the* Investigator, *and tend to bring it into general use. [Flinders, 1814, vol. II, p. 297]*

The three ships sailed from Port Jackson on Wednesday, 10 August, and steered north-east for Torres Strait. The weather was mostly fine, the winds favourable, and the ships made good progress up the east coast. On the 17th they were about 50 leagues (150 nautical miles) from the coast when the *Cato* reported land on the port side. It was a dry sandbank, small and without vegetation, and surrounded by breakers. 'Some apprehensions were excited for the following night by meeting with this bank', but when no other lands was seen for the next 35 miles 'it did not seem necessary to lose a good night's run by heaving to'.

The *Porpoise* signalled the two merchantmen 'to run under easy working sail during the night', and took her usual station ahead. At eight o'clock the lead was cast and found no bottom at 85 fathoms. Thirty minutes later the lookouts saw breakers ahead, and the helm was put down; but with only three double-reefed top sails set, the ship scarcely came up to the wind.

Fowler, Flinders and the other gentlemen are at their ease in the gun-room when a commotion breaks out above. Fowler springs up and hastens on deck, but Flinders stays below talking to the others, supposing the sound to be caused by the tiller rope carrying away. On going up, he finds the sails shaking in the breeze, the ship in the act of paying off, and high breakers not a quarter of a cable-length away to leeward. Before the ship can come round she is carried among the breakers and strikes hard upon a coral reef.

Fowler orders a warning gun fired to alert the other vessels, but the ship's violent motion and the surf flying across her decks prevents this being done. Before lights can be brought up, the Bridgewater *and the* Cato *have hauled to the wind across each other; their bows go to meet, and a collision seems certain. At the very last moment they open off, and pass side by side without touching. The* Cato *steers north-east, and the* Bridgewater *to the south.*

The men on the Porpoise, *their safety dependent upon the preservation of the ships, rejoice at the narrow escape. Their relief is short-lived – some two cable-lengths away the* Cato *runs full upon the same reef, the rending, rasping crash of keel on coral clearly audible above the roar of the breakers. Through the murk and surf the horrified watchers see her fall over on her broadside, and her masts go overboard; soon they can hear the cries of her crew. On the other side, the* Bridgewater *has cleared the reef, the lights at her masthead showing her to be safe. Flinders and Fowler congratulate each other on her survival – Captain Palmer will surely come to their rescue at daybreak.*

The damage reports filled the two officers with dismay. The *Porpoise*'s foremast had been carried away, her bottom planking was stove in, and the hold full of water. There was no chance the ship could be saved. Fortunately she had heeled towards the reef on striking, so that the boats could be put out in the smooth water under the lee. A small four-oared gig was launched successfully, but a second boat, a six-oared cutter, jerked against the sheet anchor as she was lowered, 'and being stove, was filled with water'.

Judging that rescue boats could safely approach from the leeside, Flinders jumped overboard and swam to the gig, intending to row to the *Bridgewater* and discuss with Captain Palmer the best way of saving the survivors of both vessels. On hauling himself in, he found to his dismay only two mismatched oars instead of four, and nothing to bale out the water in her; as well, the usual crew of four men was one short, although three others – the armourer, the cook, and a marine – were hiding under the thwarts. Setting the stowaways to bale with their hats and shoes, and two of the regular crew to row as best they could with the odd oars, he steered towards the *Bridgewater*'s lights on the southern horizon.

The ship is standing away from them, and clearly it will be impossible to get close to her before she tacks. Turning back, Flinders comes across the damaged cutter; the men in her have patched her up enough to stay afloat, and are rowing aimlessly in the lee of the reef. He orders them to stay close to the wrecks till daylight, so that if either breaks up in the surf some at least of the people on board may be saved. The bottom here is coral rock, and the water 'so shallow, that a man might stand up … without being over head'.

The wind blows fresh and cold, and the gig's crew are quite drenched. Flinders keeps up their spirits by promising that the Bridgewater *will rescue them in the morning, and they will soon be homeward bound for England. He keeps the lights of the* Porpoise *in view throughout the night, but the darkness and the distance prevent any communication with her before dawn.*

At low water the ship lies much quieter on the reef, and there seems less risk of her going to pieces in the night. In preparation for the next flood tide, Fowler keeps the crew busy making a raft of the spare topmasts, yards, and other timber, with short ropes all round by which the men may hold on in the sea. Casks of water, a chest full of provisions, and Flinders's irreplaceable logbooks from the Investigator *are secured on the raft.*

At dawn the Porpoise *still holds together, and on climbing aboard, Flinders finds the crew in good spirits, with not a man lost in the wreck. The* Cato *is in a far worse state, with only the bowsprit and the fo'c'sle above water; these are crowded with men waving and crying for help. About half a mile away, a dry sandbank is visible, large enough to hold both ships' companies, together with all provisions that might be salvaged. More encouraging still, the* Bridgewater *can be seen under sail in the distance, standing towards the reef. Flinders is about to push off in the gig when she goes upon the other tack and disappears from view.*

Captain Palmer had consulted his officers on the possibility of going to the aid of the wrecked ships during the night. It had been generally agreed

> that from the state of the weather, which was now much aggravated by the increase of the wind and sea, and which [appeared] likely to continue with increasing violence; as also the surf upon and near the reef which a boat could not approach without certain destruction; all these taken into consideration it was concluded impossible to yield any assistance that night; but it was determined if possible to be with them by break of day.[31]

At sunrise the anxious men clustered on the *Bridgewater*'s deck could see clearly the full extent of the disaster. The weather threatened, and boisterous seas surged over the two wrecks stuck fast on the coral. The *Cato* lay on her side, her masts gone, her bottom exposed to the full fury of the sea. Further off they could make out the *Porpoise*, half hidden in the surf. The third mate, Williams, rejoiced with every man on board 'that they should have it in their power to assist their unfortunate companions; as there was every probability of our going to within two miles of the reef'.

The wind blew hard from the south-east with a heavy sea, and Captain Palmer had difficulty weathering the reef. Failing at his first attempt, he reluctantly concluded that, with the wind blowing as it was, it would not be possible to rescue any survivors who might have escaped the wrecks. More concerned now to find a safe passage for his ship through the patches of reef, he ordered the helmsmen to put her on the other tack. The order shocked Williams and his shipmates:

What must be the sensations of each man at that instant? Instead of proceeding to the support of our unfortunate companions, to leave them to
the mercy of the waves, without knowing whether they were in existence or had perished? … there was every probability of their existing;
and if any survived at the time we were in sight, what must have been
their sensations on seeing all their anxious expectations of relief blasted.[32]

By late afternoon the *Bridgewater* had passed the westernmost part
of the reef, and Palmer gave orders to lay to for the night; next morning
he shaped his course for Batavia, India, and home. On arrival at Bombay he sent Williams ashore with his report on the loss of the two ships
and their crews. Williams, in his own words, 'neglected his duty' and
gave a contrary account, convinced that the captain had callously abandoned the survivors of the *Porpoise* and *Cato*. He quit the ship in disgust, forfeiting his pay. The action also saved his life, for the *Bridgewater*
was lost with all hands in the Indian Ocean on the homeward voyage.

Years later, in London, Flinders received from Williams a copy of
his journal. Reading it, he could not help but wonder 'How dreadful
must have been [Captain Palmer's] reflexions at the time his ship was
going down!'

Wreck Reef

Wreck Reef, Friday, 19 August 1803.

> The Porpoise *was lost beyond a possibility of hope, and the situation of
> the commander and crew thereby rendered similar to that of their pas-
> sengers; I therefore considered myself authorised and called upon, as the
> senior officer, to take command of the whole; my intention being com-
> municated to Lieutenant Fowler, he assented without hesitation to its ex-
> pediency and propriety, and I owe to Captain Park a similar acknowl-
> edgement. [Flinders, 1814, vol. II, p. 305]*

Suspecting that Palmer had abandoned them, Flinders set about organising the rescue of the *Cato*'s crew; the ship was fast breaking up
and unlikely to last another day. He rowed across to the sandbank
and found that it lay above the high-water mark, thus offering a safe
haven for both ships' companies. The boat was then sent across to the
Cato, 'and Captain Park and his men, throwing themselves into the
water with any pieces of spar or plank they could find, swam to her

through the breakers'. They were taken to the *Porpoise*, and given food and clothing.

Several of the men were bruised against the coral rocks, and three young lads drowned in the surf. One of these boys,

> who, in the three or four voyages he had made to sea, had been each time shipwrecked, had bewailed himself through the night as the persecuted *Jonas* [sic] who carried misfortune wherever he went. He launched himself upon a broken spar with his captain; but having lost his hold in the breakers, was not seen afterwards.[33]

The reef near the ship was dry at low tide, and officers and men together spent the day bringing up water and provisions from the hold and transferring the salvage to the sandbank. Before dark, five half-hogsheads of water and quantities of flour, salt meat, rice and spirits had been landed, along with several sheep and pigs that had escaped drowning. Apart from the three boys, every man from both ships reached the bank safely. Some of the *Cato*'s men appeared in officers' uniforms, given them on the *Porpoise*, and Flinders was pleased to see that their situation 'was not thought so bad ... as to hinder all pleasantry upon these promotions'. Those who had managed to save greatcoats or blankets shared them with the less fortunate; tired out, all lay down to sleep 'in tolerable tranquillity'.

Next day the ship still held together, though thrown higher on the reef; the *Cato* had gone to pieces overnight. The *Bridgewater* had not been sighted, but a flagstaff was hastily put together from a topsail yard, and a blue ensign hoisted with the Union Jack downwards, as a signal of distress should she return.

Flinders assembles the ninety-four survivors – officers, passengers and crews of the two ships – at the top of the bank. As the senior officer present, and 'for the better preservation of discipline', he is assuming command. In the present circumstances, he declares, it is expedient that all 'should be put on the same footing and united under one head'. He warns a few of the Cato*'s seamen, who have shown signs of discontent at being ordered to work, that they must exert themselves if they expect to be fed. A large party is sent off to the wreck to bring off everything still salvageable.*

The winds moderate, the weather turns fine, and most of the men work hard, bringing ashore in two days the remaining stores of food and stocks of fresh water – enough to last them for three months. Each mess of officers and men is given a private tent, and 'their manner of living and working assumed the same regularity as before the shipwreck'. The Cato*'s men, who have saved nothing, are quartered in the* Porpoise*'s messes in the proportion of one to three. One man, a former convict, is charged with disorderly conduct and found guilty; the men are called together, the articles of war read, and the culprit is flogged at the flag-staff. The example, Flinders notes, will serve 'to correct any evil disposition, if such exists'.*

On 24 August Flinders convened a council of officers to discuss 'the precarious situation in which our misfortune, and Captain Palmer's want of energy and humanity', had placed them. The *Bridgewater* had abandoned them, and the responsibility for salvation now rested in their own hands. He proposed to dispatch the large cutter to Sydney to request the governor's help in rescuing the castaways; it was unanimously agreed that Flinders should lead the mission and that Fowler should take over command at the bank. In case the cutter should meet with some accident, two decked boats, large enough to carry all those remaining at the reef, were to be constructed at the wreck site from salvaged timber. Two days later Flinders, with Captain Park of the *Cato* and a picked crew of twelve, began the long voyage south.

Three rousing 'hurrahs' from the men on the bank cheered them on their way, and were answered as lustily by the boat's crew. One seaman, more assertive than his shipmates, ran to the flagstaff, hauled down the signal of distress, and rehoisted the ensign with the Union in the upper canton. This symbolic expression of contempt for Palmer's betrayal was also loudly cheered.

Fowler put the castaways to work, salvaging more materials from the wreck, erecting tents made from reclaimed sails and spars as store-houses, and adding to the food supply by daily fishing expeditions. The ship's guns were brought ashore and placed round the camp, primarily to impress on the men that this was a disciplined naval base. The carpenters began building the first of the two boats that would carry them to Port Jackson if the captain's rescue mission failed.

Midshipman John Franklin, with the buoyant confidence of youth, takes time to draft a letter to his father:

> *We live, we have hopes of reaching Sydney. The* Porpoise *being a tough little ship hath, and still does in some measure, resist the power of the waves,*

and we have been able to get most of her provisions, water, spars, carpenter's tools, and every other necessary on the bank … on which 94 souls live. Captain Flinders and his officers have determined that he and thirteen men should go to Port Jackson in a cutter and fetch a vessel for [us]; and in the meantime [we are] to build two boats sufficiently large to contain us if the vessels should not come. Therefore we shall be from this bank in six or eight weeks, and most probably in England by eight or nine [months].[34]

Franklin and his fellow midshipmen have lost none of their high spirits. William Westall, the only member of the scientific team to take passage home with Flinders, watches in disbelief as the youngsters drive the surviving sheep over his precious paintings, laid on the sand to dry. The mindless prank further embitters the painter, long disillusioned with the voyage.

The sailors' frantic efforts of the first few weeks settled into a steady routine – Fowler was well aware they must be away from the bank before the onset of the summer cyclones. Anything serviceable was taken off the wreck and stored for possible use. When the weather permitted, the cutter was sent to an island some 10 miles distant, well stocked with birds, eggs and turtles. Meanwhile the construction of the first of the new boats went on apace – already it was taking on the appearance of a rakish schooner.

Rescue!

Wreck Reef Bank, Friday, 26 August 1803.

On 26 August, the largest cutter being ready for her expedition, was launched and named the Hope. *The morning was fine, and wind light from the southward; and notwithstanding the day [Friday], which in the seaman's calendar is the most unfortunate of the whole week to commence a voyage, I embarked for Port Jackson with the commander of the* Cato. *[Flinders, 1814, vol. II, p. 315]*

From the bearings taken at the sandbank, Flinders calculated that the nearest land lay about 150 miles due west. He set a westward course, intending to reach the coast first and then follow it south to Sydney. The cutter was loaded with three weeks' provisions and two half-hogsheads of water, and lay so deep in the water that on the second day, with a fresh wind and a cross sea, so much water came aboard she seemed likely to sink. The men were set to bailing, and as a last resort a cask of fresh water was emptied overboard, followed by the stones of their fireplace, the firewood carried for cooking, a bag of pease, and whatever else could be spared. The boat then rode more easily, the following day the coast was sighted, and they turned south.

The fourth day out, with the need for fresh water becoming urgent, Flinders made for shore and anchored at the southern entrance to Glasshouse Bay (now Moreton Bay). Some twenty natives were seen on a nearby hill, and amused the boat's crew with dances mimicking kangaroos; when the men in the boat made signs of wanting water, they pointed out a small stream falling into the sea. The surf was too rough to run into the land, but two sailors leapt overboard and hauled an empty cask ashore. They were able to fill it from the stream without incident, the natives remaining aloof the whole time.

Continuing south, the mouth of the Hunter River came abreast on 6 September, the eleventh day after leaving Wreck Reef. Gusty southerly winds and squally weather made it unsafe to remain at sea, and Flinders sought shelter in a shallow cove. It was the first time they had slept ashore, but with the stormy weather, soaking rain, and cold, and the boat's sail their only cover, the night was far from restful.

From here it was a short run to Sydney. Broken Bay was passed on the morning of the 8th, a sea breeze set in and, crowding on all sail for Port Jackson, 'soon after two o'clock [we] had the happiness to enter between the Heads'.

> The reader has perhaps never gone 250 leagues at sea in an open boat, or along a strange coast inhabited by savages; but if he recollect the eighty officers and men upon Wreck-Reef Bank, and how important was our arrival to their safety, and to the saving of the charts, journals, and papers of the *Investigator*'s voyage, he may have some idea of the pleasure we felt, but particularly myself, at entering our destined port.[35]

The cutter had covered about 750 miles in less than fourteen days. Apart from one man with dysentery, the crew were all in excellent health, 'notwithstanding our cramped-up position in the boat and exposure to all kinds of weather'.

Flinders and Park hurried to Government House, where they found the governor at dinner with his family. King was shocked by the sudden appearance of the two sea-weary men, bearded and unkempt, their clothes torn and salt-stained, whom he supposed to be well on their way to England:

> but so soon as he was convinced of the truth of the vision before him, and learned the melancholy cause, an involuntary tear started from the eye of friendship and compassion, and we were received in the most affectionate manner.[36]

As a naval officer, King knew the hazards of shipwreck, and lost no time in planning with Flinders the rescue of the castaways on Wreck Reef. He arranged with the captain of the *Rolla*, a 430-ton merchantman then in port, and about to sail for China, to call at the reef and take on board those of the survivors who wished to take that route home. For those preferring to return to Sydney, the colonial schooner *Francis* would accompany her. To Flinders he offered another colonial-built schooner, the *Cumberland*, 29 tons, and suggested that he might sail via Torres Strait – thus arriving in England some months before the *Rolla*.

The *Cumberland* had the reputation of being a good sea-boat, and Flinders, anxious to report to the Admiralty and secure another vessel to complete his survey of the Australian coasts, agreed. The little flotilla sailed from Sydney harbour on 21 September, the *Cumberland* in the lead. Her crew consisted of Flinders, the boatswain, and ten seamen.

Friday, 7 October. Lieutenant Fowler is out with some seamen in the newly built schooner, named the Resource, *putting her through her sea-trials. Suddenly one of the men cries out 'Damn my blood, what's that?', pointing to a speck of white on the far horizon. The others take it at first for a seabird, but soon it is seen to be a sail, and is shortly followed by another – and another. The men shout with joy, and slap each other on the back; Fowler hastens back to the bank with the good news.*

Samuel Flinders is in his tent calculating some lunar observations when a midshipman rushes in, shouting 'Sir! Sir! A ship and two schooners are in sight!' Samuel completes his calculations before saying that he supposes it is his brother come back. He enquires if the vessels are near, and when told 'not yet' desires to be informed when they have reached the bank. He returns to his calculations.

It is early afternoon when the ships anchor under the lee of the bank, in 18 fathoms, to be welcomed by a salute of eleven carronades brought ashore from the Porpoise – *the courtesy due to a commodore. 'Every heart', writes Seaman Smith, is 'overjoyed at this unexpected deliverance'.*

When Flinders sets foot on the bank he is loudly cheered, the men crowding round to shake his hand. The pleasure of rejoining his companions 'so amply provided with the means of relieving their distress' makes the day one of the happiest of his life. It is six weeks to the day since the Hope *sailed from Wreck Reef Bank.*

Flinders called the castaways together at the flagstaff and outlined his plans. First he chose the *Cumberland*'s crew for the passage home; all were from the *Investigator*, and included John Aken as master, Edward Charrington as boatswain, and seven picked men. John Franklin volunteered, but to his great disappointment was not chosen 'on account of the smallness of the vessel and badness of accommodation'.

The rest were given the choice of continuing on the *Rolla* to Canton, or returning to Port Jackson in the *Francis*. Lieutenants Fowler and Samuel Flinders, with John Franklin and most of *Investigator*'s former crew, sailed with the *Rolla* and reached England safely the following year. Fowler carried with him, for delivery to the Admiralty, four charts of the east and north coasts, 'which there had been time to get ready'.

The *Cumberland*

The Sydney Gazette, *Sunday, 18 September 1803.*

> *This vessel is only 29 Tons burthen; but … there is no doubt of her continuing as good a sea boat as experience has shewn her to be in very tempestuous weather off Norfolk Island and in Bass's Straits, and in every way equal to carry a sufficiency of Provisions and Water for Captain Flinders, the officers, and nine men who are appointed to navigate the first Vessel built in this Colony to England—May her Voyage be safe and expeditious!*

Flinders must have wondered whether he was aboard the same vessel! 'From her want of breadth,' he wrote to Governor King from Wreck Reef,

> the *Cumberland* is exceedingly crank, so that if there is not room to run before the sea, she must lie to in a double-reefed topsail breeze. She has always been leaky, and in one hour-and-a-half's cessation from pumping, the water washes on the cabin floor.[37]

Her standards of hygiene and comfort were if anything worse than her seaworthiness. 'Of all the filthy little things I ever saw,' he continued, this schooner, for bugs, lice, fleas, weavels, and mice, rises superior to them all. I have almost got the better of the fleas, lice, and mosquitoes, but in spite of boiling water and daily destruction amongst them, the bugs still keep their ground.[38]

To rid himself of the 'vile bug-like smell', he and his clothes 'must undergo a good boiling in the large kettle'. As for the mice, he had set his friend Trim to work upon them.

Despite the schooner's obvious unsuitability for a long voyage, Flinders seems not to have considered any significant change of plan. After calling at Timor, he told King, he proposed to sail straight for the Cape of Good Hope – although on arrival there he hoped he might find passage on a larger and more comfortable vessel bound for England.

The ships prepared to sail from Wreck Reef Bank on 11 October. Flinders boarded the *Rolla* to take leave of his brother, Mr Fowler and the other officers, and to farewell his men. The crews exchanged cheers, and at noon the *Cumberland* parted company, 'the *Rolla* steering north-eastward for China, whilst my course was directed for Torres Strait'. In ten days, Flinders reached Pandora's entrance, and in a further three days had found a new and intricate passage through the strait (still known as Cumberland Passage).

The schooner is leaky, more so than before, and one of the pumps begins to fail; rarely is the cabin free of water. With the stubbornness of his Flemish forebears, Flinders presses on, determined to complete the survey begun in the Investigator. *Hurrying to beat the monsoon, he enters the Arafura Sea, and reaches Timor on 10 November – it has taken just thirty days from Wreck Reef in a vessel 'something less than a Gravesend packet boat'. Without a doubt the route he has pioneered through the strait promises a far quicker passage to Batavia than that to the north of New Guinea – in a fast-sailing ship perhaps no more than thirty-five days from Port Jackson to Java Head.*

At Fort Concordia there is a new governor, Mynheer Giesler having died a month before. His successor is no less helpful, but the colony lacks the resources to repair the pumps or even provide pitch to pay the seams in the upper works after they are caulked. Flinders learns Baudin had visited the port shortly after his own departure from Kupang in April, and had sailed in early June to the Gulf of Carpentaria:

... and I afterwards learned, that being delayed by calms and opposed by south-east winds, he had not reached Cape Arnhem when his people and himself began to be sickly; and fearing that the north-west monsoon might return before his examination was finished, and keep him in the Gulph beyond the extent of his provisions, he abandoned the voyage and steered for Mauritius in his way to Europe.[39]

Flinders sailed from Kupang on 14 November, and set his course for the Cape of Good Hope. Though King had left the ports to be touched at on the homeward voyage to Flinders's discretion, the governor had advised him not to call at Mauritius,

> both on account of the hurricanes in that neighbourhood, and from not wishing to encourage a communication between a French colony and a settlement composed as is that of Port Jackson.[40]

As the *Cumberland* crossed the Indian Ocean, the wind was unsettled, showers of rain were frequent, and it seemed they were only just in time to save their passage. The schooner laboured in the choppy seas, the leaks 'were augmented so much that the starboard pump, which was alone effective, was obliged to be worked ... day and night'. Fearing the tiny vessel could not make the Cape, Flinders decided to put in at Mauritius to have the upper works caulked and the pumps rebored and fitted. On 6 December he 'altered the course half a point for that island, to the satisfaction of the people'.

Notes

1. Flinders, 1814, vol. II, pp. 12-13.
2. Ibid., p. 47.
3. Ibid., p. 88.
4. Ibid., p. 96.
5. Ibid.
6. Austin, 1964, p. 155.
7. Known today as Flinders Passage.
8. Flinders, op. cit., p. 104.
9. Seaman Samuel Smith's journal, MS, Mitchell Library.
10. Flinders, op. cit., p. 132.
11. Ibid., pp. 141-142.
12. Ibid., p. 143.

13. The trunk is now exhibited in a Brisbane museum.
14. Flinders, op. cit., p. 153
15. Seaman Samuel Smith's journal, op. cit.
16. Flinders, op. cit.., p. 161.
17. Ibid.
18. Ibid., p. 172.
19. Brown, in press.
20. Ibid.
21. Flinders, op. cit., p. 189.
22. Seaman Samuel Smith's journal, op. cit.
23. Flinders, op. cit., pp. 247-248.
24. Mack, 1966, p. 148.
25. Seaman Samuel Smith's journal, op. cit.
26. Good, 1981, p. 122.
27. D'Entrecasteaux's ships had completed a circumnavigation of the continent from Batavia to Batavia, but sailed to the north of New Guinea, not through Torres Strait.
28. Flinders, op. cit., p. 275.
29. Scott, 1914, pp. 277-278.
30. Ibid., pp. 279-280.
31. *Orphan* newspaper, Calcutta. 3 February 1804. National Maritime Museum, Greenwich.
32. Flinders, op. cit., p. 308.
33. Ibid., p. 304.
34. Sutton, 1966, p. 52.
35. Flinders, op. cit., p. 321.
36. Ibid., p. 322.
37. Ingleton, 1986, p. 246.
38. Ibid.
39. Flinders, op. cit., p. 348.
40. Ibid., p. 351.

CHAPTER NINE

⚓

The Shadow of a Captain on a Ghost Ship,
December 1802 to August 1803

King Island capers

Sea Elephant Bay, King Island, Wednesday, 8 December 1802.

> *Since the* Naturaliste *was to leave on the morrow, I went aboard her with Citizen Guichenot, our gardener, to examine the condition of the plants. We found them all healthy and the live quadrupeds and birds like-wise, except that we had the misfortune to lose our last remaining kanga-roo. It died as a result of an abscess on the thigh. However, I have no doubt that all the surviving animals will arrive safely. After dinner I took leave of Captain Hamelin and wished him a good journey. [Baudin, 1974, p. 441).*

Sea Elephant Bay at King Island, charted by Murray in the *Lady Nelson* in January 1802, was the appointed rendezvous for the French ships after leaving Port Jackson. The *Géographe* and the *Naturaliste* sailed into the bay – no more than a shallow inlet offering little shelter in a heavy sea – on 6 December, followed hours later by the *Casuarina*, which had been buffeted by storms in Bass Strait. As soon as his caulkers had repaired leaks to *Casuarina*'s deck seams, Baudin dispatched Frey-cinet back through the strait to survey the Hunter Islands – 'the geog-raphy of which the French government wants to be studied with the greatest exactitude …'.

On the 8th, Baudin boarded the *Naturaliste* to carry out a final inspec-tion of the ship and to take his leave of Hamelin. He presented his second-in-command with a large wombat, a gift from an English captain met on the passage south from Port Jackson, while Hamelin in return handed over several stowaways found hidden on board after leaving the port. Both men felt real regret at their parting. For the commandant,

the moment of separation was extremely painful ... I was truly fond of
Captain Hamelin for his personal qualities, and when one has shared the
same dangers for two years it was natural to feel as I did at his departure.[1]

Hamelin too was touched, writing in his journal that night:

Commandant Baudin gave us another and final proof of the generosity
and kind-heartedness that characterises him, by sending a pig, three sheep,
and some chickens for the invalid Depuch.[2]

Baudin, however, was less charitable towards Hamelin's subordinates:

I had the pleasure of seeing only one of his officers, and even then I
think that he would not have appeared if Captain Hamelin had not invited him to dine with us. As I left, I had them thanked for their politeness
and told that I was very happy to have no farewells to make to them.[3]

An hour after Baudin returned to his ship the colonial schooner
Cumberland headed into the anchorage, dropping anchor a few lengths
astern of the *Naturaliste*. Hamelin sailed an hour later, but not before
he had been visited by the schooner's captain, Lieutenant Robbins, and
a companion whom he recognised as Surveyor General Charles Grimes
from Sydney. Their news was unsettling – the *Cumberland* was under
orders to find suitable sites for settlement in Van Diemen's Land. 'Thus
there remains no doubt,' he recorded in his journal,

that the English are about to take from us the D'Entrecasteaux Channel,
where it would interest the French government to have a settlement, as
I shall try to prove in [my] report to the Minister.[4]

Next morning, Robbins and Grimes boarded the *Géographe* and asked
to see the commandant. Surprised by the *Cumberland*'s appearance
in the bay, Baudin invited his visitors to lunch, during which Robbins
handed over a letter from Governor King and a copy of his instructions. The contents 'threw enough light on the matter for me to realise
the purpose of his voyage, which was quite simply to watch us'. The
lieutenant's orders restated the British government's claims on Van
Diemen's Land, leaving Baudin in no doubt

that Governor King was afraid, from whatever he had been told, that I
was going to put someone on that land so as to occupy it first, and that
was his sole [reason] for despatching a ship to observe us.[5]

The two men also solicited the Frenchman's help. The *Cumberland* had left Port Jackson in such haste that she lacked many essential supplies – they had only the one set of sails, very worn, no canvas to repair them, and neither thread nor needles. One anchor was broken, and there was no powder for the guns. Their requests were so many and varied that Baudin asked for a list. He provided everything required – a sounding lead, a 60-fathom cable, one length of canvas measuring 108 English yards, a pound of sail thread, six needles, six padlocks and the nails to go with them – and added 12 pounds of powder from his limited store. The *Géographe*'s forge was set to work repairing the schooner's broken anchor. It was, he felt, some recompense for Governor King's generosity at Port Jackson.

Meanwhile, shore parties set up the observatory tents and prepared a camp site for the scientists and their servants. When the time came for the large dinghy to ferry them to the beach, however, it was so cluttered with the cooks' paraphernalia that not all of them could find a place. Shaking his head about their 'pomp and magnificence', Baudin withdrew to his cabin, 'dissatisfied that the whole lot of them had not left on the *Naturaliste*'.

A few days later the naturalists were startled to see Robbins and Grimes, escorted by the *Cumberland*'s three marines, march into their encampment. Hoisting the Union Flag to the top of a nearby tree, the lieutenant posted the marines at its foot and ordered them to fire a volley (doubtless with French powder) – after which the British party gave three hearty cheers, and Robbins completed the ceremony by reading out the 1788 act of possession. Péron and his friends chose to treat it as a joke, inviting their visitors to lunch with them, while Petit sketched a 'complete caricature' of the proceedings.

Baudin was not amused when he arrived at the camp an hour later. Noting the flag flying head downwards he thought at first

> it might have been used to strain water and then hung out to dry, but seeing an armed man walking about, I was informed of the ceremony which had taken place that morning.[6]

The incident, he wrote to King, was 'childish'. There was no basis to the story of a proposed French settlement in the area,

> nor do I believe that the officers and naturalists who are on board can have given cause for it by their conversation ... I took great care in mentioning [this] to your captain.[7]

Baudin gave no details in his letter, but rumour in Sydney had it that

the French commodore … observed with much pleasantry, that the English
were even worse than the Pope, for His Holiness had the moderation to
divide the world, whereas the English were grasping at the whole of it.[8]

*In their excursions, the naturalists come across a 'colony of eleven
wretched sealers' living on the island. They are a rough gang, mostly
former convicts, among them two Irishmen deported for their political
opinions. According to their leader, a man named Cooper, they have
been there for thirteen months, hunting seals and preparing cargoes of
their oil and skins for the China trade. But the oil casks have been full
for months, the ships are overdue, and the men are restless.*

*Soon it is the naturalists' turn to find themselves 'abandoned'. A wild
storm drives the* Géographe *from the bay, and she makes for the safety
of the open sea. The same gale tears their tents to shreds, leaving them
without shelter from the torrential rain. Drenched to the skin, their
provisions lost, unable even to collect shellfish because of the violence
of the waves upon the shore, they make their way to the sealers' camp
some 6 miles distant.*

*The Frenchmen are overwhelmed 'with demonstrations of concern
and kindness' by the sealers, who willingly share their scanty rations
with their visitors. Cooper takes them into his hut, a ramshackle hovel
where he lives with a Sandwich Island woman he has brought here 'as
wife and housekeeper'. She serves them an excellent and savoury meal,
containing 'masses of different meats, well-cooked in their own juices
… though we had to eat them without bread or biscuit'. Their diet, Péron
learns, consists largely of wombat, emu, and kangaroo meat. The wombats are domesticated, while the kangaroos and emus are hunted by
dogs, who first chase them to exhaustion, then kill them by tearing at
their throats.*

The *Géographe* returned on 23 December, eight days after her hurried
departure from the anchorage. The ship had lost her anchors and a
boat, been carried across Bass Strait and beyond Wilson's Promontory
by the storm, and narrowly escaped being wrecked among the offshore islets. Expecting to find the *Casuarina* in the bay, Baudin worried what accidents 'might have befallen her, if she was amongst the
islands when the squall began'.

At daybreak he sent a boat ashore to bring off the scientists. All were aboard by 9 a.m. 'except for M. Péron, who seeing nothing but molluscs at every step, had amused himself by missing the first boat'. Cooper came on board to seek replacements for the provisions supplied to the marooned men, and was rewarded with rum, six bottles of red wine, and sugar, tea, and biscuit. Baudin gave him a letter for Governor King, which he was to send to Port Jackson by the first ship to call. He also bought from his men several live emus, a kangaroo, three wombats, and one of their trained hunting dogs.

The *Casuarina* returned on the 27th from her survey of the Hunter Islands. She had twice run aground, said Freycinet, and he had been obliged to empty his water casks to refloat her. Anxious to be off, Baudin indicated an anchorage on the east coast of Flinders's Kangaroo Island as their next rendezvous.

Borda's Island

On board the Géographe, *Friday, 31 December 1802.*

> *An immense scarf of vapour lying along the horizon looked so exactly like land that everybody aboard both ships was deceived by it. On all sides men thought that they could make out the capes, peaks, and various indentations which constitute a long stretch of coast; but after running towards these fantastic shores for several hours, we realised our error and hurried to resume the course we had so inopportunely altered. [Péron and Freycinet, vol. II, unpublished translation]*

Cape Hart, the south-eastern extremity of Kangaroo Island, came into view on 2 January 1803. Aware that Flinders had not surveyed the south coast, Baudin sailed west, ordering Freycinet in the *Casuarina* to follow, coasting the land as closely as he could. During the day they covered about 10 leagues (30 nautical miles) of featureless shoreline, apart from one high hill with a sandy summit, sporting a few scattered tufts of vegetation. It was Flinders's Prospect Hill.

Rounding the island's southernmost point (Cape Gantheaume) next morning, the ships entered a large but shallow bay 4 or 5 leagues across (Vivonne Bay). The coast appeared dreary and forbidding, sandy dunes alternating with rocky plateaux, bleak hills from 200 to 300 feet tall; close to the sea they rose like ramparts above the waves, which everywhere broke savagely on the shore, making a landing impossible. Their

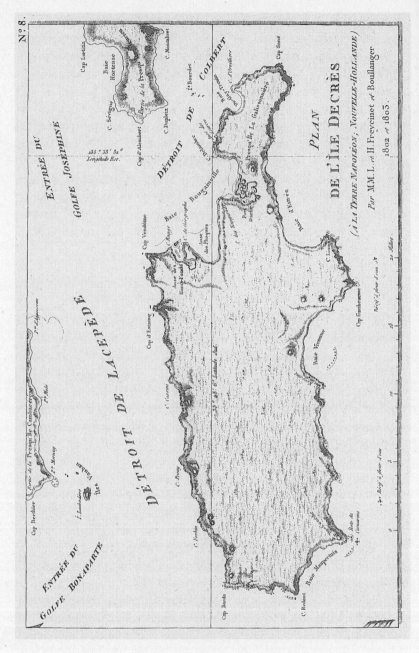

Kangaroo Island, 1802 and 1803. Chart by Freycinet and Boullanger.
[Courtesy of Scotch College, Hawthorn]

colours were melancholy and wild, grey, brown and black, occasionally softened by a yellowish ochre.

Cape succeeded cape, mostly guarded by fearsome reefs thrusting out to sea.

At 3 p.m. the *Géographe*'s lookouts reported two small islands (Casuarina Islets) ahead, protected by a double line of reefs over which the sea broke with great force. Rounding a rocky cape (Cape du Couedic), they found the land ran to the north-west. After an uncomfortable night searching for land seen to the west – another mirage – the flagship reached the position near Cape Borda, where the previous April Baudin had broken off his survey of the island's north coast. It was, he noted smugly, more than Flinders had achieved:

> We had [now] finished our geographical work on the whole of Kangaroo Island, so that if the English have the advantage over us of having reached it a few days earlier, we have the advantage over them of having circumnavigated it and determined its geographical position in a way that leaves nothing to be desired for the safety of navigation.[9]

To commemorate the French exploit, Baudin renamed the island Île Borda, after the mariner and mathematician Jean-Charles de Borda (1733-1799).[10] The *Géographe* entered Bougainville Bay (Flinders's Nepean Bay) on 6 January and dropped anchor off Cape Delambre (Kangaroo Head). Freycinet arrived next day and moored nearby.

Baudin wastes no time, sending boat parties off to explore the eastern and western shores of the bay, and above all to search for fresh water in the ravines running down to the beach. A third boat leaves with the carpenters, who must find and cut timber for the new longboat under construction on board the ship.

The boats return at dusk. Not a drop of drinkable water has been found, nor have the carpenters located any usable timber – all the fifty or so trees cut down were quite rotten at the heart. So numerous were the kangaroos that they seemed like flocks of sheep, and were as tame. None of the boat parties report any sign of man's presence on the island.

On the days that follow, the Géographe*'s forge is kept busy repairing the* Casuarina*'s anchors. Work gangs are sent across to the schooner to overhaul her masts and rigging, recaulk her topsides, and prepare her for her next task – completing the survey of the two large gulfs on the mainland, left unfinished the previous year. At Freycinet's request,*

Baudin grudgingly transfers four casks of fresh water to his consort, but makes it clear he can expect no more: 'I was unable to run the risk of lacking it myself just for the pleasure of giving him some.'

When Freycinet sails after nightfall on 10 January, he carries strict instructions to return to the bay within twenty days. If he fails to do so, Baudin warns, 'I shall sail for the St Peter and St Francis islands ... The impossibility of replacing my water does not allow me to wait any longer'.

The *Géographe* remained at the anchorage to the end of January. Detailed surveys were made of the large coves on the western side of Bougainville Bay (Western Cove and Bay of Shoals) and of the inlet at its head (American River and Pelican Lagoon). Bernier, the astronomer, set up his instruments on Cape Delambre, taking more than 250 measurements to establish its exact position.

Work continued on the rigging and other shipboard repairs, while fishing and hunting parties were dispatched to gather fresh food for the ship's company. Surprisingly, the waters of the bay yielded few species of fish; barely a dozen species were procured, 'new ones it is true, but five or six of which are not commonly eaten'. The answer most probably lay in the monstrous sharks, 15 to 20 feet long, which prowled round the ship 'numbing with terror all those watching'.

There was no shortage of red meat. With no predators, kangaroos had multiplied on the island; in some parts the ground was so trampled that not a blade of grass was to be seen. Hunting was so easy that twenty-seven kangaroos were captured alive and taken aboard, quite apart from those killed and eaten by the crew. The hunters did not need to exert themselves at all, or to use their ammunition:

> One single dog, called Spot, was our provider. Trained by English sealers in this type of hunting, he chased the kangaroos; and when he had caught up with them, he killed them on the spot by tearing open their jugular arteries. Nothing less than the presence and shouts of the hunter was required to rescue the victim from certain death.[11]

Spot had been brought aboard at King Island. Much as Péron admired his prowess, he feared that 'the weak and harmless race of kangaroos' would soon be wiped out if others of his breed were introduced to the island.

The naturalists, meanwhile, made numerous excursions around the bay, examining the island's singular flora and fauna. Péron delighted in the great flocks of land and sea birds – the former including 'the beautiful golden-winged pigeon, the pretty ultramarine ringed tomtit, the red-rumped bullfinch, and the white New Holland goshawk'; and among the latter, 'yellow-necked pelicans with half-black, half-white wings, seagulls … terns, sea pies, a great sea eagle, teals', and many others.

Baudin's prime concern was the well-being of the living animals on board – it was his responsibility to ensure their survival on the long voyage to France (and their safe delivery to Mme Bonaparte). Appropriate food, fresh water and security on the ship were all essential. Over the officers' bitter protests, he ordered several cabins turned into sheltered pens for the kangaroos.

Water has been found in a sandy bay east of Cape Delambre, which they name Anse des Sources (now Hog Bay). Three pits dug in muddy ground adjoining the beach yield enough drinkable water for their daily needs. To while away the time, one of the watering party chisels an inscription on a nearby rock to commemorate their visit.

The carpenters find some timber suitable for the longboat's planking. To hurry the work along, Baudin accompanies them, but is nearly crushed when a tree falls askew, knocking him to the ground. Though trapped for some time in the branches, he escapes with cuts and bruises to his head and chest.

On the ship, saws are set up to cut the wood into planking. Hoping to set an example the commandant saws the first piece, then hands over to Engineer Officer Ronsard. None of the other officers shows up.

Baudin shrugs off the accident, as he ignores his worsening health. Since his near-fatal illness at Timor, he has suffered recurrent bouts of fever; now he has a persistent cough and is spitting blood. The convict girl who shares his quarters, Mary Beckwith, worries him more. Clearly he was mistaken to bring her on board – her presence has caused tensions among the crew and, he suspects, within his staff. He resolves to disembark her at their next port of call.

Mary has put behind her the drudgery of her life at Port Jackson. Though no beauty, she is young and headstrong, with a trim figure, and enjoys feeling the men's eyes following her as she walks the deck. There is no shortage of offers to take her ashore – it will be a welcome change after the close confines of the ship. She has no reason to refuse. In January 1803 she becomes the first European woman to set foot on South Australian soil.

The twenty days allowed for the Casuarina*'s return have run out by 31 January. Baudin sails early the following day, 'leaving [the schooner] to her good fortune, as she has chosen to leave us to ours'.*

Gulf mirages

Aboard the Casuarina *in Gulf St Vincent, January 1803.*

> *The entire stretch of coast that we had in view is low, swampy, and covered in small trees. A line of high hills could be seen inland and ran perceptibly in a north-south direction. Access to the shore is prevented by a multitude of sandbanks, below and above water-level, which formed a very long barrier; six miles offshore we had only two fathoms of water.*
>
> *Such are the obstacles which had halted us on our first campaign [April 1802]; and if the reader recalls now those dark, stormy nights, during which we were reduced to beating about in the middle of this gulf with a great ship loaded with sail, he will undoubtedly shudder at the perils to which we were exposed and whose complete extent we were then far from suspecting. [Péron and Freycinet, vol. II, unpublished translation]*

Freycinet sailed from the anchorage during the night of 10 to 11 January and headed across Backstairs Passage for Gulf St Vincent.[12] Passing Rapid Head at dawn he coasted the eastern shore, the leadsman calling the depth every few minutes. Although 6 or 7 miles offshore, for much of the day the *Casuarina* sailed in shallows of 4, 3, and even 2 fathoms of water.

Arriving finally at the head of the gulf, they found it ended in low, sunken land, with not even a creek emptying into it. Many fires in the interior denoted the presence of 'several tribes of savages', but none were to be seen along the shore. The western coast posed fewer problems for navigation; below Black Point the lead rarely showed less than $4\frac{1}{2}$ fathoms. The terrain too was more interesting, with stretches of reddish cliff displaying 'a rather beautiful vegetation'.

On 18 January the *Casuarina* rounded Troubridge Point at the gulf's western exit and sailed rapidly along the southern coast of Yorke Peninsula, only to run aground on an extensive sandbar. She floated off without damage and by nightfall was off the Althorpe Islands, at the entrance to Spencer Gulf.

Doubling Cape Spencer next morning, Freycinet again headed north, and two days later was off Cape Elizabeth (south of the present Moonta

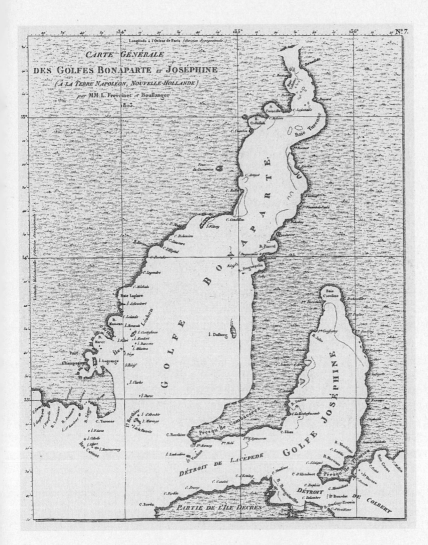

Spencer Gulf (Golfe Bonaparte) and Gulf St Vincent (Golfe Josephine), 1803.
Chart by Freycinet and Boullanger.
[Courtesy of Scotch College, Hawthorn]

Bay). Here too the land was low-lying, and the sea obstructed by shoals that prevented him sailing close to the shore. Above Port Germein there was an abrupt change of scenery; the gulf narrowed and the land rose, turning each coast into a formidable rampart:

> everything seemed to guarantee the existence of a great river, that deep, vast bed, those imposing shores ... the extraordinary prolongation of the gulf, its direction, its very shape – everything helped to complete the illusion ... They crowded on sail, and almost immediately the river vanished. Countless sandbanks were squeezed into a bed, four to five miles in extent, the shores drew closer and closer together.[13]

Faced with impenetrable swamps connecting the east and west shores, with no waterway visible between them, Freycinet turned towards the western coast. A succession of sandbanks drove the *Casuarina* away from the land, as the soundings dropped to fewer than 3 fathoms. Contrary winds obliged him to tack about for almost sixty hours in the dangerous waters; worrying as the delay was, it gave Freycinet and the hydrographer Boullanger time to chart accurately the position and extent of the shoals.

On 25 January, off what is now Whyalla, the entire ship's company suddenly found themselves amid what seemed to be a vast basin, enclosed by fantastic shores. Even Freycinet was beguiled:

> The mistake was so general and so complete, that if we had not been quite certain that there could be no land in the areas where we had just been sailing, it would not have been possible to disclaim such a marvel. Some of my sailors were so perfectly deceived, that they believed they could make out trees on this would-be coast; only the lifting of the mist could convince them of their mistake.[14]

Time was now running out for Freycinet. Baudin had given him only twenty-one days (half the time taken by Flinders) to complete his survey of the two gulfs. His water and provisions would last barely a week. Allowing himself a fleeting examination of Flinders's Port Lincoln, 'capable of harbouring all the navies of Europe', he headed south-east for the rendezvous with the *Géographe*. Calms and contrary winds delayed him, and he was still well short of the anchorage when, at two in the afternoon of 1 February, he saw the corvette steering towards him under full sail.

Bay of Saints

On board the Géographe, *Tuesday, 1 February 1803.*

> *At two o'clock we sighted the* Casuarina *running on the easterly leg. I expected that as soon as she saw us, she would go on the same tack as us and follow us. Consequently, when we were within range of each other, I furled the mainsail so that she should have less trouble in keeping up with us. But that disturbed her extremely little and she continued running East, and so rapidly, that by half past three she was out of sight. It is undoubtedly difficult to explain this manoeuvre on the part of Citizen Freycinet ...* [Baudin, 1974, p. 472]

On board the Casuarina, *Tuesday, 1 February 1803.*

> *At ten past three [the* Géographe*] was abeam of us to windward, a league distant. I went about again towards her; I was expecting to see her heave to or bear towards me; it was in vain: no change was made in her course or her sails. Unable from then on to keep up her fast pace I soon lost sign of her.* [Freycinet, 1815, vol. III, cited in Horner, 1987, p. 281]

Péron and his companions aboard the flagship were relieved to see the schooner tacking towards them; they had feared the worst when Freycinet had failed to make the rendezvous the previous day. 'Only a few moments were needed to effect the reunion of the two vessels' when the unthinkable happened:

> not the slightest change was ordered in the *Géographe*'s course. This ship had all sails set – not one was lowered. Incapable of matching our speed, pinned down by the wind, the *Casuarina* remained astern and soon disappeared from view ...[15]

The *Géographe*'s logbook, signed by Bonnefoi as officer of the watch, contradicts Freycinet and Péron. It records that the corvette did shorten sail, almost halving her speed from $7\frac{1}{2}$ knots to less than 4. Nevertheless, it was not enough to allow the *Casuarina* to catch her before dark, and Baudin spent the night hove to off the western end of Kangaroo Island, with a lantern burning on the poop. Next morning he retraced his tracks eastwards, but without sighting the schooner. Freycinet, continuing his westward course, had passed the flagship in the night, then headed north-west for the next rendezvous at the St Peter and St Francis islands.

The *Casuarina* reached the St Francis group on 5 February, and for two days cruised among the islands without sighting the *Géographe* or discovering any suitable anchorage. Freycinet's position was desperate; only four days' water remained, and he was 300 leagues (close to 1000 miles) from King George Sound – the only part of the coast where he knew it could be obtained. He headed west for the sound, halving the crew's water ration, and setting all the sail that the schooner could carry. Incredibly, a fresh easterly breeze blew continuously for the next six days, driving them before the wind to safety:

> We finally made it on the afternoon of the 13th February … At that stage the *Casuarina* was so damaged that it was necessary to beach her immediately. There were only a few bottles of water left … Thus without that truly extraordinary occurrence – of stiff winds for six days – the cruellest death would have been the result for us of a separation as incomprehensible and, so to speak, as deliberate as this one was.[16]

Baudin for his part had sighted the St Peter Islands, some 25 miles north of the St Francis group, at about the same time on 5 February. Unaware that Freycinet was close by, he doubled the islands to the north and dropped anchor two days later, midway between the mainland and the islands, in a large inlet he called the Bay of Saints. Next day, the 8th, he dispatched Bonnefoi and Ransonnet to explore the shores of the bay (named Denial Bay by Flinders the previous year), giving each explicit instructions 'to avoid in all circumstances using on the natives the firearms which are aboard for the security of those accompanying you'.

The boats returned the following day, both with negative reports. The head of the bay offered no resources at all, 'being entirely without water and barely able to be visited by small craft'. What appeared at first to be rivers were mere arms of the sea, disappearing into the swamps. Though each party reported smoke rising inland, and one man claimed to have spotted natives at a distance on the beach, no contact had been made. As for the land, it seemed so parched and arid that even in the rainy season the water would be 'soaked up before being able to run in any direction'.

The naturalists – 'who until then', complains Baudin, 'remained peacefully on the ship through fear of not being comfortable in a boat that might spend one or two nights ashore' – join a 'pleasure party' on an excursion to Eugène Island, one of the group.

Once ashore the scientists separate, Leschenault to study the island's botany and Lesueur to join the hunters; Péron, 'avid for what these shores have to offer for their being new to science', makes for a great sandbank stretching almost to the mainland. With a sailor named Lefebvre, he crosses to the bank, much of it exposed by the low tide, to examine its deposits of shells and molluscs; their richness surpasses even his expectations.

Hours pass as Péron collects his precious specimens, until Lefebvre draws his attention to the fast-rising tide. Retracing their steps, they find the sea surging across the bank, cutting them off from their companions on the island. Deceived by the whiteness of the sand, and suspecting no difficulty in the crossing, the two wade into the water. Soon the water is up to their waists, and still rising; it is impossible to turn back.

> A stranger to the art of swimming and exhausted by my efforts, I found myself in imminent danger. It was the zeal of the good Lefebvre that saved me. He alone guided me in the midst of the rising waves; in the deepest places his arms supported me. In a word, this brave man did everything to prevent my drowning. But, despite all his care, it would certainly have occurred, had he not managed by dint of groping and testing, to discover a sandbank over which there was not much water, and which led us close to the shore that we were trying to reach.[17]

The shore party returned on board at sunset. Leschenault had collected a number of new plants, and the hunters brought back several birds not seen before, five short-tailed possums (two of them alive), and two kangaroo rats, which, Baudin noted, 'seemed to be different from the ones we knew, at least in their colouring'. All was made ready to sail in the morning.

Few regretted leaving the anchorage in the Baie des Saints. Years later, Péron used a quotation from the Roman writer Seneca to sum up his impression of the mainland; translated, it reads 'the barren desert of the abysmal fields lies all untilled, and the foul land lies torpid in endless sloth'.[18] The quote contrasts oddly with the eminent names he and Freycinet bestowed on this western outpost of Terre Napoléon on the latter's maps – among them Îles Josephine (empress of the French); Baie Murat (marshal of the empire, king of Naples, and Napoleon's brother-in-law); Baie Louis (king of the Netherlands, Napoleon's brother); Îles Jerome (king of Westphalia, Napoleon's brother); and Île Eugène (viceroy of Italy and Napoleon's stepson).

Black hunters, white sealers

At anchor in King George Sound, Friday, 18 February 1803.

> *At daybreak ... we put our boats out. One was [sent] to reconnoitre the watering-place and the other was employed in the establishment of our observatory on a reasonably pleasant island lying North-West by North of the ship. When the boat returned from the watering-place I was informed that it was very good and easy of access. Ten casks were found there belonging to the* Casuarina, *which we knew to be in one of the two ports, for she had informed us of her arrival ... by a flag placed on the summit of Seal Island ... we had told her of our own arrival by a cannon shot fired during the night. She had replied to it with several flares which had indicated that she was in the port de la princesse Charlotte [Princess Royal Harbour]. [Baudin, 1974, p. 483]*

The *Géographe* anchored on the western side of the sound at sunset on 17 February. Her six-day passage from the Bay of Saints had been uneventful, apart from one near-disaster when the ship, running at 6 knots, narrowly avoided a small, flat rock, almost invisible in the setting sun, at sea level. At breakfast next morning Freycinet came on board to report to the commandant. The reunion was frosty, each man blaming the other for their separation off Kangaroo Island. 'I pointed out to him,' wrote Baudin,

> the bad manoeuvre he had made and which seafaring men will be able to judge, but he remained none the less convinced that his had been better than ours, which was the sole cause of the separation ... He objected, moreover, that having only very little water, he had thought it wiser to come to this port rather than continue looking for me, because of the uncertainty of knowing if he would find me.[19]

It was Freycinet who broke off the argument, saying that he needed help in repairing his ship, still beached in Princess Royal Harbour. Baudin grudgingly agreed, sending a caulker and a carpenter to work on the schooner. He himself set off in his boat to inspect the watering-place. Satisfied, he 'straightaway established our tents there, one for our sick men, then numbering four, and the other for the naturalists'.

On the 20th he dispatched Midshipman Ransonnet to explore the coast to the east: 'Generals D'Entrecasteaux and Vancouver only saw [it] from a distance; you are to examine it closely and with scrupulous attention.'

At midday Ransonnet's party is abeam of Mount Gardner, at the sound's eastern limit. Rounding Cape Vancouver they are astonished to find a two-masted ship at anchor in the bay beyond. From her master, Isaac Pendleton, Ransonnet learns she is an American brig, the Union, *four months out from New York, in search of sealskins for the China market. Pendleton sails at once for the sound to meet the French commandant.*

Leaving Two People's Bay, Ransonnet continues eastwards. The coast consists mostly of high, granitic cliffs, nearly sheer and inaccessible. Tired after several days of blustery south-westerlies, his party seeks shelter in a small bay behind Bald Island. Here, finally, the French meet the Aboriginal people of Nuyts Land.

Eight natives emerge from the bush, five men and three women. The latter retreat, but the men – first hurling their spears away, to signal their peaceful intentions – come forward and help their visitors to land. The sailors offer presents, which are accepted 'with an air of satisfaction, but without eagerness'; soon the trinkets are discarded, left lying on the ground.

Several large and handsome dogs accompany the natives. Ransonnet tries to bargain for one of the animals, offering everything he has to give, but without success; 'their will was unshakeable'. The dogs are used for hunting kangaroos, and are indispensable. The French share coffee, biscuit, and salt meat with their new-found friends, but cannot tempt them to try the bacon also offered, which is left untouched on some stones.

The warriors refuse to allow the visitors to approach the place where their womenfolk are concealed, consenting only to lead one sailor to a well they have dug close by; the water is fresh and sweet. As night falls, Ransonnet farewells them and anchors safely out to sea, 'ready to sail with the first good wind'.

Later, on the Géographe, *Ransonnet describes his meeting with these peaceful people to Péron. The expedition's 'anthropologist' finds himself yet again reporting an absorbing meeting with mainland Aborigines at second hand. He listens enviously to the midshipman's account:*

These men are tall, thin and very agile. They have long hair, black eyebrows, a short, flat nose – sunken where it begins, deep-set eyes, a large mouth, protruding lips, and very beautiful, very white teeth. The inside of their mouths appears as black as the outside of their bodies. The three eldest, who could have been forty to fifty years old, wore big, black beards;

their teeth looked as though they had been filed down; the septum of their noses was pierced; their hair was trimmed and was naturally curly.

The other two, whom we judged to be from sixteen to eighteen years old, had no kind of tattoo at all; their long hair was gathered back into a knot, powdered with ochre, such as the old men had rubbed over their bodies. For the rest, they were all naked and wore no ornament, apart from a sort of broad belt made of many little strands of kangaroo skin. They talk volubly and sing from time to time, always on the same note and accompanied by the same gestures.[20]

At dawn on the 21st, Baudin set off in his dinghy, determined to tour the two 'ports' in the sound – Princess Royal and Oyster harbours. Calling first on Bernier to inspect his island observatory, he then made an unheralded visit to the *Casuarina*, beached in shallow water at the head of the harbour. Not unexpectedly, the schooner to his eyes 'was in complete disorder and singularly dirty ... Most of the sailors were off hunting or fishing. I stayed for only half an hour ...'.

Continuing along the coast, 'much pleasanter than all I had seen so far', he came to Oyster Harbour late in the evening, where he found Faure and Bailly encamped. Next morning, while the sailors were occupied fishing for oysters, Baudin and the two scientists made their way up a small stream running into the harbour. After travelling for a league and a half they came upon two 'peculiar and interesting structures' on its banks:

The first was 7 or 8 feet from the stream, on a piece of bare ground that was 3 feet in circumference and surrounded by finely-tapered spears painted red at the tip. There were eleven in all. Parallel with this trophy, on the other side of the stream, was a plot of ground similar in shape and with the same number of spears. But these seemed to be guarding the passage to the right bank from the left, just as those on the left seemed to be guarding it from the right.[21]

Debating the purpose of the structures with his companions, Baudin guessed they were probably graves containing the bodies of

two warriors of different tribes, buried there either after a private battle between themselves or after some more general fighting, and seeming still to defy one another after death.

He forbade any of his men to desecrate the graves or remove the spears.

On his return to Princess Royal Harbour, Baudin found the observatory island a blackened desert, and Bernier's tent set up on the mainland. A conflagration started inadvertently by the cooks had swept across the island in a matter of minutes; the astronomer 'considered himself extremely lucky that the instruments and the tent had escaped'. Bernier also reported the arrival of an American brig, which Baudin had seen at a distance but mistaken for the *Casuarina*. He was still with the astronomer when Captain Pendleton made his appearance at the tent.

The American is invited to dine on the Géographe *next day. He comes aboard a worried man, deceived, he tells Baudin, by Vancouver's exaggerated report of seas teeming with countless herds of seals. Needing 20 000 skins for a profitable cargo, he has succeeded only in catching a few individuals here and there. He lacks information about the best sealing grounds, and has no maps showing the hazardous coasts to the east.*

Baudin seeks to set his visitor's mind at rest. Pendleton has chosen the wrong time of year for his voyage – Vancouver had visited in the southern spring, when the herds would have been abundant. Now, in late summer, they have sought refuge in cooler, more southerly waters. However, he adds reassuringly, he should still find enough skins to fill his holds at the St Francis and St Peter islands, and at Île Borda further to the east. He presents Pendleton with copies of Beautemp-Beaupré's charts and Flinders map of Bass Strait, and points out the best anchorages and the most likely sealing grounds.

As his visitor prepares to leave, Baudin particularly recommends a secure anchorage at the head of Bougainville Bay, on the north coast of Île Borda (Kangaroo Island), and requests the American, if he should rest there, 'to prevent his men from killing the pigs and poultry I left there for the use of future navigators in those parts'. Pendleton promises no harm will be done to them, and returns to his ship to prepare for an early departure. He sails on 27 February, and heads eastwards.

Baudin too was anxious to continue his voyage. On 28 February, after a stay of ten days, the boats were sent ashore for the last time to bring off 'everybody and everything'. At half past four next morning 'we hoisted the topsails and got under way with all sails set'.

West coast revisited

On board the Géographe *in King George Sound, Monday, 28 February 1803.*

> *During this rest I made up the* Casuarina's *provisions to a five-month supply, so that if she should become separated from me and be unable to rejoin me at the meeting-place indicated, she will still have all that is necessary to make Isle de France …*
>
> *The botanists were very pleased with the collections they made. From what he has told me, Citizen Leschenault appears to have gathered roughly two hundred new species of plants that were unknown to him, and as many varieties within the species. Guichenot, the gardener, is no less satisfied, and has even more specimens.*
>
> *I think that Citizen Péron, too, will be able to write a volume on worms and molluscs. He has one or two cases of broken shells [which], he claims, should help him to establish the period at which New Holland must have risen from the floor of the sea. [Baudin, 1974, pp. 493-494]*

During their stay in the sound, Baudin grumbled often about the naturalists' 'nonchalant' commitment to their collecting. In similar mood, he reprimanded Louis de Freycinet for feasting with his fellow officers when he should have been supervising his men on a work detail:

> Citizen, I should not be obliged to reproach you with such thoughtless behaviour, the slowness that it causes in your operations, the resulting loss of time, and finally, the drunken state of the men in the longboat when they returned from your vessel.[22]

He hoped the scolding would not be forgotten by the schooner's commander, 'as much for the good of the Service as for himself'.

Nothwithstanding the commandant's complaints, the *Géographe* was fast becoming congested with living plants and live animals. The former too required secure storage to protect them from the hazards of the ocean voyage, and Baudin took some of the finer specimens into his cabin,

> so as to put off for a few more [weeks] the lamentations that I shall have to listen to when it becomes absolutely necessary to take the cabins of those still in occupation for housing the additional objects of Natural History that we may collect during the remainder of the expedition.[23]

Guichenot's collection, Baudin noted with approval, was the result of 'work and not wit'. Péron and Leschenault, by contrast, could be trusted to produce '60 pages of writing which, for a different reason, will be all wit and no work'.

Before sailing, Baudin drew up Freycinet's sailing instructions, and sent them across to the *Casuarina* by a midshipman. Her commander was forbidden to open them until he was beyond the entrance to the sound.

It is his intention, writes Baudin, to revisit Géographe Bay and complete the survey of that part of the coast, and then sail for Rottnest Island; from there he will proceed to the roadstead off Dirk Hartog's Island to obtain some turtles, before heading further north; he will halt for a while at North-West Cape to settle some problems with the old Dutch maps. This is to be the rendezvous should the ships again become separated – it is 'the only one I can indicate to you', he writes, adding 'as I do not want to stop anywhere there are Europeans, it is only too likely that we shall not meet again'.

Freycinet is to wait for ten days, after which, if the Géographe has still not appeared, he is to head directly for Isle de France, where the vessel is to be handed to the colonial government for use in the colony's service. 'I think that you will retain command until my return ... but if that does not suit you, you are free to return to France whenever you like.'

The closing paragraph adds to Freycinet's confusion:

...it rests with you alone to accompany me or not. At no stage did I overlook anything that was necessary to prevent the two separations that we have already experienced and that you could easily have prevented ... And so, you will have to be personally answerable to the authorities in France for the expenses incurred in the equipping of your ship, since they will have grown burdensome for the government and pointless for the expedition.

Your fellow-citizen, N.B.[24]

What, he puzzles, is the purpose behind these strange instructions? At first glance the commandant has given him virtual carte blanche *– he is neither ordered to keep company with the* Géographe *nor to assist in the west coast survey, merely to attempt a rendezvous at North-West Cape. Baudin acknowledges that they are unlikely to meet again, and has given him five months' supplies to ensure the* Casuarina *will make Isle de France in comfort.*

Self-interest, duty and ambition all point Freycinet in the same direction. There is the barely concealed threat in his instructions that he will be held liable for all costs incurred in the event of another separation.

On the other hand, the past months have demonstrated the value of his
survey work for the expedition and for French science. He will keep
Baudin company – at least to the rendezvous at North-West Cape.

'Thwarted by strong south-easterly winds, assailed by heavy squalls
and dense mists, [and] worn out by a perpetually stormy sea', the ships
tacked off the coast for four days. When the weather at last moderated
on 5 March, Baudin sent Freycinet to examine the inshore waters between Cape Howe and Point Nuyts, hoping that they might offer 'some
ports … of use to navigators'. For another two days he awaited the
Casuarina's return, then made for Cape Leeuwin, annoyed at the time
wasted 'waiting for her to carry out a reconnaissance that was the work
of three hours'. He wrote angrily:

> Until the present time, this officer (who is insubordinate on principle)
> has not been able to carry out a single task nor even obey my orders to
> him. And that is why, if he rejoins me at Rottnest Island (which I indi-
> cated to him as a meeting place), there is a strong possibility that I shall
> take his command from him to see if someone else will execute more
> thoroughly the orders that circumstances will cause me to give.[25]

The *Géographe* doubled Cape Leeuwin on the 9th. From the position
of St Allouarn Island nearby, Baudin realised that the point believed to
be the cape at their landfall in May 1801 'was not the one the name
belonged to'.[26] As a result the charting of this part of the coast was
begun afresh, 'judging what had been done earlier to be inaccurate'.

The running survey from Cape Leeuwin to Rottnest Island took just
four days. Baudin entered Géographe Bay, intending to anchor off the
beach where Vasse and the longboat had been lost almost two years
previously, but they passed it by in the evening mist. Beyond the shoreline the glow of native fires reminded all on board, according to Péron,
'of the fate of our compatriot who had been lost in the sea, and whose
name had been given in his memory to the river'. The ship was 3 leagues
past the anchorage before it was recognised; rather than retrace his
steps and lose perhaps a day in tacking against the wind, Baudin
continued on.

Next day the *Géographe* anchored in the eastern reaches of the bay;
here the entire coast was a continuous series of dunes, behind which
could be seen a range of mountains inland. Sent with a boat's crew to
examine an inlet 'that seemed to promise the entrance to a port',

Bonnefoi returned before sunset, reporting that it was suitable for small vessels only, and bringing back some new plants and casuarina branches for the kangaroos. The inlet was named after Leschenault, the botanist.

Running before a strong south-westerly breeze, they made rapid northing, and sighted Rottnest Island at dusk on 12 March. At daybreak Baudin headed into the anchorage between the island and the mainland, intending to await the *Casuarina*'s arrival, but to his surprise she was already there. 'This unexpected encounter decided me to go on, for the place, already explored by the *Naturaliste*, was not worth the trouble of stopping there.' He put down a boat so that Freycinet could report on his explorations, and explain how this latest separation had come about:

> I did not tell him of my resolution to remove him from [his] command, but confined myself to asking him why he had once again carried out my orders so badly, his disobedience having caused a fresh separation and the loss of the two days that I spent in waiting for him.
>
> Citizen Freycinet replied that he had himself been most astonished by this separation, for he had taken only five hours to explore the coast that I had indicated, and had immediately returned to join me. One must attribute it solely to the afternoon mist and the weather on the following day. This seemed fairly reasonable to me.[27]

With no reason to delay further, Baudin ordered Freycinet to follow him from the anchorage, and to keep in sight of the flagship. Though the winds were favourable the *Géographe* could take little advantage from them, 'being obliged to carry very little sail in order to wait for the *Casuarina*'. North of the Abrolhos Islands on the morning of 15 March they changed course, heading east of north to stand in for the land, planning 'to enter Shark Bay by the same pass as Dampier, and collect some turtles for the crew's food'. At 4 p.m. land was sighted from the masthead, stretching from east to north-east as far as the eye could see.

The giants of Shark Bay

At the anchorage in Shark Bay, Thursday, 17 March 1803.

> *In returning to these shores, we had as our main goal the capture of as many as possible of those great turtles which, at the time of our first sojourn, covered the vast sandbanks of Hamelin Pool. To this end, we came to occupy the* Naturaliste*'s anchorage in Dampier Bay, and on the morning*

of the 17th we dropped anchor there in fathoms, bottom of fine sand. [Péron and Freycinet, vol. II, unpublished translation]

The *Géographe* entered Shark Bay by the middle entrance first used by Dampier a century before. Worried by the shallows, Baudin put two boats down 'with orders to go on ahead of the ship and take soundings, one to port and the other to starboard'. He anchored in five fathoms off the northern tip of Dampier's 'Île du Milieu' (Péron Peninsula), still some 2 leagues offshore.

Freycinet left in the *Casuarina* to sound the waters north of the anchorage, with instructions to discover whether the northern channel leading out of the bay could be navigated safely. He spent five days on the task,

> fixing the positions of all the banks and determining all the passages, thus ... preparing for our expedition one of the most important charts with which it has enriched nautical science.[28]

Ransonnet, meanwhile, was dispatched with two boats' crews to the eastern shores to catch turtles, 'our stay in this bay having no other purpose than that'. He was also ordered to collect 'any objects of curiosity and Natural History' that he might happen to find, and hand them to Baudin upon his return. The ship's two dinghies were lowered, with four men in one and three in the other, 'to fish and find an easy landing place' on the peninsula; the men had orders to return at sunset.

The first dinghy returns a few hours later, fear etched on the men's faces. At least a hundred uncommonly tall, strong and robust savages, they gasp, had opposed their landing. When they attempted to step ashore these giants, their long black beards covering their chests, rushed towards them as though berserk, brandishing their spears and uttering loud, piercing cries. The fishermen had fled precipitately back to the boat.

Their comrades on the ship scoff at the unlikely tale, but fall silent when the second boat arrives in great haste, its crew also panic-stricken. They were already on the beach, they said, when the giants appeared and made to attack them; they count themselves lucky to be alive.

The commandant listens attentively to the sailors' garbled tale. He thinks it probable that the sudden appearance of the natives 'so scared our men, that they were unable to judge the number properly'; some putting it as high as 200, while those less frightened thought there were

thirty or forty. Hoping to make contact with this singular tribe, he or-
ders the longboat to be made ready for an excursion the following day.

Command of the expedition was given to Ronsard, the first time the engineer had led an armed party ashore. His mission, wrote Baudin, was

> to become acquainted with [the natives], using every means that may help
> you in this, whether the normal gestures for such situations, or the sight
> of the presents that you are taking to them.[29]

He and his men must take all possible care to avoid bloodshed; if the
need arose to open fire, only blanks were to be used. Should the natives launch an attack, he must re-embark 'in order to avoid being forced
to extremities harmful to [them]'.

From the ship the officers followed the longboat's every movement.
Ronsard and his men landed without difficulty, and disappeared into
the sparse scrub behind the beach. Baudin remained on tenterhooks:
the wait 'made the day seem very long, even though I spent the whole
of it on geographical work for drawing up our charts'.

The longboat returned at six o'clock, Ronsard reporting that his party
had seen none of the giants encountered the day before. They had, however, come across a small native village, made up of some twelve to
fifteen huts, in several of which fires were still burning. He had left some
presents behind, including beads, metal buttons, mirrors, and a few small
snuffboxes bearing the portrait of a negro, in the hope that these might
'perhaps persuade [the natives] to place more confidence in us'.

Next day Baudin again sent the longboat ashore, this time with Bonnefoi in command. His orders were to obtain salt for preserving the
large hauls of fish being caught in the bay, and to bring back fresh
plants and foliage to feed the animals aboard. He was to return no
later than the evening of the second day. Péron, Petit and Guichenot,
also in the party, were given the same instruction.

Petit's task is to sketch the natives' camp, and Guichenot's to gather new
plants to add to the Géographe*'s collections. Péron has other plans;*
encouraged by Dampier's account[30] of the wealth of curious and beau-
tiful shells found around the bay, he persuades his two companions to
abandon their assigned work and follow him to the sheltered sandbanks
on the peninsula's eastern shore; it is, he assures them, no more than a
league distant.

They set off at dawn, not telling Bonnefoi of their intentions, and with neither food nor water. They have weapons, two pistols and a musket, taken ashore in defiance of the commandant's orders. Hot and thirsty after four hours' march under a scorching sun, they reach the far coast about ten o'clock. With Péron in the lead, they wade out onto the flats, cooling themselves in the placid waters. It is enough to plunge their hands into the sand to amass a treasure-trove of the most beautiful shells, speckled with red and green and yellow; shoals of fish swim fearlessly in the shallows around them.

Abruptly the idyll is shattered. A large fish brushes Petit's leg; taking it for a shark he panics, pulling out his pistol and firing. The crack of the shot echoes up and down the deserted shore. Péron reacts at once, calling to his friends to leave the water and take refuge in the bush. He has noticed native tracks on the beach, and is sure the sound of gunfire will bring them back. Guichenot hurries to join him, but the foolhardy Petit stays in the water, jeering at their fears. In a moment his rashness turns to terror, as a band of naked warriors rush down a nearby dune and advance with menacing cries along the beach. He scrambles ashore, and the three Frenchmen prime their weapons, prepared to sell their lives dearly.

There is nothing extraordinary about these natives – they are of medium height, perhaps smaller than average. Realising that any retreat will embolden their attackers, Péron and his friends walk straight ahead, guns at the ready. The Aborigines hesitate, then turn their backs and walk slowly, and with no signs of fear, along the beach. The French follow at a short distance, suiting their pace to that of the natives and making no attempt to catch them. The latter scale the dune and turn towards their pursuers, with many signs and gestures beckoning the French to follow them. 'We bade them some sort of farewell,' Péron recalls later, 'and peacefully continued on our way.'

Their troubles were far from over. Loaded down with the rich collection of shells, they set off to return to the boat, but shortly lost their way in the dunes. The sun was now at its height, the heat, reflecting off the sand, was unbearable, and they had neither food nor water with them:

Our bodies ran with constant and excessive sweat; our weakness was soon at its height. In vain did we fill our mouths with small pebbles to induce a few drops of saliva – the supply appeared to have dried up. A feeling of desiccation ... and an unbearable bitterness made our breathing difficult and somehow painful; our trembling legs could no longer support us. One or other of us fell at every moment ...[31]

Returning to the shore, the exhausted men found some relief by submerging themselves in the sea: 'this salutary bath probably saved us from death', wrote Péron. Discarding the shells 'acquired at the cost of so much devotion and danger', along with their shoes and most of their clothes, they staggered on, dragging their feet through the water in an effort to keep cool.

At sunset a light breeze sprang up off the sea; they left the water to move a little faster along the beach, but could barely drag themselves from point to point. At last they made out a reddish glow in the darkness ahead – a great fire lit by their companions as a signal to guide their return. The sight gave them fresh heart, and they struggled on towards the encampment:

> At that moment our strength was at its lowest ebb; less than two hundred paces away we fell as though senseless on the sand. Our good friends hurried to our side; they lifted us up and supported us and, lighting several fires around us, managed to rekindle the flame of life that was at the point of extinction.[32]

Shark Bay – in parenthesis[33]

Extracts from François Péron's Voyage de Découvertes aux Terres Australes, *vol. II, 1816*[34]

[Our companions'] eagerness was all the greater for their having already given up hope of finding us. My friend, Mr [Bonnefoi de] Montbazin, had sent search parties in all directions, and the futility of these efforts had led him to believe that we had been killed by the natives. His generous friendship had not stopped there; he had refused to obey the thrice-repeated order that the Commander had given him to return aboard by firing the ship's cannon three times. The reasons for his disobedience were so pressing, that he had no doubt of its being approved by the Commander ...

No food or drink of any kind remained in the longboat; we had to spend the whole night stretched out on the sand in our clothes that were completely soaked in salt water. And to crown all, a thick mist rose next morning from the water's surface and prevented us (for lack of a compass!) from rejoining the ship until two o'clock in the afternoon. We then found ourselves reduced to the most deplorable state. We had had neither food nor drink for the last 48 hours and we had walked for fourteen of them. Pale and trembling, our eyes hollow, our countenances lifeless, we could scarcely stand. I could barely distinguish objects, my hearing was almost gone, and my withered tongue could not form words.

Upon seeing us in this weakened state, everybody was filled with compassion and concern; it was a matter of who could show us the kindest attention, the most affectionate solicitude. Only our Commander maintained towards us the attitude that he had thus far maintained towards all his miserable companions ... In vain did Mr Montbazin bring back twelve to fifteen hundred pounds of salt from an expedition which was not expected to yield ten pounds. The Commander made it a crime for him not to have abandoned all three of us *(these are his very words; he condemned him to pay ten francs for each cannon-shot fired for his recall ...). Wretched man! In order to save his life in Timor, I had shared with his doctor the small supply of excellent quinine that I was saving for myself ...*

Extracts from Captain Nicolas Baudin's Journal, *March 1803*

Sunday, 20 March:

At eight o'clock ... I gave the order for the firing of a cannon-shot to summon this boat back to the ship. At ten o'clock, seeing no sign of it, we fired a second one with no greater success, and so I lost hope of being able to make the tour I had planned to do myself the following day ... Since the absence of my longboat constitutes formal disobedience of the orders that the commanding officer received from me, I have had thirty francs added to his account to pay for the cannon-shots that he had obliged me to expend ...

Monday, 21 March:

By half-past one [p.m.] the longboat was back and moored alongside the ship. The officer in command informed me that he had 600 pounds of very good salt that had been collected from a salt-water pool, which, for the moment, was almost dried up ... I [then] asked him why he had not obeyed my orders to be back at sunset. He told me that he had found himself in a very awkward position through the absence of Citizens Péron, Guichenot and Petit, who had left him in the morning without telling him anything of their plans, and who had not returned in the afternoon. However, since they had set off with neither food nor water, he had presumed that they could not be far from the meeting-place, unless they had become lost.

When they had not returned by four o'clock, he had immediately dispatched four armed men who had searched for them in various directions until dark ... Citizen Bonnefoi, fearing my reproaches if he did not bring everyone back, yet knowing well that he would be reprimanded for not returning, chose to spend the night still waiting. Finally, at about nine o'clock, the three missing people appeared, exhausted and weak with hunger and thirst. Two of them lay down 12 to 15 feet from the tent and had to be carried to it.

... Citizen Péron told me that he was not in a state to be able to give an account of anything and begged me to allow him, before making his report, the long rest that he was undoubtedly in great need of, since he could hardly talk and remain standing ... This is the third escapade of this nature that our learned naturalist has been on, but it will also be the last, for he shall not go ashore again unless I myself am in the same boat. And the limits that I shall set to his excursions will not be broad enough to allow him to delay the boat's departure or to stray too far.

The north-west coast

Aboard the Géographe, *Tuesday, 22 March 1803.*

Few objects of Natural History were brought back from this excursion – merely a few common shells like those of which we already had large numbers. Nevertheless, they fetched good prices from our naturalists and scientists, who bought them from the sailors in return for rum; so that in the evening sixteen people were fighting in every corner of the ship and I was myself obliged to strike several of them in order to obtain peace. It was not the first time that such a disturbance had occurred, but I thought I had remedied the matter by prohibiting the payment of anything in rum. However, any prohibition is worthless for scientists like ours, and since they consider themselves independent, I gave a special order to the ship's steward not to deliver a drop of rum in future without an order signed by me ... [Baudin, 1974, pp. 510-511]

Ransonnet returned from his turtle-hunting expedition with only twelve, all but three weighing less than 40 pounds. The turtles had been difficult to catch, he reported to Baudin; the men had been obliged to take them in the water, going out over the shoals

> where they were found in depths of up to four feet. This made the operation very difficult and even dangerous, for one Lefevre was attacked by a large shark that he could not protect himself from until he had knocked it senseless and harpooned it.[35]

The excursion nonetheless provided the solution to a perennial puzzle in natural history. Among the shells and other objects brought back by Ransonnet's crew were several large teeth, taken from a long-dead animal found half-buried in the sand. They described its skin as being 'dark grey, rough to the touch and with a few traces of sparse hair of the same colour'. They took it to be a fish, but because the carcase was badly decomposed and gave off a foul smell, they had not attempted to see whether it had feet or flippers.

Baudin had always regarded Dampier's claim to have found part of the head of a hippopotamus in a shark's stomach as 'extraordinary, even improbable'. Now, listening with growing interest to the men's story, he had second thoughts:

> I thought I recognised in it a perfect resemblance to the hippopotamus, which I am well acquainted with, having killed several ... on the African coast. Above all, the position of its teeth seemed absolutely the same. The scientists of Europe will decide, after seeing them, whether this conjecture is valid or false.[36]

Before commenting in his journal, Baudin should perhaps have sought the views of his scientists. Péron had trained in anatomy with Professor Georges Cuvier in Paris, and later identified the teeth as those of a dugong – which tallies with the sailors' sketchy description of the rotting carcase.

The *Casuarina* returned from her survey of the bay's northern reaches on the afternoon of the 22nd; next morning the two vessels sailed through the Naturaliste Channel and headed for North-West Cape. Baudin's instructions to Freycinet were simple: 'keep near us overnight on the new course ... as it [will] be the only way to avoid separation'. Doubling the cape three days later, they changed course to the east, 'bearing north as much as necessary to sail along the coast'. On this second visit to the north-west, Baudin planned a more thorough survey than he had been able to make in 1801 – 'the least-advanced ... and the most poorly done' of his earlier work in New Holland.

For a thousand miles the two ships follow the treacherous coast, from
the North-West Cape to Cassini Island (the northernmost of the cluster
Freycinet later calls the Îles de l'Institut – each named after a member
of the Institut National). Much of it is a labyrinth of archipelagos, a
succession of shoals, islands, islets, rocks and coral reefs. Beyond lies
the mainland, remote and inaccessible, often difficult to distinguish
from the chain of offshore islands; to penetrate this barrier is at best
dangerous, and sometimes impossible, protected as most are by outlying
reefs, banks, and shallows. The old Dutch charts are worse than use-
less, showing only a continuous coastline with few islands; their 'pen-
insulas' mostly prove to be islands mistakenly linked to the mainland.

To Freycinet's frustration, Baudin keeps a prudent distance offshore
during the day – from 5 to 15 nautical miles – for fear of being carried
on to the shoals. At evening he anchors out to sea, rather than run the
risks of night navigation. The schooner's captain is impatient to utilise
his newly developed skills in inshore surveying:

> *Though I am a long way from land [he writes in his journal on 29 March]*
> *I have begun to survey it. I don't know the commandant's intentions ... If*
> *Mr Baudin wants to survey the coast we have had in view since [the 26th],*
> *it is quite ridiculous to keep at such a prodigious distance from it. How*
> *can one hope to fix the position or the form of the land when one can*
> *hardly see it? ... If Mr Baudin is frightened of compromising the safety of*
> *his ship by approaching the land too close why doesn't he make use of the*
> Casuarina, *whose shallow draught makes it suitable for navigating among*
> *the banks and in low soundings.*[37]

The ships came to anchor off Cassini Island on 24 April. Freycinet's survey work, carried out in difficult circumstances, had been competent,
and his charts contributed significantly to European knowledge of the
north-west coast when published a decade later. Some eighty placenames
given by the expedition are retained on modern maps of the region.

Baudin's health had visibly worsened during the passage up the west
coast; by 12 April he was so weak that he could barely stand:

> Apart from a general lassitude in all the limbs, I was suffering greatly in
> the chest, and a dry, almost continual cough decided me to try and get
> rid of it with the usual remedies ...[38]

It was a rare concession to his illness; a week later he was 'beginning
to improve', and resumed his duties.

Before sailing for Timor and a rest period for the crew, Baudin put down a boat to reconnoitre the island – the first landing since Shark Bay. He gave the command 'to the skipper alone, being very certain that he would carry out my orders much better than any of the officers, for whom such tasks are infinitely disagreeable'. Doubtless for the same reason the gardener Guichenot was the only naturalist allowed ashore. The decision angered Péron; no-one in the shore party, he wrote, had the education to profit from what took place.

The skipper, 'an intelligent man with a will to perform his duties thoroughly', returns next day with astonishing news. At sunrise his men had sighted four large proas, or canoes, leaving a sandy cove near where the boat was anchored. They set sail in pursuit, but with the aid of their paddles the proas moved much faster and were soon far away. Rounding the point of the cove, the Frenchmen came across two more proas preparing to sail. One was able to escape, but the second was stopped by the show his men made of firing upon it. When they came alongside, the boat's crew were surprised to see the five men manning it were Malays dressed in Timorese style; all were armed with a dagger and a kris. Several bows, 4 or 5 feet long, hung in the stern end of the canoe, with some wooden boxes containing arrows.

Though the Malays were frightened at first, the skipper continues, they soon put their fears to rest – one even climbed into the boat. His own men had no presents to give them, but the Malays exchanged some turtles' eggs for biscuit, and also offered his oarsmen fresh water, from jars like those found in Timor. Before leaving, they handed Guichenot a betel-leaf and some finely chopped tobacco.

Hoping to discover the reason for the Malays' presence, Baudin manned the boat with a fresh crew and instructed the petty officer in charge – Fortin, the yeoman of signals – to reconnoitre the island to which the proas had sailed. Freycinet followed in the *Casuarina* to protect the boat in case of attack. The wind died and Freycinet turned back, but the boat continued under oars. Suddenly Fortin and his men found themselves amid a flotilla of some two dozen proas. Perhaps they had a Malay-speaker with them, for they were well-received and brought back much useful information.

The flotilla was one of several to visit the coast each year in search of trepang, an aphrodisiac much prized by the Chinese. It was commanded by an old rajah, the only man in the fleet with a compass; the instrument, 'only two inches in diameter and extremely poorly made', was nevertheless 'adequate for him to direct the course of all the ships under his command'. More importantly, they learnt that fresh water could be found on several of the islands; there was also a small freshwater river on the mainland, but the natives there were very savage, and often attacked parties sent to procure some drinking water.

At this point Baudin seems to have fallen ill again and discontinued his daily journal entries, not resuming them until the expedition's departure from Timor in early June. Writing to the geographer Jean-Nicolas Buache from Kupang, he offered the encounter with the Malay flotilla as proof that the 'Grande Terre' discovered by seamen from the Moluccas at the start of the 16th century could only have been New Holland, and speculated that it was they who guided the early Dutch navigators to its shores.

Return to Kupang – the crocodile hunt

On board the Géographe *at Kupang, Saturday, 7 May 1803.*

> *After the customary salute, and while the crews were busy with the task of mooring our ships, the naturalists and officers of the expedition hastened with the Commander to carry out a formal duty: [to pay a call on] the current Governor of Kupang, Mr. Johannes Giesler … He told us of the death of the former Governor, Mr. Lofstett, who had been carried off in three days by an acute fever. Mr. Giesler had replaced him [and] was himself dangerously ill … The climate of this part of Timor is such that Europeans cannot survive there long … [Péron and Freycinet, vol. II, unpublished translation]*

The passage from North-West Cape had taken seven weeks. Though the ships had remained far offshore for much of the time, Baudin believed the expedition had corrected many faults in the Dutch charts; he was not to know that through his caution they had missed many of the great inlets along the coast – among them Exmouth Gulf, Roebuck Bay, and King Sound. Writing to the Minister of Marine from Timor, he stressed the hazards of navigation in the region: 'A proper exploration of this coast would take ten years, and even then I could not be sure it would be properly done in this time'.

From Governor Giesler, the French learnt of Flinders's visit just one month earlier. News of the *Investigator's* decayed state, and the sickness ravaging her crew, distressed Baudin, but he realised that now his rival would be in no position to replicate his work on the west and north-west coasts. Perhaps it was this, suggests Dr Horner, that prompted the French navigator to change his mind and return to Cassini Island, where he had abandoned his survey.

The Dutch again provided their visitors with comfortable lodgings ashore. Baudin moved into the former house of Governor Lofstett, while the naturalists were given rooms in Fort Concordia. Péron and Lesueur, 'having gathered a multitude of specimens of various zoological objects', decided to add the skeleton of a crocodile to their collections.

The two naturalists set off at daybreak, attended by five Malays on horseback and four on foot as guides and escorts. The Frenchmen are forced to ride bareback, for the pamalis *(a prohibition on the use of the saddle) operates between Kupang and Babau, their journey's end. The road passes through groves of coconut palms and tamarind trees, crosses a vast plain bordered to the north by a curtain of mountains, then skirts ancient mangrove swamps half-submerged by the sea. Moving inland they venture through sweet-scented woods that become denser and darker as they proceed. At one village an old man comes forward to offer them refreshments of coconut, milk and rice. The Malay escort, dressed only in gracefully draped loincloths, reminds Lesueur of 'the progress of those journeying patriarchs of which the Bible tells'.*

Reaching Babau towards evening, they alight at the house of the rajah to whom the governor has recommended them. He receives them courteously and promises his support for their crocodile hunt. But for the local Malays the animals are at once an object of veneration and of terror; they quiver with horror at the request, and not even the rajah's authority can prevail over superstition and fear. The most the two Frenchmen can obtain is two guides who will go with them to point out the monsters' usual retreats.

At first light the four men move off towards a vast swamp penetrated by several deep rivers; here at the slightest false step 'one's leg sinks up to the knee ... in the mire'. Searching among muddied lagoons for crocodiles, they are often threatened with abandonment by their guides. At length they see a crocodile about twenty-five paces off, partly submerged and apparently asleep. Lesueur, a good shot, fires in such a way as to

*break its backbone. Unable to throw itself into the water the animal
thrashes about with fury, bleeding freely. Certain that it cannot escape,
the hunters put off the task of skinning it to the following day.*

*Returning to the spot with a dozen Malays they have engaged as por-
ters, Péron and Lesueur stand waist-deep in water to dissect the 10-foot-
long crocodile. The Malays watch fearfully from a distance; at first they
refuse to touch any part of the remains, but are finally persuaded with
fresh promises to carry the load strung from two long bamboo poles
resting on their shoulders.*

*They reach Babau at 4 p.m., having marched through the heat of
the day. The rajah greets them with his usual kindness, but will not allow
them to come close until they have been ritually cleansed. With the
Malays gathered around, the two Frenchmen undress and are doused
twenty times in a trough cut from a tree trunk; their clothes are washed
in a nearby spring. Only then are they invited to attend a feast and
dance given by the rajah in their honour. The celebrations last far into
the night.*

*Their journey back to the ship begins before dawn. The horse carry-
ing the remains is led by a slave, holding a rope some 60 feet long for
fear of contamination; similarly the Malays along the route flee into
the woods at the approach of the procession. Exhausted by weariness
and heat, they reach Kupang in the afternoon. Despite all their precautions the crocodile's skin has begun to decompose, and is thrown
into the sea, leaving its skeleton as the naturalists' only reward for the
expedition. They set about cleaning the various parts before sending
them aboard.*

On this second visit to Timor the expedition stayed only four weeks.
Again the visitors enjoyed 'the most delicious and varied fruits and the
most wholesome animals', but again the unhealthy climate served 'as
a reminder of the miseries from which man is inseparable, and com-
pensates with actual diseases for so many riches ...'. Indeed, the French
received a clear warning within a week of their arrival when an American three-master, the *Hunter*, was sighted flying distress signals at the
entrance to the bay. The *Géographe*'s longboat, sent to offer assistance,
helped bring her to the mooring place.

The ship was from Dili, in Portuguese East Timor, and carried a cargo
of wax and sandalwood. Her crew were suffering from 'the most frightful epidemics, dysentery and fever, and reduced to three men in a state

to work'. The others were dead or dangerously ill, including the captain, who was at the point of death. Dr Lharidon and his colleagues had the sick men brought ashore, but their treatments proved useless 'against a sickness which, once it has developed to a certain point, is irrevocably fatal'. In the end the ship was abandoned in the roadstead.

Crew members of both ships came down with dysentery, but it had yet to reach 'that frightening aspect of malignancy' that had proved so fatal on their earlier visit. Among the scientists, the botanist Leschenault, the artist Petit, and Bernier the astronomer all fell seriously ill. Leschenault declared himself unfit to continue, and disembarked.[39] Bernier decided to follow him, but at the last moment, believing that 'his astronomical observations were of the greatest importance, determined to follow the fortunes of the voyage'. He died at sea three days after they sailed.

Baudin grows weaker by the day. He is worn out with fatigue and again spitting blood, bringing up sputum so thick 'that one would have said that it was pieces of lung coming away from my body'. Confronting at last the seriousness of his illness, on 13 May he writes to his two potential successors, lieutenants Ronsard and Henri de Freycinet, that the time has arrived for a decision that

> *I would rather had not been taken until after my death. Tomorrow ... I shall assemble the crew, in order that they should tell me which of you they prefer to have as leader, and their choice will be irrevocable, for it is appropriate that men should be commanded by those who suit them best and in whom they have most confidence.[40]*

The vote runs sixty to twelve in Freycinet's favour.

Before he sails from Kupang, Baudin faces another hard decision. Whatever his motives for bringing Mary Beckwith aboard, he can no longer tolerate her presence on the ship. During the voyage her affairs with several members of the crew – including perhaps his staff – have not only affected shipboard discipline but also, he believes, his health. As the ships make ready to sail, he confronts her at his residence in the town, telling her she must remain behind. Lesueur describes the scene in his journal:

> *In no way would she agree to this, but rushed off headlong like a fury; she ran toward the bridge crossing the little river in Kupang, and threatened to throw herself from it. Reassured by two blacks sent after her by the*

Commandant, she allowed herself to be brought back to him, and it was agreed between them that she could continue …

The same evening she was embarked, but meanwhile she had drowned her sorrows in drink, and had to be carried to the boat. She was in a truly horrible state; distraught, her hair dishevelled, her clothing in disorder, out of her mind … They had to haul her up [over the side], and she was carried to the Commandant's cabin in front of all the crew.[41]

The last campaign

On board the Casuarina, *Friday, 3 June 1803. Lieutenant Louis de Freycinet records his departure from Kupang roads:*

> *At nine o'clock in the morning, as soon as the breeze had sprung up, the Commander gave the signal to weigh anchor and then set sail an instant later. The* Casuarina *was preparing to follow him, when four of the seven pigs we had taken aboard jumped into the water and swam towards land. This loss of the greater part of my refreshments was a heavy blow, but nothing could be done about it. It was impossible for me to delay any longer; the* Géographe, *moving off under full sail, repeated the formal order to weigh anchor with all speed, and so I quickly followed her. [Péron and Freycinet, vol. II, unpublished translation]*

The commandant had no wish to prolong his stay at Kupang. It was no place for sick men – venereal diseases as well as dysentery and fevers were making inroads among the crew. Already six seamen had deserted; four were discovered in a hurried last-minute search, but seeing no possibility of finding the other two he left them to their fate.

After another night of spitting blood, Baudin was woken at 5 a.m. on 6 June and told of Bernier's death. The news saddened him deeply, for 'of all the scientists given me, it was he who worked the most, whether to acquire knowledge or to carry out the government's design'. Unlike his colleagues, the young astronomer (he was only twenty-three) regarded science as 'both his work and his pleasure, and [he] turned it to account by assiduous and unceasing labour'. At ten o'clock the commandant and his staff 'paid their last respects and his body was cast into the sea'.

Baudin resumed the survey of the north-west coast at the point where it had been broken off in late April, ignoring advice from other seamen that 'it would be impossible to make easting with the South-East monsoon' at this time of year. Sighting the coast on 12 June, the two ships set their course eastwards; the *Casuarina*, 'heavily laden and encumbered', lagged behind.

Slowly and with difficulty the two vessels work their way into the vast Joseph Bonaparte Gulf. Light breezes alternating with calms hold them back, the Géographe *constantly forced to shorten sail so as not to lose her consort. The Dutch charts are as useless here as on the north-west coast. Detailed surveys are impossible; offshore shoals and contrary winds and currents keep them far out to sea. In the sick bay the numbers increase daily 'without our knowing why'.*

Following four days of unprofitable tacking, Baudin heads north-east across the gulf; to the south, unseen and unsuspected, lies the mouth of the great Victoria River. The ships make landfall on the far side at Cape Dombey, then follow the coastline northwards to Péron Island; now the winds and currents carry them north-west away from the land. The weather turns threatening, with drenching rain and squally easterly winds. Worried by irregular soundings, Baudin keeps the deck for most of the night of 26 June; he is bone weary, and again spitting blood.

Next day they double Cape Van Diemen, on Melville Island, and confront the Arafura Sea to the east. The weather resembles 'the worst winter days in Europe ... full of mist and big dark clouds, black and dense'. Baudin's goal is the Gulf of Carpentaria, but the monsoonal winds are too great an obstacle; he sets course for Cape Vals in New Guinea, intending to sail south-east from there to the gulf. He drives his wasted body as hard as his exhausted crew, but for six days they make almost no easting against the monsoon. He holds the poor sailing qualities of the Casuarina *largely to blame:*

> *This ship helped in no small way to cause us not only a strong drift (with the short sail we had to carry to wait for her), but also [a loss of time] through the westing that we had to make each night to rejoin her, and which I estimate to be 2 leagues a day.*[42]

The crews are at breaking point. On the Géographe *twenty men are in the sick bay, with many others barely fit to work the ship. Shaken by the continuous heavy swell, the animals refuse to eat, and seem likely to die from hunger. The emus are force-fed with pellets of rice mash, the kangaroos kept alive with wine and sugar. On 2 July Baudin writes:*

Top: 'Study ... A View ...
Island of Timor', by William
Westall, circa 1808. [By
permission of the National
Library of Australia]

Above: A Malay horseman
on Timor, by N.-M. Petit.
[Collection Lesueur, Muséum
d'Histoire Naturelle, Le Havre]

Top: 'Wreck of the Porpoise,
Flinders Expedition', by
William Westall, circa 1803.
[By permission of the National
Library of Australia]

Above: 'View of Wreck Reef
bank taken at low water, Terra
Australis', by Westall, circa
1803. [Rex Nan Kivell
Collection; by permission of
the National Library of
Australia]

'if the weather does not turn fine after the full moon, I have decided to make for Isle de France rather than lose them all'.

Five days later the decision can no longer be deferred. Most of the surviving animals are on the brink of death, the living plants too are endangered by the shortage of fresh water. Baudin takes to his bunk, overcome by another haemorrhage. Shortly before midnight on the 7th, after waiting two hours for Freycinet to rejoin him, he gives the order to turn back and set course for Isle de France:

> It was not without regret that I decided upon this step. A thousand rea-
> sons should even have made me take it earlier; but without listing them
> all, I shall limit myself to saying that we no longer had anything more
> than a month's supply of biscuit, at the rate of 6 ounces per man, and two
> months' of water, as a result of the birds' and quadrupeds' consumption
> of their supply. Twenty men were ill: several with dysentery, and others
> unfit for duty because of severe venereal diseases contracted at Timor.
> Nobody to replace me ...[43]

The commandant's decision, for which, says Louis de Freycinet, 'each of us had long been yearning ... produced a joy that was as keen as it was natural'. Despite the stormy weather, many of the men spent the entire night on deck, and 'went frequently to consult the compass for fear they had not heard aright'. With the easterlies now favouring them the two ships made good speed, and soon the high peaks of Timor were in view. On 13 July they passed through the Roti Strait, south of Kupang, for the last time, but Baudin, fearing for the animals, did not revisit the port. Two of the four emus died before they reached Isle de France.

'On the journey home,' wrote the French historians Bouvier and May-nial,[44] 'Baudin was no more than the shadow of a captain on a ghost ship.' Daily more distressed by his illness, and as surely depressed by his failure to complete the mission, Baudin kept to his cabin. On 30 July he wrote to the expeditions' officers and scientists, directing them to hand in all their journals, memoirs and other papers for delivery under seal to the Minister of Marine in Paris.

This formality was carried out on 4 August. Two days later the *Géographe* docked in Port North-West (the former Port Louis) at Isle de France. The *Casuarina*, buffeted by fierce gales in her crossing of the Indian Ocean, arrived a week later and anchored alongside. On the 29th her crew were paid off and transferred to the *Géographe*.

Notes

1. Baudin, 1974, p. 438.
2. Horner, 1987, p. 264.
3. Baudin, op. cit., p. 441.
4. Horner, op. cit., p. 264.
5. Baudin, op. cit., p. 442.
6. Ibid., 446.
7. *Historical Records of New South Wales,* vol. IV, p. 1009.
8. Turnbull, 1813, p. 487.
9. Baudin, op. cit., p. 460.
10. On Louis de Freycinet's maps, published a decade later, the island is named Île Decrès in honour of Admiral Denis Decrès, Napoleon's Minister of Marine.
11. Péron and Freycinet, vol. II, unpublished translation by C. Cornell.
12. The French at the time were unaware of Flinders's nomenclature (with a few exceptions such as Kangaroo Island). They assigned their own names to the two gulfs – Bonaparte (Spencer) and Josephine (St Vincent) – and other features on the South Australian coast. For ease of reference, Flinders's names have been used.
13. Ibid.
14. Ibid.
15. Ibid.
16. Ibid.
17. Ibid.
18. Somverville, 1947.
19. Baudin, op. cit., p. 483.
20. Péron and Freycinet, op. cit..
21. Baudin, op. cit., p. 486-487.
22. Ibid., p. 491.
23. Ibid., p. 492.
24. Ibid., pp. 492-493.
25. Ibid., p. 499.
26. Probably the point seen in 1801 was Cape Hamelin.
27. Ibid., p. 503.
28. Péron and Freycinet, op. cit..
29. Baudin, op. cit., p. 506.
30. Dampier, 1981.
31. Péron and Freycinet, op. cit..

32. Ibid.

33. Péron wrote his version of events back in France with Baudin's *Journal* in front of him; the translation by C. Cornell of volume II of his account of the voyage remains unpublished. Baudin was writing at the time, and died six months later; his *Journal* was first published in 1974 (in English translation), and has yet to appear in French.

34. Péron and Freycinet, op. cit.

35. Baudin, op. cit., p. 510.

36. Ibid., p. 513.

37. Horner, op. cit., p. 301.

38. Baudin, op. cit., p. 532.

39. Leschenault remained for some time at Kupang before moving on to Java, where he botanised extensively. He did not return to France with his collections until 1807.

40. Horner, op. cit., p. 309.

41. Collection Lesueur, Muséum d'Histoire Naturelle, Le Havre [no. 17076-1, p. 62].

42. Baudin, op. cit., p. 558.

43. Ibid., p. 560.

44. Bouvier and Maynial, 1947, p. 210.

⚓

CHAPTER TEN

Nemesis – Isle de France:
Death, Detention, Deliverance, 1803 to 1810

'M. Baudin ceased to exist'

Port North-West,[1] Isle de France, Saturday, 17 September 1803.

> *At the time of abandoning the exploration of the southern lands our Commander was dangerously ill. From the time of our arrival in the colony his condition had grown much worse. All hope of recovery had already been long since lost, and the doctors' efforts had as their sole aim the prolongation, by a few days, of a life whose end had been decided by the very nature of the illness. This last moment finally arrived; and on 16th September, 1803, around the middle of the day, M. Baudin ceased to exist.*
>
> *He was buried on the 17th, with all the honours due to his rank in the Navy. All the officers and the scientists from the expedition took part in the funeral procession, at which the principal authorities in the colony were also present. [Péron and Freycinet, vol. II, unpublished translation]*

The *Géographe* reached Port-North West, the principal harbour of Isle de France, on 7 August 1803; she was joined five days later by the *Casuarina*, which had become separated during a fierce gale while crossing the Indian Ocean. On arrival, Louis de Freycinet reported the loss of the master sail-maker, carried overboard by a sudden lurch of the sloop in the gale:

> He had been on deck, attending to the ship's handling. I immediately had the life-buoys thrown out, and since the state of the wind and sea prevented me from reaching the place where the unfortunate fellow had fallen, I went about and positioned myself to leeward of him. If this man had known how to swim he could easily have returned to us; but he did not, and we soon saw him raise his hands to heaven and then sink.[2]

After an absence of more than two years the men of both ships were overjoyed to be reunited with friends and compatriots at Isle de France.

Letters from families and friends in Europe awaited them, adding to their happiness. 'This alternating of sorrow with joy, weariness with repose and scarcity with abundance, to which seamen are so often exposed,' mused Louis de Freycinet, 'must accustom them to withstand more easily than other men the vicissitudes of life.'

Here at last they received news of their friends in the *Naturaliste*. Hamelin had been forced to call at the island to obtain medical help for his many sick men, as well as fresh food and water for the animals on board. In a letter left for Baudin, he reported that only one of the latter had died – a Samoan turtle; the trees and plants were also mostly doing well. The mineralogist Depuch, left on the island by Hamelin in the last stages of dysentery, had died soon after – 'a touching victim of our common disasters', wrote Freycinet.

The health of both crews, the care of the live animals, birds and plants carried on the Géographe, *repairs to the ship for her voyage back to France, and the replacement of provisions and supplies all demand attention. The most pressing need is the first, and the sick men are immediately sent to the town's hospital – though for a few it is too late.*

Baudin is terminally ill, but does not admit it. He takes lodgings with his friend Mme Kerivel, a widow, and arranges for the animals and plants to be cared for in her gardens until he is ready to sail in December. He writes to the Minister that the voyage will be 'full of honour for the French people', and to Governor King at Sydney trusting they will always remain friends. In a month he is dead.

Few of his staff show any distress or regret at the commandant's death. 'His funeral,' wrote Charles Baudin decades later, 'was a dismal event; he was universally detested.' Only Capitaine de frégate *Milius acknowledges Baudin's unswerving dedication to his task: 'The Commandant was so determined that he resolved … to sacrifice all the time necessary and even his life to fulfil entirely the object of his mission.'*

The records make no mention of the young convict girl Baudin had taken on board his ship at Port Jackson. Doubtless Mary would not have been allowed to attend the ceremony, but one likes to think she may have witnessed it at a distance. Afterwards, perhaps, she went into service with Mme Kerivel or with one of Baudin's friends or relatives on the island.

The ship's officers expected that command of the *Géographe* would pass to the first lieutenant, Henri de Freycinet: 'his labours throughout the voyage with regard to geography and astronomical observations', wrote his brother Louis, 'seemed to strengthen the rights that he had as a naval officer'. Since their arrival in the colony, however, the political and military situation had changed significantly. Within a week, Admiral Linois's squadron, including the battleship *Marengo*, three frigates, a corvette and two transports, had sailed into the port, carrying on board the colony's new governor, General Decaen, and 800 troops. The admiral, exercising his authority as senior naval officer, appointed *Capitaine de frégate* Pierre-Bernard Milius, Hamelin's former first lieutenant, as the expedition's new commandant.

With Baudin's consent, Milius had left the expedition at Port Jackson due to continuing ill health, and taken passage on an American ship to Portuguese Macao, on the China coast; there he found a merchant ship bound for Isle de France. Linois, however, considered him still a member of the expedition's staff; 'from then on', noted the younger Freycinet, 'his superiority in rank assured him of the command'. Linois had judged well – Milius had proved a very capable deck officer on the outward voyage before his illness, and Baudin had promoted him at Timor.

Milius received his appointment as commandant on 29 September, and immediately boarded the *Géographe* to carry out a thorough examination of the ship and its crew, cargo, and accounts. The accounts and the ship's muster roll were in a state of disorder, he reported to the shore authorities. His first concern, however, was for the well-being of the men, still restricted to shipboard rations of biscuit and salt meat – fresh food was officially reserved 'to Admiral Linois' division, [and] my men would have continued eating salt meat, if I had not had recourse to other means of procuring fresh meat and vegetables'. Next he proceeded with an inventory of all items belonging to the government:

> All natural history objects gathered and cared for by M. Baudin, both for the collection of Mme. Bonaparte and for the Museum, were placed by my orders in the care of M. Péron.[3]

Like Baudin before him, Milius found himself faced with the ill will of the local administration. Throughout the refitting and reprovisioning of the ship for her voyage to France 'obstacles without number' were placed in his way by the colonial prefect:

> All my transactions were delayed or otherwise impeded under various pretexts. My requests were always badly received, and in the end I found it necessary to ask General Decaen for his protection.[4]

By mid-December the *Géographe* was at last ready for sea – remasted and rerigged, her masts repositioned to improve her steering. The animals and plants all stowed safely aboard, Milius sailed on 16 December.

The captain general

Paris, June 1802. General Decaen writes to First Consul Bonaparte:

> *Even if it were necessary to spend ten years of my life awaiting a favourable opportunity of acting against the English, whom I detest because of the injury they have done to my country, I would undertake that task with the utmost satisfaction.*

Paris, January 1803. The first consul's instructions to General Decaen:

> *Aided by the information received as well as by the exact observance of instructions pertaining to military and political affairs, I will some day be able to place the Captain-General on the road to fame outlasting the memories of men and the passing of centuries. [H. Ly-Tio-Fane Pineo, 1988, pp. 45-49]*

At thirty-three, General Charles-Mathieu-Isidore Decaen was the same age as Napoleon Bonaparte. His exploits on the European battlefields had brought him to the notice of the first consul, who, despite his inglorious exit from Egypt, still nursed ambitions to be the new Alexander and conqueror of the East. He saw in the former *général de brigade* his chosen instrument for extending French power on the subcontinent and throughout the Indian Ocean, and in June 1802, during the Peace of Amiens, appointed Decaen captain general of all French territories east of the Cape of Good Hope. His headquarters were to be at Pondicherry, one of the French settlements then under British occupation that were to be returned to France by the terms of the peace.

Decaen's promotion did not unduly worry the British government at the time. 'He is a young man,' reported Lord Whitworth, British ambassador in Paris,

> and bears a very fair character in private life, but possesses no very shining talents either as a general or a statesman. We may therefore conclude that, as far as he is concerned, it is intended rather to improve what possessions they already have in India than to extend them by conquest or intrigue.[5]

His Lordship had clearly been misinformed.

The appointment was not so well-received in Paris. Admiral Denis Decrès, Minister of Marine and Colonies, took it as a personal affront, an infringement of his ministerial prerogatives. The first consul might select colonial governors and issue instructions, but it was the Minister who implemented them, and Decrès – himself a survivor of the ill-fated Egyptian expedition – made skilful use of the bureaucratic machinery to impede the project. So successful were his delaying tactics that, despite Decaen's appeals to Bonaparte, the squadron of six ships under the command of Rear Admiral Charles Linois did not sail until March 1803 – nine months after the general's appointment and only two months before the final collapse of the peace.

In a further blow to Decaen's authority, Decrès separated the naval command from the military – perhaps to ensure that the more cautious Linois was not forced to hazard his irreplaceable squadron at the direction of the aggressive general. As a result the passage out was marked by increasingly bitter quarrels between the two, jeopardising the chances of success from the start.

Linois's squadron carrying Decaen and his troops reached Pondicherry in July. The general was incensed to find British troops still in occupation; even as negotiations began for the transfer of power a superior British fleet, with four ships of the line and five frigates, sailed into the harbour and anchored nearby. To the suspicious French it bore all the signs of a carefully engineered British trap. Linois and Decaen, for once forgetting their differences, managed to extricate the ships overnight (although forced to abandon a company of French troops already on shore) and sailed for Isle de France, arriving at Port North-West on 16 August; Decaen landed at the head of his troops the following day. It was not until the end of September, however, when dispatches from Paris confirmed the renewal of hostilities with Great Britain and his appointment as captain general of Isle de France and Réunion, that Decaen formally took over from Governor Magallon.

At once the general set about reorganising the colony's administration on quasi-military lines, concentrating executive power in his hands. The civilian Colonial Assembly was disbanded and new institutions of government set up in its place. The duties of mayors were downgraded, existing justice tribunals abolished, and the island divided into *quartiers*, each with a military commander. Isle de France would become the centre of action for France's war against Great Britain in the East, and the captain general would be the mastermind directing operations.

The general's forced retreat from Pondicherry has further embittered him against the British. 'Les Anglais sont les tyrans des Indes,' Bonaparte had written, and the first consul's orders are that Decaen should support the princes and peoples of the subcontinent in their struggle against oppression. Already his agents are at work seeking to build an alliance with the Maratha princes; they carry a message promising, in General Bonaparte's name, French support if they rise against the common enemy. Success depends on an expeditionary force from France coming to their aid.

The first consul must be informed directly of the situation in India. Here the captain general faces a quandary – a single ship risks capture by the Royal Navy, while Linois's squadron is sailing for Batavia to lie in wait for the East Indiamen of the China tea fleet (a potentially devastating blow against Britain's eastern trade). Fortunately a simple and safe alternative is close at hand – the discovery ship Géographe is about to sail for France; moreover, she is protected by a British safe-conduct.

General Decaen calls in her commander, Captain Milius, who briefs him on the military capabilities of the British colony of New South Wales and the defences of Port Jackson. The naturalist François Péron provides a lengthy report on the English establishment at Port Jackson – 'this freshly set trap of a great Power' that will shortly 'extend over the continent of New Holland, Van Diemen's Land, New Zealand, and the numerous archipelagos of the Pacific Ocean'. Péron adds that the English navigator Flinders is at present on 'an expedition of discovery which is calculated to last for five years', and is at the present moment 'traversing the region under discussion'.

When Milius sails on 16 December, he carries a new 'geographer' on his staff – Major Barois, Decaen's brother-in-law and aide-de-camp, who carries dispatches for personal delivery to the first consul. The same day, the general receives word from Commandant Bolger at the harbour of Baie du Cap, on the island's south coast, that an armed English vessel has chased a colonial vessel into the anchorage. Her commander claims to be the navigator and explorer Captain Flinders, and has requested assistance to continue his voyage.

'You are imposing on me, sir!'

Aboard the Cumberland, *Port North-West, Saturday, 17 December 1803.*

At four in the afternoon ... we got to an anchor at the entrance of Port Louis, near the ship which I had hoped might be the Géographe; *but Captain Melius [sic] had sailed for France on the preceding day, and this proved to be the* L'Atalante *frigate.*

The peculiarity of my situation, arising from the renewal of war and neglect in the passport to provide for any accident happening to the Investigator, *rendered great precaution necessary in my proceedings; and to remove as much possible, any doubts or misconceptions, I determined to go immediately with my passport and commission to the French governor, and request his leave to get the necessary reparations made to the schooner; but learning from the pilot that it was a regulation of the port for no person to land before the vessel had been visited by the officer of health, it was complied with. [Flinders, 1814, vol. II, p. 359]*

Flinders sighted the jagged peaks of Mauritius at first light on 15 December. Wind and tide drove the *Cumberland* southwards, and by mid-morning she had doubled the south-east point and was running along the south coast, in clear view of the shore. He carried no charts of the island, and followed a local schooner into a small harbour at the Baie du Cap (now Cape Bay), hoping to find a pilot to take him to Port Louis. To his astonishment the crew abandoned the vessel and hastily clambered up a steep hill behind the anchorage; shortly after, several armed men appeared on the hilltop, leading him 'to apprehend that England and France [must be] either at war or very near it'.

Two French officers, Major Dunienville and the area commandant, Etienne Bolger, came aboard and confirmed that the war had been resumed. They inspected Flinders's commission and passport, and after arguing between themselves agreed to provide a pilot to take the ship round to Port North-West (the former Port Louis) next day. When they left, Flinders sat down to study his passport closely – the first time he had done so, he wrote later. Its 'general purport' had been explained to him in England, but since then he'd scarcely glanced at it. He knew little French, but with the aid of a dictionary made out that it seemed to be solely for the *Investigator*. 'The intention,' he wrote in the *Voyage,*

no doubt was to protect the voyage generally ... but it appeared that if the governor of Mauritius should adhere to the letter of the passport and disregard the intention, he might seize the *Cumberland* as a prize.[6]

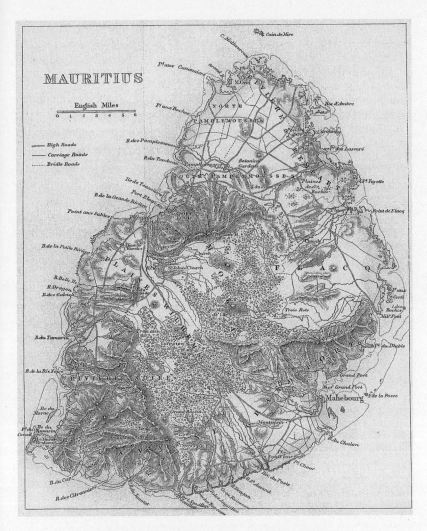

A 19th-century map of Mauritius, or Isle de France.
[Flemying, Mauritius, 1862, from the Royal Geographical
Society of South Australia Library]

Flinders considered whether to sail at once for the Cape of Good Hope, but decided against it – the schooner was unseaworthy; the Dutch, allies of the French, again held the Cape, and he had no passport from the Netherlands; whereas in support of his French passport he had a letter from Governor King for Governor Magallon. Besides, the French officers had told him that the *Géographe* was still at Port North-West, although expected to sail for France any day. And although Captain Baudin was dead, his successor was Captain Milius, whom he had known well at Port Jackson. Dismissing his concerns about the passport as groundless, he

> determined to rest confident in the assurance ... that the conduct of a governor appointed by the first consul Bonaparte, a professed patron of science, would hardly be less liberal than that of two preceding French governments ... and to banish all apprehension as derogatory to the governor of Mauritius and to the character of the French nation.[7]

The *Cumberland* arrived off Port North-West on the afternoon of 17 December. The mastheads of a large ship seen within the harbour proved to be those of the frigate *Atalante* and not the *Géographe*, which Flinders learnt had sailed the previous day. He was not worried by the news that Governor Magallon had been replaced by a General Decaen. Knowing nothing of the man he was about to meet, Flinders went ashore to present his passport and commission to the new governor.

Arriving at Government House about six o'clock, he finds the captain general is at dinner. Told he must wait for an hour or more, Flinders joins several officers lounging in the square outside. Some of them speak a little English, and by way of passing the time,

> *they asked if I had really come from Botany Bay in that little vessel ... and if I had not sent a boat on shore in the night? Others asked questions of M. Baudin's conduct at Port Jackson, and of the British colony there; and also concerning the voyage of M. Flinedare, of which, to their surprise, I knew nothing; but afterwards found it to be my name which they so pronounced.[8]*

At seven Flinders returns to Government House with the Frenchmen. His companions are called in to report on their conversations with him, and he is kept waiting a further half hour. Far from well (he is still recovering from a bout of fever), tired, and thirsty, his resentment grows.

Ushered at last into the captain general's office, he finds himself facing two senior officers – one, a shortish thickset man, is obviously General Decaen; the other is taller and more genteel-looking, and 'his blood seemed to circulate more tranquilly'; this is the general's aide-de-camp, Colonel Monistrol. Both men remove their hats to receive him.

Stepping forward to introduce himself, Flinders, perhaps unthinkingly, keeps his own hat on, and is reminded by the interpreter to remove it. The general's temper flares, and he never forgets the slight. He demands his visitor's papers, then fires a volley of questions at him:

'Why has Captain Flinders undertaken a voyage in such a small vessel?'
'What is his motive for coming to Isle de France?'
'To what place does Captain Flinders intend to go to from this island?'
'What reason led him to chase a boat in sight of the island?'
'Was he informed of the war?' and so forth.[9]

Decaen, with his experience in military intelligence, found Flinders's story implausible. What experienced navigator would choose to sail from Port Jackson to England, or even the Cape, in such a decrepit vessel of only 30 tons? Had not the *Géographe*'s officers told him this same schooner, the *Cumberland*, commanded by an arrogant young officer, Lieutenant Robbins, had spied on them at King Island just a year ago? The naturalist Péron had said the explorer Flinders was voyaging in the Pacific. The man before him must be an impostor, perhaps Robbins himself: 'You are imposing on me, sir! It is not probable that the governor of New South Wales should send away the commander of an expedition on discovery in so small a vessel!'

Flinders was escorted back to the ship by two French officers, who informed him they had orders to impound his books and papers. Fatigued, and in a foul temper, he made no attempt to hide his feelings about his treatment, adding 'that the Captain-General's conduct must alter very much before I should pay him a second visit, or even set my foot on shore again'. They replied, apologetically, that he had no choice – he was under arrest.

The Café Marengo

Port North-West, Sunday, 18 December 1803.

We were conducted to a large house in the middle of the town, and through a long dark entry, up a dirty staircase, into the room destined for us; the

aide-de-camp and interpreter then wished us a good night, and we after-
wards heard nothing save the measured steps of a sentinel, walking in
the gallery before our door … It seemed to me a wiser plan to leave the
circumstances to develop themselves, rather than to fatigue ourselves with
uncertain conjectures; therefore, telling Mr. Aken we should probably know
the truth soon enough, I stripped and got into bed; but between the mus-
ketoes [sic] above and bugs below, and the novelty of our situation, it was
near daybreak before either of us dropped asleep. [Flinders, 1814, vol. II,
p. 362]

Flinders and the master, John Aken, were taken to lodgings ashore,
the officers explaining that their stay would probably be for a few days
only, while the ship's papers were examined. In the meantime, the
general had said they should want for nothing. Flinders woke at dawn,
to find an armed grenadier pacing up and down between their beds.
Unable to sleep in such company, he woke his companion, 'who, seeing the grenadier and not at first recollecting our situation, answered
me in a manner that would have diverted me at any other time'.

In the darkness, they had taken their room for 'one of the better
apartments of a common prison'. Now, peering through the grimy window, they found themselves in a tavern, 'though a very dirty one', dignified with the name Café Marengo. Breakfast was brought at eight,
followed by dinner at twelve, and they both ate heartily; good bread,
fresh meat, fruit and vegetables were rarities after shipwreck and the
months at sea.

In the afternoon, Flinders accompanied Colonel Monistrol to Government House, where a clerk repeated Decaen's questions of the night
before, recording his answers in French and English. He also dictated
a full account of the shipwreck and of the *Cumberland*'s voyage, and
signed a declaration that he had no knowledge of the contents of the
dispatches he was carrying for Governor King. Worn out after five hours
of interrogation, he was about to return to the café when Monistrol re-
entered the room.

The colonel brings an invitation for the prisoner to join His Excellency
and Mme Decaen for dinner, which is about to be served. Incredulous,
Flinders refuses: 'the invitation accorded so little with my previous treat-
ment, that I thought it to be a piece of mockery, and answered that I
had already dined'. Monistrol presses him to accept, or at the least to go

to the table. Flinders, feeling 'grossly insulted both in my public and private character' by the events of the past twenty-four hours, is in no mood to listen:

> *My reply was, that 'under my present situation and treatment, it was impossible; when they should be changed – when I should be at liberty, if His Excellency thought proper to invite me, I should be flattered by it, and accept his invitation with pleasure.'* [10]

The general has planned (so he later claims) to discuss his visitor's plight in pleasant surroundings over dinner, which doubtless 'would have brought about a change favourable to his position' – his interrogation of the young man has convinced him that he is indeed the navigator he claims to be. Instead, his conciliatory gesture is met with this calculated insult – directed at his wife as much as himself. 'Tell him,' he orders Monistrol icily, 'that I will invite him when he is set at liberty!'

The privations of the past eighteen months had, it must be assumed, affected Flinders's judgement. Ill and confused, almost paranoid, he saw the invitation as a trap, an experiment 'to ascertain whether I were really a commander in the British Navy; [and] I could not debase the situation I had the honour to hold by a tacit submission'. His rash refusal had grim consequences – it was this action, the records suggest, that triggered the general's lasting hostility; he refused to see Flinders again during his six-and-a-half years on the island.

Heedless of the antagonism he had provoked, Flinders still expected their early release. Conditions improved at the café: the sentry was removed, they were given use of the billiard room, and a second bedroom was provided. Aken was permitted to board the *Cumberland* to collect the ship's timekeeper and navigational instruments, and Flinders's servant John Elder joined them, bringing Trim with him. Flinders penned the first of a series of letters to the general, requesting that repairs be carried out to the ship 'so that I may be able to sail as soon as possible after you shall be pleased to liberate me from my present state of purgatory'.

The *Cumberland*'s boatswain, Charrington, brought bad news from an unexpected quarter. The men 'were committing many irregularities, taking spirits out of my cabin and going on shore as they pleased'; the French guards took little or no notice of their actions. Discipline had to be restored, and Flinders wrote again to Decaen, requesting that Aken be allowed to return to the schooner 'to correct the disorders'.

The general had no more wish than Flinders to have a band of drunken English sailors carousing in the port's streets and taverns, but neither was he prepared to release their young commander from what Flinders insolently called his 'state of purgatory'. Decaen ordered that any crew members found ashore should be locked up in the guardhouse, and the remainder transferred to the prison ship; the corporal of the guard was dismissed. Next day Monistrol called with fresh orders from the captain general that for the first time led Flinders to consider the possibility of a longer confinement.

The documents he had handed over for inspection, explained the colonel, were found to contain incriminating evidence. His Excellency believed that Captain Flinders 'had absolutely changed the nature of the mission for which the First Consul granted a passport' – in no sense did this allow him to acquaint himself 'with the periodical winds, the port, and the present state of the colony', as recorded in his logbook. Further, he had disobeyed his instructions from the British Admiralty, which required him 'not to take letters or packets other than such as you may receive from this office or His Majesty's Secretary of State'; yet he admitted carrying dispatches from the governor of New South Wales.

Judging such conduct to be a 'violation of neutrality', continued Monistrol, General Decaen had ordered him to accompany Flinders on board the *Cumberland*, and to collect in his presence 'all other papers which might add to the proofs already obtained'. He was to make a full inventory of everything on board the schooner, after which the stores were to be sealed and removed to the arsenal. Flinders was to remain at the inn 'where my suspicious conduct had made it necessary to confine me from the instant of arriving in port'. He and Aken would be held in custody there until further notice.

'I have found a prison!'

The Café Marengo, Wednesday, 21 December 1803. Captain Flinders writes to Captain General Decaen from his 'place of confinement':

> *Now, Sir, I would beg to ask you whether it becomes the French nation …*
> *to stop the progress of such a voyage, and of which the whole maritime*
> *world are to receive the benefit? How contrary to this was her conduct some*
> *years since towards Captain Cook! But the world highly applauded her*

Top: *Elephant seals on King Island, by C.-A. Lesueur. [Collection Lesueur, Muséum d'Histoire Naturelle, Le Havre]*

Above: *The French camp on King Island, with the Union Flag flying after it was raised by Lieutenant Robbins. Sketch by Lesueur [?]. [Collection Lesueur, Muséum d'Histoire Naturelle, Le Havre]*

Top: *Frenchman's Rock, now in the Island Gateway Information Centre at Penneshaw, Kangaroo Island. The inscription, 'Expedition de decouverte par le* Commendant Baudin sur le Géographe 1803', *was carved by a member of a watering party sent ashore at the island by Baudin. [Photo by Jill Gloyne]*

Above: *Native huts on Péron Peninsula, Shark Bay, WA, by C.-A. Lesueur. [Collection Lesueur, Muséum d'Histoire Naturelle, Le Havre]*

conduct then; and possibly we may sometime see what the general senti-
ment will be in the present case.

I sought protection and assistance in your port, and I have found a
prison! Judge for me as a man, Sir – judge for me as a British officer em-
ployed in a neutral occupation – judge for me as a zealous philanthro-
pist, what I must feel at being thus treated. [Flinders, 1814, vol. II, p. 370]

Flinders had little option but to comply with the general's orders. His remaining books and papers, including private letters from family and friends, were impounded. Despite his anger, he acknowledged that Monistrol and the interpreter, M. Bonnefoy, 'acted throughout with much politeness … and the Colonel said he would make a representation to the Captain-General, who doubtless lay under some mistake'.

Back at the tavern, Flinders wrote an irate letter to Decaen, protesting long and bitterly against his arbitrary detention, and recklessly contrasting his reception with that of captains Baudin and Hamelin by Governor King:

> The governor in chief at Port Jackson knew too well the dignity of his own nation, either to lay any prohibition upon these commanders, or to demand to see what their journals might contain.[11]

He would have been still more irate had he known then what he discovered later – that the *Géographe* had sailed with secret dispatches for the French government (including Decaen's report on the political situation in India and requesting that a French expeditionary force be sent to support the princes in a revolt against the British). Writing in the *Voyage,* Flinders claimed

> this transaction being contrary to the English passport, and subjecting the ship to capture, if known, it was resolved to detain me for a short time, and an embargo was laid upon all neutral ships for 10 days.

He continued:

> The first motive for my detention therefore arose from the infraction of the *English* passport, by sending despatches in the *Géographe*; and the probable cause of its being prolonged beyond what seems to have been originally intended, was to punish me for refusing the invitation to dinner.[12]

The conjecture is reasonable enough, so far as it goes. Flinders's position was not helped, however, by the suspicion already implanted in Decaen's mind by Péron's comment that his voyage had military and strategic objectives in the Pacific – a suspicion apparently confirmed when he came to read Governor King's dispatches taken from the *Cumberland*. These sought an increase in the colony's military establishment to defend it against any possible attack from Isle de France, and to 'annoy the trade of the Spanish settlements on the opposite coast'.

Ignorant of the geopolitical background, Flinders made a bad situation worse by writing further 'insolent and arrogant' letters to the general – or so they appeared to the recipient. As commander-in-chief of French territories in the East, responsible directly to the first consul, Decaen's patience rapidly wore thin. When Flinders wrote on Christmas Day,

> I cannot think that an officer of your rank and judgement, could act either so ungentlemanlike, or so unguardedly, as to make such a declaration [that is, that Flinders was a spy] without proof; unless his reason had been so blinded by passion, or a previous determination that it should be so ...[13]

Decaen would take no more. His response was abrupt – and final:

> I was far from thinking that after having seriously reflected upon the causes and circumstances, you should take occasion from a silence so delicate to go still further; but your last letter no longer leaves me an alternative. Your undertaking, as extraordinary as it was inconsiderate, to depart from Port Jackson in the *Cumberland*, more to give proof of an officious zeal, more for the private interests of Great Britain than for what had induced the French government to give you a passport ... had already given me an idea of your character; but this letter overstepping all the bounds of civility, obliges me to tell you, until the general opinion judges of your faults or of mine, to cease all correspondence tending to demonstrate the justice of your cause; since you know so little how to preserve the rules of decorum.[14]

Flinders brushed aside the general's comments as 'embarrassment sheltering itself under despotic power' – at least they removed any lingering illusions as to his current status. Abandoning for the time being any further attempt to regain his freedom, he decided to resume work on his charts of the Gulf of Carpentaria, and wrote to Decaen requesting the return of the charts, logbooks and other papers impounded after his arrest. He also raised a complaint from his own seamen in the prison hulk, of being shut up at night in a place where not a breath of air could come to them; which in a climate like this must be … very destructive to European constitutions. Also, that the people with whom they were placed were affected with that disagreeable and contagious disorder the itch; and that their provisions were too scanty, except in the article of bread, the proportion of which was large, but of a bad quality.[15]

Decaen did not reply, although Monistrol delivered the trunk holding Flinders's books to the tavern, and informed him that the men's complaints were being investigated. He let it be known, politely, that in his view the prisoner's letters to the general were couched 'in a style so far from humble' that he thought they 'might rather tend to protract than terminate my confinement'. In no mood to take the hint, Flinders replied tartly:

that I demanded only justice, so I did not think an adulatory stile [*sic*] proper to be used … It was not the custom in England when justice only was the object in view to apply for it in the supplicating stile of a criminal, and although I was now to the east of the Cape of Good Hope, yet I retained too much of my native good manners to address the captain-general as if he were an eastern monarch from whom I had favours to ask.[16]

All was now set for the long-drawn-out denouement of Flinders's tragedy. He helped to lay the framework. Writing yet again to Decaen, he requested: that he be permitted to depart with his vessel, papers, stores, and so on, on his word of honour not to reveal anything concerning the Isle de France for a prescribed time; that he be sent to France to have his case heard there; or that if his continued detention were considered necessary, that Aken and his crew be allowed to leave on parole.

This letter too was ignored. For Decaen the Flinders affair was no more than an irritating distraction – his time was fully occupied in placing the island's administration on a war footing, strengthening its defences against a possible enemy attack, and developing his plans for taking the war to the subcontinent. There was no point in sending the navigator to France – in all probability he would be rescued on the voyage by an English cruiser. Instead he referred the matter to Paris for adjudication – the Ministry could weigh up the legal and diplomatic issues, leaving him free to pursue the great task given him by the first consul. Thus Flinders became a 'prisoner of state'.

January 1804. Visiting Government House to collect his private letters and papers, his logbooks and journals, and 'such charts as are necessary to completing the Gulph of Carpentaria', Flinders is dismayed to learn that the third logbook, recording his reasons for calling at Isle de France, has been confiscated by the captain general. Worse by far, it contains his irreplaceable observations made in Torres Strait and the gulf, information that cannot be made good. Confiscated too are the dispatches he carried from Governor King and Colonel Paterson.

In the tavern's squalid surroundings, he begins to work, with Aken's help, on his charts of the gulf. It serves to ease his mind, but does nothing to lessen his resentment against the man responsible for his imprisonment. He does not forget his men, sending them once a week a large basket of fruit and vegetables, which Elder buys at the market and takes to the prison hulk.

Now high summer, the rainy season sets in, the sultry heat raising still further the humidity in the stifling room. The warm rain falls at random, tumescent storm-clouds form seemingly from nowhere in a clear sky and burst directly above the little town. Rain falls in billowing silvery sheets, then as suddenly stops. After the rain the sun beats down as before, dazzling rainbows colouring the sky from horizon to horizon.

Nightfall brings no relief. The prisoners lie shrouded in mosquito nets, their sweating bodies the target of endless insect armadas hunting for fresh blood. Moths blanket the solitary lantern, spiders the size of a man's palm tenant the corners of the room, scorpions, cockroaches, mice and more anonymous vermin scurry across the dirty floor and scratch behind the walls. A good night's sleep is unknown.

Since independence, the United States had opened a promising trade with the French merchants at Isle de France. Its South Sea sealers and whalers visited regularly for supplies, relaxation for their crews, and to trade their products. Flinders contrived to smuggle copies of the *Sydney Gazette* reporting his shipwreck, along with details of his imprisonment, on board one such ship. Thus it was through the Americans that the British government first learned of the *Cumberland*'s arrival at Port North-West, and of her captain's arrest by Captain General Decaen.

Prisoner of state

The Café Marengo, Tuesday, 1 March 1804.

I asked M. Bonnefoy to give me his opinion of what was likely to be done with us? He replied that we should probably be kept prisoners so long as the war lasted, but might perhaps have permission to live in some interior part of the island, and liberty to take exercise within certain limits. This opinion surprised me; but I considered it to be that of a man unacquainted with the nature of a voyage of discovery, and the interest it excites in every nation of the civilised world, and not the least in France. [Flinders, 1814, vol. II, p. 385]

Stress, depression and lack of exercise brought about a breakdown in Flinders's health; scorbutic ulcers again broke out on his legs and feet. A surgeon was sent to treat him, but he was refused permission to walk out in the open air. A note to Decaen requesting an interview went unanswered, but according to the interpreter Bonnefoy the general had remarked: 'It is needless for me to see him, for the conversation will probably be such as to oblige me to send him to the tower'.

In February, and again in early March, Captain Bergeret of the French Navy visited Flinders at the tavern and promised to intercede for him with the captain general. He regretted having been at sea when Flinders arrived, 'believing it would have been in his power to have turned the tide of consequences'; though doubtful whether anything could now be done to obtain their release, he hoped they might be permitted to live in the country.

Bergeret was as good as his word. On 23 March, Flinders heard from Bonnefoy that his crewmen had been taken ashore from the hulk and sent to a prison camp at Flacq, on the eastern side of the island. On the 31st, 105 days after their arrival, Flinders and Aken were moved to a large house, the Maison Despeaux, standing in its own grounds at the edge of the town. Known as the Garden Prison, it was used by the French authorities to intern captured British officers and the masters of merchantmen taken at sea. In the healthier surroundings, and in the company of other officers, Flinders's health and spirits quickly recovered, and his work on his charts progressed.

In his private journal, begun on the day of his arrival at Port North-West, Flinders records his daily routine at the Maison Despeaux:

Before breakfast my time is devoted to the Latin language, to bring up what I had formerly learned. After breakfast I am employed making a fair copy

of the Investigator*'s log in lieu of my own which was spoiled in the shipwreck. When tired of writing, I apply to music, and when my fingers are
tired with the flute, I write again until dinner. After dinner we amuse ourselves with billiards until tea, and afterwards walk in the garden till dusk.
From thence till supper I make one of Pleyel's [flute] quartettes; afterwards
walk half an hour, and then sleep soundly till daylight when I get up
and bathe.*[17]

*A domestic tragedy intrudes on this ordered life – Trim the circum-
navigator is reported missing. Left in the care of a French lady as a
companion for her small daughter, he has disappeared. The island's
public gazette offers a reward of 10 Spanish dollars for his return to his
grieving young mistress, but without result.*

*It is only too likely, thinks Flinders, that the faithful, intelligent Trim,
his affectionate companion for five years, the survivor of a dreadful
shipwreck, has been stewed and eaten by some hungry black slave.
Sorrowfully he writes his epitaph:*

*Never, my Trim, to take thee all in all, shall I see thy like again; but never
wilt thou cease to be regretted by all who had the pleasure of knowing thee!*

As Ernest Scott noted, Decaen at this time was simply waiting for an
order from Paris to release Flinders. In March he had advised Bergeret
'to have a little patience, as he should soon come to some determination of the affair'. Flinders wrote to King in August that he had been
told 'I must wait until orders were received concerning me from the
French government'. In 1805, and again the following year, he wrote:

General Decaen, if I am rightly informed, is heartily sorry for having made
me a prisoner, [but] he remitted the judgement of my case to the French
government, and cannot permit me to depart, or even send me to France,
until he shall receive orders.[18]

Meanwhile the *Géographe*, with Major Barois on board, arrived in
France in mid-1804. Napoleon, having declared himself emperor in May,
was devoting his prodigious energies on the one hand to the comprehensive reorganisation of government needed to change a republic into
an empire, and on the other to the vastly detailed preparations for the
cross-Channel invasion of England.

When Barois at last managed to obtain an interview, the Emperor
appeared to approve of Decaen's plans, but gave no directions for their

implementation. At the time, all ministries were working full-time on far more pressing matters than the affairs of a distant colony.

Admiral Decrès was inspecting installations at Boulogne when he received the general's dispatches reporting the detention of the *Cumberland* and her commander. He had taken this action, said Decaen, because of the British government's unprincipled conduct towards France; the seizure of the *Naturaliste*[19] by the Royal Navy; and third, Captain Flinders's intention, confirmed by his logbook, to 'acquire knowledge' of the colony. As proof he enclosed copies of official correspondence carried by Flinders in contravention of his passport. In a second dispatch, Decaen proposed an early attack on India, both to check British expansion and to forestall an enemy invasion of Isle de France.

These were distractions the admiral could well do without. Apart from his ministry's responsibilities for the construction and deployment of the invasion armada, he was not one to disregard the importance of the change to an imperial form of government to high-ranking officials like himself. The new emperor too was fully occupied with these matters, and approved Decrès' bureaucratic solution to *l'affaire Flinders* – to refer it to the Council of State for advice. The Minister wrote to the captain general in cautious terms:

> I have placed before H.M. the statement of motives that led you to detain the English Schooner *Cumberland* commanded by Captain Flinders. My opinion is that there is a case for approving the course you have taken, and I have put this on record in my report; but in view of the importance of the facts under discussion and in order that the British government may know that a question of this nature has not been disposed of lightly, I have proposed to H.M. that my report be sent to the Council of State; as soon as a definite decision has been reached ... I shall hasten to inform you of it.[20]

It was twenty months later that Decrès wrote to advise the general of the Council of State's decision on Flinders. And many more months passed before Decaen received the dispatch on his blockaded island.

'One advantage of being confined in the Garden prison,' Flinders writes while counting the slow passing of the days, 'is the frequency of visitors.' Through them he learns all the public news, and also the colonists' views on his own imprisonment. Many think Decaen's conduct has been severe, impolitic, and unjust; one correspondent cites his

confinement as an example of the captain general's many tyrannical acts as their governor.

Among his visitors is the captain of a Danish merchant ship, who introduces himself as Augustin Baudin, Nicolas' brother. He is 'exceedingly kind and civil in offers of service', and shows the prisoner a letter sent by his brother from Port Jackson in 1802, praising the 'attention and assistance' given him and his officers by Governor King; he deprecates the great difference in Flinders's treatment by General Decaen.

Captain Baudin comes from Tranquebar, a Danish fort and trading settlement on the Coromandel Coast, south of Madras. He seeks the Englishman's advice 'concerning the propriety of taking a young woman to India whom his brother had brought hither from Port Jackson'. Flinders, unfortunately, does not record his reply.

Visitors to the Garden Prison performed another invaluable service for prisoners – arranging for the safe dispatch of unauthorised mail. Ships of neutral nations, the United States among them, were regular callers at Port North-West, and it was an easy matter to bypass the nominal French censorship. Flinders took full advantage of this laxity; during his long detention, notes Ingleton, 'he wrote innumerable letters, [and] it is a remarkable fact that most reached their destination safely'. Letters poured forth, not only to his wife and friends, but also to persons of influence – in France as well as England and her colonies – pleading the justice of his cause, complaining of his harsh and invidious treatment by General Decaen, and entreating their help in regaining his freedom.

When the recipients responded officially with direct appeals to the captain general for Flinders's release, as did the governor-general of India, Marquess Wellesley, and Governor King of New South Wales, the effect was to fuel Decaen's stubborn determination not to free his most important prisoner until it suited him to do so. He could not comply with their request, he replied, because the case had been referred to the French government for a decision.

Letters home

From the Garden Prison, July to December 1804. In July Flinders writes to Sir Joseph Banks about his hopes for promotion:

> *I hope, Sir Joseph, that even from the charts which I have sent Home you will think we did as much as the lateness of the season with which we first*

came upon the [north] coast, and the early rottenness of the Investigator, *could well allow; and I think our labours will not lose on a comparison with what was done by the* Géographe *and* Naturaliste. *No part of the unfortunate circumstances that have since occurred can, I believe, be attributed to my neglect or mistakes; and therefore I am not without hope, that when the Admiralty know I am suffering an unjust imprisonment, they will think me worthy to put upon the post-captains' list. My age now exceeds the time which we judge in the Navy a man ought to have taken there who is to arrive at anything eminent.*

It is to you only, Sir Joseph, that I can address myself upon this subject. I have had ample testimonies of your power, and of the strength of your mind in resisting the malicious insinuations of those who are pleased to be my enemies; nor do I further doubt your willingness to give me assistance, than that I fear you do not yet think me worthy of it. But I will be. If I do not prove myself worthy of your patronage, Sir Joseph, let me be thrown out of the society of all good men. I have too much ambition to rest in the unnoticed middle order of mankind. Since neither birth nor fortune have favoured me, my actions shall speak to the world.[21]

To a friend he reveals his private impression of General Decaen:

The truth I believe is that the violence of his passion outstrips his judgment and reason and does not allow them to operate; for he is instantaneous in his directions, and should he do an injustice he must persist in it because it would lower his dignity to retract. His antipathy, moreover, is so great to Englishmen, who are the only nation that could prevent the ambitious designs of France from being put into execution, that immediately the name of one is mentioned he is directly in a rage, and his pretence and wish to be polite scarcely prevent him from breaking out in the presence even of strangers. With all this he has the credit of having a good heart at the bottom.[22]

On 24 August, Matthew writes to his 'beloved Ann':

I yesterday enjoyed a delicious piece of misery in reading over thy dear letters. Shall I tell thee that I have never before done it since I have been shut up in this prison? I have many friends who are kind and much interested for me, and I certainly love them. But yet before thee they disappear as stars before the rays of the morning sun. I cannot connect the idea of happiness with anything without thee. Without thee, the world would be a blank. I might indeed receive some gratification from distinction and the applause of society; but where could be the faithful friend who would enjoy and share this with me, into whose bosom my full heart could unburthen itself of excess of joy? I am not without friends even among the French. On the contrary, I have several, and but one enemy, who unfortunately, alas, is all-powerful here; nor will he on any persuasion permit me to pass the walls of the prison; although some others who are thought less dangerous have had that indulgence occasionally ... [23]

The following day, Flinders writes to congratulate his brother Samuel on his part in the action off Pulo Auro, in which the China tea fleet (carrying the survivors of the Wreck Reef shipwreck home to England) repulsed an attack by Admiral Linois's ships. He informs Samuel that the hydrographical record of the Investigator's **voyage should soon be completed, and goes on:**

Much other useful work I have also done, and shall continue to do until I am permitted to depart. Their rage is disarmed of its sting, for they cannot make me very unhappy whilst I have useful employment. To external appearance they are doing me the greatest imaginable injury, but I will so contrive it, that it shall prove to have been of the greatest advantage to me in the end. They do indeed keep me from my wife and friends, from the regulation of my pecuniare [sic] and family concerns, from the credit due to the exertions I have made fulfilling the objects of the voyage, and they keep me very probably from promotion, but the enjoyment of these will be all increased in their degree from this suspense, and in a few months I shall triumph.

Although I use the expressions they and them, there is but one man here who I believe is my enemy. The governor both hates and fears me, for he has deeply injured me; but he is too proud to alter his measures, although he has acknowledged that of myself I have not done anything to forfeit the passport. I have offers of service from many individuals here, and have received both attention and kindness from some, but they cannot procure me liberty ... Upon a review of all probabilities I think we shall be at liberty in December and in England 4 months afterwards. [24]

By midyear the last three charts in a series mapping the coast from Cape Leeuwin to the Gulf of Carpentaria are complete, and Flinders turns to the most important of all – the general chart of the continent. Two fellow internees permitted to return to England by way of America agree to deliver duplicate copies of the charts to Sir Joseph Banks and the Admiralty. With the general chart – the first authoritative map of the

entire coastline of the southern continent – he encloses an explanatory letter to his patron:

> *… The propriety of the name* Australia *or* Terra Australis *which I have applied to the whole body of what has generally been called New Holland must be submitted to the approbation of the Admiralty and the learned in geography. It seems to me an inconsistent thing that Captain Cook's New South Wales should be absorbed in the New Holland of the Dutch, and therefore I have reverted to the original name Terra Australis or the Great South Land, by which it was distinguished even by the Dutch during the 17th century; for it appears that it was not until some time after Tasman's second voyage that the name New Holland was first applied, and then it was not long before it displaced T'Zuydt Landt in the charts, and could not extend to what was not yet known to have existence. New South Wales, therefore, ought to remain distinct from New Holland; but as it is required that the whole body should have one general name, since it is now known that it is certainly all one land, so I judge that one more acceptable to all parties and on all accounts cannot be found than that now applied.[25]*

On New Year's Eve, Matthew unburdens himself once more to Ann, this time with a rare introspection:

> *… I shall learn patience in this island, which will perhaps counteract the insolence acquired by having had unlimited command over my fellow men. You know, my dearest, that I always dreaded the effect that the possession of great authority would have upon my temper and disposition. I hope they are neither of them naturally bad, but when we see such a vast difference between men dependant and men in power, any man who has any share of impartiality must fear for himself. My brother will tell you that I am proud, unindulgent, and hasty to take offence; but I doubt whether John Franklin will confirm it, although there is more truth in the charge than I wish there was. In this island these malignant qualities are ostentatiously displayed, and I am made to feel their sting most poignantly. My mind has here been taught a lesson in philosophy, and my judgement has gained an accession of experience that will not soon be forgotten.[26]*

Last days at the Garden Prison

The Maison Despeaux, Sunday, 18 August 1805.

> *Rose at half past six. Slipped on my shoes and morning gown, and went down to walk in the garden. Met the sergeant and bid him bon-jour. Think*

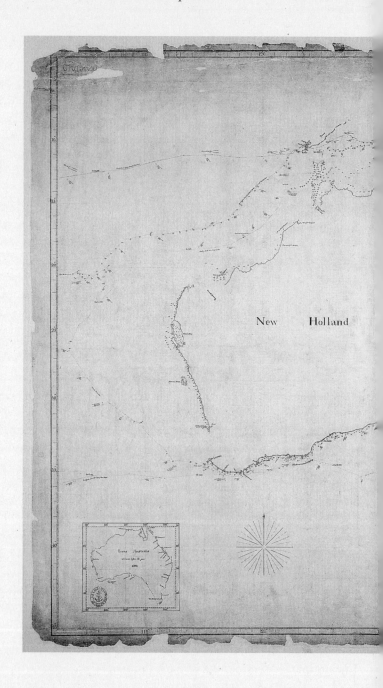

The general chart of Australia, or Terra Australis, completed in 1804 by Flinders while he was still held captive on Isle de France. [British Crown Copyright © 2000. Published by permission of the Controller of Her Majesty's Stationery Office and the UK Hydrographic Office]

the old man looks a little melancholy at the prospect of his last prisoner leaving the house, for he will lose his situation. The dogs came running after me. Meditated during my walk on the extreme folly of General De Caen keeping me prisoner here, for it can answer no good purpose either to him or to the French government; and some expense, and probably odium, must be incurred by it. This injury that is to me is almost incalculable – but this will not bear to be dwelt upon, it leads almost to madness. [Flinders, Private Journal, 1803-1814, MS, Mitchell Library; see also appendix C]

At the Garden Prison, Flinders met a young French merchant, Thomi Pitot, a frequent visitor. This soon ripened into 'the most agreeable, most useful, and at the same time durable' friendship. Pitot was 'well-informed, a friend to letters, to science, and the arts', and spoke English well. Widely respected in the colony, he was not intimidated by Decaen and spoke his mind freely. Yet Flinders found to his surprise that his friend considered his conduct towards the general undiplomatic and needlessly antagonistic. Writing to his brother Samuel, he confessed that he might have been 'too obstinate to sacrifice one tittle … either of the honour of my country or of myself'.

It was Pitot who persuaded Flinders to make a fresh appeal to Decaen in December 1804, on 'the anniversary of that day on which you transferred me from liberty and my peaceful occupations to the misery of a close confinement'. In his letter, Flinders requested that he either be 'fully released with my people, books and papers … or that we shall be sent to France', where, if the government's decision proved favourable, they could then return immediately 'to our country, our families and friends'. The general's response, as before, was silence.

Summer returns, Flinders's second on the island. Steamy tropic heat blankets the little town, hemmed in by craggy mountains. The stone-walled compound of the Garden Prison is alternately burnt by the sun, washed by rain. Mosquitoes swarm in puddles of rank water. Flinders falls ill, a recurrence of a 'constitutional gravelly complaint' that first affected him in Sydney,

> *to which confinement had given accelerated force, and by a bilious disorder arising partly from the same cause, from the return of hot weather, and discouraging reflections on our prospects.*[27]

His eyes are affected, and 'almost every evening' he reproaches him-self 'for not having made greater progress' in his labours. Dr Laborde, the principal physician of the medical staff, certifies that a move to the country is needed to restore to health 'a man whose work serves the progress of science and is of such utility to his colleagues'. Flinders is later told 'that Dr. Laborde had received a message from the General, desiring him not to interfere with matters which did not concern him'; this, it seems, is 'the sole mark of attention' paid to the doctor's certificate.

By August 1805 Flinders was alone at the Maison Despeaux. His fellow inmates had been released, singly or in groups, and repatriated in exchange for French officers held as POWs in Ceylon and India. Aken, who had fallen ill and been taken to the military hospital, was allowed to leave in May, on giving his parole 'not to serve against France or its allies, until after having been legally exchanged' – 'that is', Flinders noted mockingly, 'as a *prisoner of war*'. Aken sailed in an American ship, taking with him sixteen original charts for personal delivery to Banks; Sir Joseph in turn forwarded them to the Admiralty's Hydrographic Office, where they were filed away until their compiler's return. The French cartographers were thus able to publish the first complete map of Australia well in advance of Flinders.

His hopes for his release boosted by Aken's repatriation, Flinders wrote to Banks in July 1805 that dispatches from Paris were expected daily, 'and then I may be either liberated or sent to France to be tried as a spy'; he looked to be in England, or France, about February or March 1806. Understandably, he had no means of knowing the intricate workings of the new imperial bureaucracy.

Also in July, two cartel ships arrived from India with French prisoners of war for exchange – among them Flinders's friend Captain Bergeret, captured in a naval action in the Bay of Bengal. Governor-General Wellesley had designated Bergeret, an officer of equal rank, as a suitable exchange for the English navigator. Decaen, however, raised his usual objection; as his secretary explained apologetically to Flinders, he was unable to accede to Wellesley's proposal, because

> the motives of your detention having been of a nature to be submitted to the French government, the Captain-General cannot, before he has received an answer, change anything in the measures which have been adopted on your account.[28]

Not for the first time, or the last, Flinders regretted the impulse that led him to suggest that his case be heard in Paris.

To celebrate the exchange, some friendly French officers call at the Maison Despeaux to invite the newly released prisoners to a theatre performance in the town. Flinders at first declines, fearing his hosts will antagonise Decaen, but is assured that the general has given his approval. He has not attended a theatre since leaving England and, apart from the wives of a couple of the internees and the few black slaves working at the prison, has been deprived of female company since his confinement. The audience holds greater interest for him than the performance:

> *The theatre was so full that, being late, I scarcely got a sight of the stage. It surprised me to see so many handsomely dressed women in the pit; the greater part of them must be the wives and daughters of tradesmen, but their dress was beyond all comparison, more expensive and gay than those of the same class in England. The younger women were some of them very pretty, and there was one that might be very well called a beauty. Her age was said to be twelve years, but she seemed very womanish, though sufficiently modest, at least by comparison.*
>
> *The older ladies seemed generally to be fat, but their dress was equally gay and bosoms equally bare with the younger. The necks of almost all, and the shoulders, and bosoms, and nearly half the breasts were uncovered, as well as the arms nearly to the shoulders. They seemed to have good clear skins, and well turned necks and bosoms for the most part, and large eyes that were by no means destitute of power. An equal number of women, equally dressed, would I think raise an uproar in one of our English theatres. The modest would be offended, the prudes would break their fans, the aged would cry Shame!, the libertines would exult and clap, and the old lechers would apply for their opera glasses!*[29]

The two cartels sailed for India in mid-August, leaving Flinders, his servant Elder, and a seaman, William Smith, as the sole remaining prisoners on the island. Gripped by depression, Flinders received a lift to his spirits on the 16th, when Captain Bergeret called on him with the news that the general had at last relented – he could move to the country once suitable accommodation was found: 'The General, it seems, objects to my residence anywhere upon the sea shore; but I understand my parole will be accepted as a security for my behaviour' – which was scarcely consistent 'with the belief of my being a spy'.

The 16th was also the feast day of St Napoleon, celebrated with all the pomp the colony could muster. The principal church was ornamented with the flags of all nations, save only an English ensign, which, as a symbol of French power, was trodden underfoot. At sunset the fort's guns fired a general salute, followed by a fireworks display that lit up the night sky. To the prisoner at the Maison Despeaux it appeared that the Corsican saint

> had more honours paid to him, and seems to be in more repute than the poor Virgin Mary; cries of 'Vive l'Empereur' were however, it is said, by no means violent among the soldiery in the morning at the review; and the threat of imprisonment extorted the greater part of what were heard from them. I know not if the officers were more sincere in their devotions.[30]

With Thomi Pitot's help, Flinders found lodgings with Mme D'Arifat, a widow, and her 'agreeable and respectable family' on their plantation at Wilhems Plains. On 19 August, Colonel Monistrol informed him that his request to move there had been approved, and next day he took leave of the kindly old sergeant of the guard, who had always treated the prisoners well.

> Finding myself without side the iron gate, I felt that even a prison one has long inhabited is not quitted without some sentiment of regret, unless it be to receive liberty.[31]

On the 23rd he went with Bergeret to the town-major's office to give his parole, insisting it be taken in writing to avoid any future misunderstanding:

> His Excellency, the captain-general De Caen, having given me permission to reside at Wilhems Plains, at the habitation of Madame D'Arifat, I do hereby promise, upon my parole of honour, not to go more than the distance of two leagues from the said habitation, without His Excellency's permission; and to conduct myself with that proper degree of reserve, becoming an officer residing in a country with which his nation is at war. I will also answer for the proper conduct of my two servants.
>
> Mattw. Flinders.[32]

The Refuge

The Refuge, Wilhems Plains, September 1805 to March 1810.

> *By signing this parole I cut myself off from the possibility of an escape; but it seemed incredible, after the various letters written and representations*

*made both in England and France, that a favourable order should not
arrive in six or eight months. I moreover entertained some hopes of Mauritius being attacked, for it was not to be imagined that either the East-
India Company or the government should quietly submit to such losses as
it caused to British commerce; and if attacked with judgment, it appeared
to me that a moderate force would carry it; upon this subject, however,
an absolute silence was preserved in my letters, for although the passport
had been so violated by General De Caen, I was determined to adhere to
it strictly. [Flinders, 1814, vol. II, p. 419]*

Flinders set off for the plantation, appropriately named *Le Refuge*, on
24 August, accompanied by Thomi Pitot and his family, who lived
nearby. The sense of liberation after twenty months of captivity was
restorative, galvanising Flinders's powers of observation:

> From the time of quitting the port we had been continually ascending;
> so that here the elevation was probably not less than a thousand feet,
> and the climate and productions were much altered. Coffee seemed to
> be a great object of attention, and there were some rising plantations of
> clove trees … A vast advantage, as well as ornament in this and many
> other parts of the island, is the abundance of never failing streams; by
> which the gardens are embellished with cascades and fish ponds, and
> their fruit trees and vegetables watered at little expense.[33]

Mme d'Arifat and her family were not in residence when Flinders arrived, but the black servant left in charge of the plantation had orders to
accommodate him 'in one of two little pavilions detached from the house,
the other being appropriated to my two men'. He quickly settled in,
bathed every morning in the stream behind the house, walked out every
fine day, and in a few weeks had almost recovered his former health:

> Those who can conceive gratification from opening the door to an impris-
> oned bird, and remarking the joy with which it hops from spray to spray,
> tastes of every seed and sips from every rill, will readily conceive the sensa-
> tions of a man during the first days of liberation from a long confinement.[34]

Mme d'Arifat arrived in early October, with two of her sons and three
daughters, and with typical colonial hospitality invited Flinders to 'par-
take of her table with the family'. He accepted readily, 'and in a short
time had the happiness to enumerate among my friends one of the
most worthy families in the island'.

Life on the plantation soon took on a pattern that in other circumstances would have been idyllic. Flinders rose every morning with the sun, bathed, walked with the ladies or read, breakfasted, then retired to his pavilion to read or write for two or three hours. Next, he would practise translating from French into English, and English into French, with the help of Madame's pretty young daughters, Delphine and Sophie.

He dined with the family at two, after which he read, or wrote, or perhaps took a nap, until five, when he rejoined the household. Tea was usually taken at half-past six, and the evening passed in reading, conversation, sometimes in singing and flute playing, or cards. Supper was served at nine o'clock, and he retired to bed at ten – 'where the agreeable employment of the day' often occupied so much of his thoughts that he could not sleep.

The first hurricane of the season strikes in the evening of 20 February:

> *the swift-passing clouds were tinged at sunset with a deep copper colour;*
> *but the moon not being near the full, it excited little apprehension at the*
> *Refuge. The wind was fresh, and kept increasing until eleven o'clock, at*
> *which time it blew very hard; the rain fell in torrents, accompanied with*
> *loud claps of thunder and lightning, which at every instant imparted to*
> *one of the darkest nights the brightness of day.*[35]

The wind blows hardest between one and three in the morning, 'giving me an apprehension that the house, pavilions, and all would be blown away together'. On neighbouring plantations the huts of the slaves, storehouses, and some smaller dwellings are unroofed or blown down. The streams swell to raging torrents, carrying away the most fertile soils from the fields.

In the morning, Flinders walks to the cascades of the Rivière du Tamarin 'to enjoy the magnificent prospect which the fall of so considerable a body of water must afford'. They descend a precipice of 120 feet, and a short distance further on another of 80 or 100 feet. His path through the forest is strewn with the broken branches and fallen trunks of great trees, the shrubs and undergrowth are so beaten down that they 'present the appearance of an army having passed that way'. The river brims its banks, the noise of the fall is so loud that he can scarcely hear his own voice; 'a thick mist rising to the cloud from the abyss admitted of a white foam only being distinguished'.

Briefly Flinders toyed with the idea that Ann might join him at the Refuge. Two letters from her, carried by an American ship sailing direct from London, had arrived in October 1805, tentatively suggesting that she would like to share his confinement. When the general refused yet another request for his release in March 1806, Flinders wrote:

> Of all things in the world, I most desire thy presence here, since I cannot come to thee; but of all things in the world I should most dread thy undertaking such a voyage without being protected and accommodated in a manner which is scarcely possible … Let the conduct of a woman on board a ship without her husband, be ever so prudent and circumspect, the tongue of slander will certainly find occasion, or it will create one, to embitter the future peace of her husband and family.

In the end, he could not bring himself to support the proposal:

> We should be very happy here, with this excellent family … but what difficulties, fatigues and risks thou wouldst have to arrive at this happiness. No, my dearest, it cannot be. I ask thee not to undertake it.[36]

Flinders kept grief at bay by working hard at his writing, completing a draft narrative of his voyages and adventures to the point of his release from the Garden Prison. This, along with various letters and dispatches for the Admiralty, was secretly taken to London by Captain Larkins, master of a captured East Indiaman who was repatriated on an American ship. Larkins also carried a scientific paper prepared by Flinders on the use of the marine barometer for predicting wind changes at sea; this was subsequently read before the Royal Society by Sir Joseph Banks.

The routine at the Refuge was often interrupted by Pitot and other friends, visiting from Port North-West, by calls on neighbouring families, joint excursions and picnics, and hunting and fishing expeditions. On one visit, in May 1806, Pitot brought with him a young officer Flinders had known in Sydney – Charles Baudin des Ardennes, formerly of the *Géographe*, now serving on the frigate *Piémontaise*. Baudin stayed at the Refuge for some days, long enough to provide a detailed account of the French explorations on the south, west, and north-west coasts after leaving Port Jackson. The two soon became firm friends.

Toussaint de Chazal, a neighbouring planter, was another close friend. He had sought refuge in London during the Terror, spoke English well, and was an accomplished artist. Towards the end of 1806 he painted

Flinders's portrait over several sittings – the only known 'natural size' painting of the navigator drawn from life. Pitot and Chazal introduced him to fellow members of the island's Societé Libre d'Emulation, a philosophical and literary association founded by some of the scientists who had deserted from Baudin's expedition five years earlier. The society enthusiastically took up Flinders's cause, and in 1806 appealed directly to the Institut National of France:

> Use then, we beg of you, the influence of the first scientific body of Europe, the Institut National, in favour of Captain Flinders, so that the error which has led to the captivity of this learned navigator may become known. In rendering this noble service you will acquire a new title to the esteem and honour of all nations, and of all friends of humanity.[37]

'Under these circumstances,' Professor Scott wrote in 1914,

> with agreeable society, amid sympathetic friends, in a charming situation, well and profitably employed upon his own work, Flinders spent over five years of his captivity. He never ceased to chafe under the restraint, and to move every available influence to secure his liberty, but it cannot be said that the chains were oppressively heavy.[38]

Decaen rarely interfered, even allowing Flinders regular visits on parole to Port North-West.

Seaman William Smith was repatriated in April 1806, and Flinders's servant John Elder – suffering a deep depression accompanied by hallucinations – followed in June. Flinders was now the only member of the *Cumberland*'s company left in the colony. His melancholia returned, triggered by the departure of Thomi Pitot for the island of Réunion, and of Captain Bergeret for France. His dejection of spirits might have proved fatal, he wrote,

> had I not sought by constant occupation to force my mind from a subject so destructive to its repose; such an end to my detention would have given too much pleasure to the Captain-General, and from a sort of perversity in human nature, this conviction even brought its share of support ... But what assisted most in dispelling this melancholy, was a packet of letters from England, bringing intelligence of my family and friends, and the satisfactory information that Mr. Aken had safely reached London, with all the charts, journals, letters and instruments committed to his charge.[39]

L'Affaire Flinders and the Indian Dream

Paris, Monday, 30 July 1804. Admiral Decrès writes to Captain General Decaen:

> *Your despatch ... contains some very interesting details of which I have given an account to H.M. Your projects on India, the plans you have communicated, the noble impatience expressed in your correspondence to break out from the state of being an observer to which you have had to confine yourself, in order to take a role more worthy of your ambition to attack and destroy the British Empire in that part of the world, are appreciated as highly as they deserve to be; they will be taken into consideration when events come to pass that merit the genius of a great man ...*
> *[Baldwin, 1964, RGSA(SA) Proceedings, vol. 65, pp. 53-67]*

Vice-Admiral Denis Decrès, Napoleon Bonaparte's long-serving Minister of Marine and Colonies, was an accomplished bureaucrat and a great survivor. His main characteristic in office, says Paul Fregosi in his study of Napoleonic campaigns outside continental Europe,[40] 'was lack of enthusiasm for any of Napoleon's distant projects' – understandable in an officer who had commanded one of the few French ships (the frigate *Diane*) to escape the holocaust of the French Fleet in Aboukir Bay in August 1798. He also seems to have nursed a strong dislike for General Decaen – an hostility that, as luck would have it, did nothing to help the cause of the general's 'prisoner of state', Matthew Flinders.

Decrès's dispatch, containing his decision to refer *l'affaire Flinders* to the Council of State, as well as his comments on Decaen's Indian projects, reached the general about the end of October. The latter were of far greater interest to him than the former; nevertheless, confirmation that the navigator's case would now be considered by the Council of State allowed Decaen to concentrate on the 'big picture' – the overthrow of British rule in India.

Following his coronation in December 1804, Napoleon turned his attention briefly to Decaen's project. Envisaging a vast invasion convoy of more than 20 000 troops, which would pick up reinforcements at Isle de France before landing on the subcontinent, he sought Decrès' advice. 'For such a project to succeed', the Navy Minister replied, in effect giving it the kiss of death, you would need ... that the man responsible for carrying it out be of such strength of intellect, will, and energy, that I do not know of such a person in your Majesty's Navy.[41]

Napoleon lost interest, and in September 1805 marched eastwards against Austria.

Cut off from the centre of power by a three-month sea voyage, Decaen sent envoy after envoy to France carrying fresh proposals for an assault on India. The fifth, his younger brother René, returned in February 1806 with the emperor's assurance that, when his campaigns in Europe were ended, he would give his full attention to Asia. Boosted by the news the general set about strengthening the island's defences, and recruited several companies of reserve troops to support the regular forces (which never exceeded 2000 men). He also increased the in-centives to the privateers based in the colony, urging them to intensify their attacks on British shipping in the Indian Ocean. The losses to shipowners and their insurers were calamitous – estimated at more than £10 million between 1803 and 1810. It was enough to keep the colo-ny's economy functioning throughout the years of blockade.

In Europe, meanwhile, Napoleon had comprehensively defeated the Austrians at Ulm (October 1805), the Austrians and Russians together at Austerlitz (December 1805), the Prussians at Jena and Auerstadt (October 1806), and the Russians at Eylau and Friedland (February and June 1807). Despite the naval defeat at Trafalgar in October 1805, the land victories left him master of Europe, and by the Treaty of Tilsit (July 1807) he gained Russian support for a French overland invasion of India. At the same time he opened negotiations with the Shah of Persia – unwisely, because Persia and Russia were traditional enemies – for a joint Franco-Persian attack through Afghanistan.[42]

Amid these great events in Europe, in July 1807 the frigate *Grey-hound* and a cartel from Madras arrived at Port North-West (now re-named Port Napoléon by Decaen) under a flag of truce. Captain Trou-bridge carried a dispatch from Admiral Sir Edward Pellew to General Decaen, enclosing 'instructions for [Flinders's] release under the au-thority of the French minister of marine, to the Captain-General of the French establishments'. Sir Edward proposed an immediate exchange with a French officer of equivalent rank.

Decaen's feelings upon receiving the Minister's dispatch via the hated enemy can readily be imagined. It was dated 21 March 1806, and three copies were sent on separate ships, not one of which made it through the blockade. The Minister wrote:

> ... I am today sending you herewith the decision of the Council [of State] approved by the Emperor and King on the 11th of this month. You will see that your conduct has been approved and that from a pure sentiment of generosity the government has granted Captain Flinders his liberty and the restoration of his ship.[43]

The dispatch concluded: 'In consequence would you be so good as to give the appropriate orders and send me a report.'

Despite the Minister's instruction, the general declined Pellew's request for an immediate exchange, and the *Greyhound* sailed without Flinders. To Flinders's inquiry about the time of his release, Decaen only answered 'so soon as circumstances will permit, you will fully enjoy the favour which has been granted to you by His Majesty the Emperor and King'. The following month he wrote to Decrès:

> the circumstances having become still more difficult, and that officer appearing to me to be always dangerous, I await a more propitious time for putting into execution the intentions of His Majesty. My zeal for his service has induced me to suspend the operations of his command.[44]

Yet again the timing had proved fatal to Flinders's hopes for freedom. Coinciding with the *Greyhound*'s arrival, dispatches came from France ordering the captain general to make contact with the French legation in Persia – no doubt in connection with Napoleon's Indian enterprise. Decaen sent his brother René back to Paris for a further briefing, and to propose a simultaneous attack on the subcontinent from Persia and Isle de France.

At this critical juncture the general could not risk releasing Flinders. He still doubted the Englishman's motives in calling at Isle de France – suspicions revived by news that the 'unseaworthy' *Investigator* had since sailed back to England. He was well aware that Flinders had blatantly evaded the censorship regulations, and questioned why Admiral Pellew should involve himself personally in the navigator's release – unless it were to interrogate him on the winds and currents, the military strength, and the defences of Isle de France. By now Flinders was a well-informed witness to the essential weakness of the island's defences, and how easily a small force might overcome them. Its capture would be a crushing blow to the emperor's Indian plans.

Plans there undoubtedly were – in the emperor's fertile imagination if not always detailed on paper. In exile on St Helena, he reminisced with Dr O'Meara that, despite Trafalgar, he still had nearly sixty ageing sail of the line; he had intended to sail to Isle de France with 30 000 men, take on 3000 blacks as colonial infantry, and proceed to India, where he would link up with the Maratha princes. Nothing came of it, of course. The Spanish ulcer began to haemorrhage in the summer of 1808; as in Vietnam a century and a half later, Spain proved a bottomless pit, sucking in more and more men and resources. To quote Fregosi again:

India stayed out of reach, while the frustrated Decaen on Mauritius raged against a fate that left him stranded on a small island in the middle of the Indian Ocean while the great actions of the war happened elsewhere.[45]

Thus, ironically, Flinders and Decaen came to share a not dissimilar fate. Flinders acknowledged this, writing that the general

saw his former companions becoming counts, dukes, and marshals of the Empire, whilst he remained an untitled general of division; he and his officers, one of them told me, felt themselves little better circumstanced than myself – prisoners in an almost forgotten speck of the globe with their promotion suspended.[46]

This latest blow – the refusal to carry out the emperor's order for his release – came close to breaking Flinders's spirit. The bouts of depression that had plagued him since his detention grew more frequent and more acute. Worse was to come; he learnt from the French papers that Freycinet and Péron had laid claim to his discoveries on the Unknown Coast, and named it *Terre Napoléon*. The southern coast of Terra Australis was now scattered with the names of French scientists, generals, and members of the imperial family. He wrote bitterly to Banks:

Whilst General De Caen keeps me prisoner here, they search at Paris to deprive me of the little honour with the scientific world which my labours might have produced me.[47]

The episode added to his gnawing resentment against the general. Despair settled on him. As the years of captivity dragged on, with no end in sight to the war, it seemed he would remain a prisoner 'for ever'. In 1806 he had written to Samuel:

though still a young man [at thirty-two] misfortunes have made me old, and if they persecute me much longer, neither my body or mind will be capable of much further exertion.[48]

Now his hair was turning grey, and age began to line his face.

Brothers in misfortune

Tamarinds, 25 July 1808. Flinders writes to his friend Charles Baudin des Ardennes, recuperating at Port Napoléon after losing his right arm in a naval action:

> *My dear Sir – I received yesterday your favour of the 22nd with a pleasure that was accompanied with painful emotions on seeing your former hand writing so changed and reflecting on the cause; but permit me to offer you my very sincere congratulations on the advance already made in your restoration to health … be assured that no one can take a more zealous part in your health, both of body and mind, than I do.*
>
> *It is an honour of which I am very sensible that the first letter written by your remaining hand was addressed to me – to your brother in misfortune. Would that you had as good a prospect of being one day reestablished in your physical powers as I have in my natural rights. The loss of liberty is severe, but unfortunately for you, my friend, it is sooner found than a limb severed from the body. How often have I regretted that you, who are so well calculated to pursue researches that might add to the mass of knowledge and happiness in the world, should have been drawn by existing circumstances to quit so truly noble a career for one which terminates for the most part in producing misfortune either to oneself or to others; the labours of Newton and Cook were beneficial, whilst those of Alexander and Caesar desolated mankind.*
>
> *Would that our two nations were convinced of this truth, and act accordingly, then might we hope the animosity which makes it a duty for one man to destroy another would become an honourable emulation for excellence in the useful arts and sciences; when the English philosopher might pay the tribute of acknowledgement to the Savant Francais with the approbation of his nation, and receive in his turn the applause due to his labours; but alas this is but a theoretical dream, which the passions of the human heart will never permit to be realized; all that we can do is to concur in producing the good and avoid joining to do the evil as far as circumstances will allow. I would, my friend, that you might wholly adopt*

my sentiments, and let the glory of contributing to the advancement of science efface the desire of military glory; to take examples from my own nation I wish that you would took [sic] Newton and Cook for models rather than Nelson. The reputations of the first are immortal as the light of the sun whilst that of the last is a flambeau, brilliant indeed for a time, but which in half a century will scarcely be remembered.

I look forward to the time when you may be able to make a journey into the country, and when I hope you will favour me with at least a few days of your society. Did not the nature of my situation present an invincible obstacle, I would not wait for that time to have the pleasure of being near you. I have made an application lately to be sent to France, despairing to obtain anything more favourable. Would that it might be granted and that you might return on board the same vessel. I would endeavour to soothe your misfortune, and to turn your mind into a channel more productive of real satisfaction than that of war. Adieu, my dear Sir, search only at present to amuse agreeably your mind and restore health to your body, and believe me to be very sincerely

> *Your affectionate friend and humble servt*
> *Mattw. Flinders.*
> *[Baker, 1962, appendix 4, pp. 134-135]*

Another of Flinders's former acquaintances from Port Jackson reached Port Napoléon in March 1809. Captain Emmanuel Hamelin arrived from France in the frigate *Vénus* to take command of French naval forces in the Indian Ocean, carrying orders from the emperor to search out and destroy all English trading vessels in the region. In a letter informing Flinders of his arrival, Pitot wrote that Hamelin desired to visit him, but feared the captain general's displeasure. He had assured Pitot (so Flinders noted in his private journal on 28 March),

> that all the officers of the *Géographe* and *Naturaliste*, but particularly MM. Hamelin and Péron, had made applications to the marine minister for my liberation, and that the minister had several times answered, that an order had been sent out for that purpose. It was known, nevertheless, in France … that I was still detained.[49]

Hamelin's career as commander of 'the Hamelin division' was brief but effective. Within the year, five well-armed and richly laden East Indiamen and numerous smaller vessels were captured and taken to Port Napoléon. The East India Company's trading post at Tappanouly on Sumatra was stormed and razed. Time was running out for the French, however – the company's directors in London had already prevailed on the Admiralty to eradicate 'that nest of pirates' on Mauritius. A full-scale land assault on the island stronghold was planned as soon as an invasion force could be assembled from India and South Africa.

The campaign opened in August 1809 with a landing on the small island of Rodrigues (about 300 miles east of Isle de France) by a mixed force of British and Indian sepoy troops from Bombay, commanded by Colonel Keating. With the island secured as a staging post, Keating joined forces with ships of Commodore Rowley's squadron from Cape Town to prepare an assault on the island of Réunion. Both men would play a part in the final act of Flinders's confrontation with Captain General Decaen.

Deliverance – at a price

Port Napoléon, Thursday, 7 June 1810. Captain Flinders hands his signed parole to Colonel Monistrol.

> *I undersigned, captain in His Britannic Majesty's navy, having obtained leave of His Excellency the Captain-General to return in my country by the way of Bengal, promise on my word of honour not to act in any service which might be considered as directly or indirectly hostile to France or its allies, during the course of the present war.*
>
> > *Matthew Flinders.*
> > *[Flinders, 1814, vol. II, p. 482]*

Deliverance, when it came, was sudden, and scarcely hoped for; Flinders had come to accept Bonnefoy's prediction six years previously that he would remain a prisoner 'so long as the war lasted'. On 13 March 1810 a letter arrived at the Refuge from Mr Hugh Hope, commissary of prisoners on the cartel *Harriet* from India, informing him that General Decaen had agreed to his release and he was free to sail on the cartel:

> This unhoped for intelligence would have produced excessive joy, had not experience taught me to distrust even the promises of the General ... for General De Caen to let me go at this time, when I knew so much of the island and an attack on it was expected, would be to contradict all the reasons hitherto given for my detention ...[50]

Flinders remained in a 'state of suspense, between hope and apprehension', until the 28th, when an express arrived from Colonel Monistrol confirming that

> His Excellency … authorises you to return to your country in the cartel
> *Harriet*, on condition of not serving in a hostile manner against France
> in the course of the present war.[51]

Three months later, Flinders was on his way to the Cape of Good Hope.

Mrs Pineo argues, convincingly in my view, that Flinders's unlooked-for release may well have resulted from secret negotiations between General Decaen and Lord Minto, the new governor-general of India. Flinders later wrote that Hope 'had been directed by Lord Minto to make an application to General De Caen for my restoration to liberty under such conditions as the General might think necessary …'. In the absence of direct evidence, suggests Mrs Pineo,

> what is most probable is that the Captain-General, an excellent strategist,
> must have demanded, in exchange for the immediate liberation of the
> navigator, assurances for his own freedom, should he … be reduced to
> capitulate to an invading force.[52]

Her thesis goes far to explain the events that followed.

The *Harriet* remained at Port Napoléon for three months, sailing on 13 June. At sea, Flinders was transferred to a British warship blockading the island, and was invited to dine with Commodore Rowley and some army officers, including Colonel Keating. He does not say whether they discussed the invasion of Réunion, now little more than three weeks away, or that of Isle de France, in both of which they took a prominent part.

'By a happy concurrence of circumstances' (Flinders's words), the same day a frigate arrived from India carrying dispatches from Lord Minto. Rowley sent HMS *Otter* with these to Cape Town, and agreed to Flinders's suggestion that he take a passage on her; this would give him the chance to catch a packet home to England. He found one at the Cape about to sail, but before he could board he received orders to report at once to Vice-Admiral Bertie, the commander-in-chief. Bertie was in the midst of planning the invasion of Isle de France, and required information on the island's topography, military strength, civilian morale, and similar intelligence.

When Flinders demurred, pointing out that this was contrary to his parole, Bertie read it through, 'conceiving with me [!] that I was under no obligation to refuse any information that might be required of me relative to that colony'. Flinders gave in; at the admiral's insistence he answered a detailed questionnaire prepared by the commander-in-chief, and drafted a map of the island, indicating potential landing places and the fortifications protecting the port. A man of integrity, he was undoubtedly embarrassed by this enforced breach of his parole. Though he gives details of the episode in his Private Journal, he omits all mention of it in the *Voyage*. He does, however, question Decaen's motives for freeing him so unexpectedly:

> If the island were attacked and he could repulse the English forces, distinction would follow; if unsuccessful, a capitulation would restore him to France and the career of advancement. An attack was therefore desirable; and as the Captain-General probably imagined that an officer who had been six years a prisoner ... would not only be anxious to forward it with all his might, but that his representations would be adhered to, the pretexts alleged for my imprisonment and the answer from France were waived.[53]

Meanwhile, Réunion had fallen almost without a struggle. Keating's troops, some 4000 strong, disembarked under the guns of Rowley's squadron on 7 and 8 July. The outnumbered French, mostly local militia, surrendered on the 10th (coincidentally the day of Flinders's arrival at the Cape). British casualties numbered fewer than 100 killed and wounded.

Made overconfident by the easy victory, the British moved on to Isle de France. In mid-August four frigates of Rowley's squadron, with a small army detachment on board, entered the harbour of Grand Port, on the island's east coast, to establish a beachhead. Surprised by a part of Hamelin's division the frigates opened fire, but in the unfamiliar waters of the harbour they ran aground; in a confused and bloody fight with an inferior force, spread over several days, all four frigates were captured or destroyed, and their captains and crews taken prisoner (in all some 1700 men). It was the Royal Navy's most ignominious defeat during the Napoleonic Wars.

The following month, Hamelin won a short-lived victory in his ship the *Vénus*, capturing the frigate HMS *Ceylon* on her way from Madras to join the invasion force. On board was General Abercromby, the commander-in-chief, and his staff, but before the *Vénus* could make the safety of Port Napoléon with the *Ceylon* in tow, Commodore Rowley and the rest of his squadron caught up and forced the Frenchman to strike his flag. Abercromby was released and Hamelin taken prisoner; he was eventually exchanged and returned to France, where the emperor created him a Baron of the Empire.

Abercromby and Rowley had learnt their lesson, and planned the final assault on Isle de France with meticulous care. The invasion fleet of some seventy ships, mostly men-of-war and armed Indiamen, carrying more than 10 000 troops, arrived off the island on 29 November. They landed on the unfortified north-east coast, where they were least expected. Against them Decaen could muster about 2000 men, including the colonial militia and several hundred Irishmen, recruited from captured British naval and merchant crews (some perhaps from the debacle at Grand Port).

Against such overwhelming force the defenders could do little. Following an honourable resistance in which he was slightly wounded, General Decaen surrendered on 3 December, accepting the generous terms offered by Abercromby. The terms of the capitulation disgusted many British officers – the regular French troops, including the general and his staff, were not treated as prisoners of war, but were allowed unconditional repatriation to France. Decaen and his entourage took their passage home on a British transport – prima facie evidence perhaps for Mrs Pineo's argument.

Napoleon commented that he had never seen better terms. Decaen was cleared of any blame for the colony's fall and appointed commander of the Army of Catalonia, serving with distinction under Marshal Suchet. Imprisoned for a while under the restored monarchy, he lived in retirement in the country writing his memoirs. He died in 1832, with the reputation of 'a man gifted with the highest qualities necessary to the soldier, the administrator, and the politician'. Many of his papers were lost when his house was ransacked by Prussian troops in the war of 1870-1871.

Notes

1. The former Port Louis.
2. Péron and Freycinet, vol. II, unpublished translation.
3. Milius, 1987, p.52. (Passage translated by J. Treloar.)

4. Ibid.

5. Fregosi, 1989, p. 230.

6. Flinders, 1814, vol. II, p. 355.

7. Ibid., p. 358.

8. Ibid., pp. 359-360.

9. Scott, 1914, pp. 324-326.

10. Flinders, op. cit., p. 363.

11. Ibid., p. 369.

12. Ibid., p. 489.

13. Ibid., p. 373.

14. Ibid., p. 374-375.

15. Ibid., p. 376.

16. Flinders, Private Journal, 27 December 1803. MS, Mitchell Library.

17. Ibid., 18 May 1804.

18. Scott, op. cit., p. 348.

19. The *Naturaliste* was intercepted in the Channel and taken into Portsmouth. After an appeal to Sir Joseph Banks, Hamelin was allowed to proceed to a French port.

20. H. Ly-Tio-Fane Pineo, 1988, p. 83.

21. *Historical Records of New South Wales*, vol. V, 1803-1805, p. 397.

22. Flinders Papers, State Library of Victoria.

23. Ibid.

24. Flinders, Private Letter Book, Mitchell Library. Quoted in Ingleton, 1986, p. 314.

25. Ingleton, op. cit., p. 311.

26. Flinders Papers, op. cit.

27. Flinders, 1814, vol. II, p. 403.

28. H. Ly-Tio-Fane Pineo, op. cit., p. 99.

29. Flinders, Private Journal, op. cit., 25 July 1805.

30. Ibid., 17 August 1805.

31. Flinders, 1814, vol. II, p. 417.

32. Ibid., p. 418.

33. Ibid., p. 421

34. Ibid., p. 438.

35. Ibid., p. 442-443.

36. Ingleton, op. cit., p. 334.

37. Flinders, 1814, vol. II, 448.

38. Scott, op. cit., pp. 364-365.

39. Flinders, 1814, vol. II, pp. 456-457.

40. Fregosi, op. cit.
41. Ibid., pp. 242-243.
42. I am indebted to Mrs Huguette Ly-Tio-Fane Pineo of Mauritius for permission to use material from her book *In the Grips of the Eagle: Matthew Flinders at Ile de France 1803-1810*. She makes sense of the complex personal and strategic factors underlying Flinders's detention by General Decaen.
43. H. Ly-Tio-Fane Pineo, op. cit., appendix 12.
44. Scott, op. cit., p. 372.
45. Fregosi, op. cit., p. 264.
46. Flinders, 1814, vol. II, p. 490.
47. Letter dated 28 February 1809, quoted in Baker, 1962, p. 100.
48. Baker, op. cit., appendix 3, p. 129.
49. Flinders, Private Journal, op. cit., 28 March 1809.
50. Flinders, 1814, vol. II, pp. 478-479.
51. Ibid., p. 479.
52. H. Ly-Tio-Fane Pineo, op. cit., p. 146.
53. Flinders, 1814, vol. II, p. 491.

CHAPTER ELEVEN

⚓

Finale

THE FRENCH EXPEDITION – RETURN TO FRANCE

The return of the *Naturaliste*, 1803

Soho Square, London, 1 June 1803. Sir Joseph Banks to Capitaine *Hamelin:*

Sir – I have much pleasure in telling you that no solicitation of any kind was wanting to procure the release of your vessel. The moment the King's Pleasure could be taken on the subject orders were issued for her liberation which will probably reach port [in France] months before this letter will arrive.

The difference of the actual tonnage of the vessel from the tonnage specified by the French Government when the Passport was required and inserted in the Passport itself, will I trust appear to you, as it does to me, a sufficient cause for the officer who detained you to suspect that your ship was not the vessel for which the Passport was granted. The dispatch with which she was released without solliciting [sic] or even application of any kind as soon as the Proofs of her being the Naturaliste were admitted is a sure proof of the good faith of the English Nation …

Yours Sir, &c.&c.
[de Beer, 1960, p. 121]

The *Naturaliste*, with her irreplaceable natural-history collections, had sailed from Isle de France on 10 February 1803, leaving on the island those of her crew too ill to continue the voyage. Among them was the mineralogist Depuch, 'reduced by dysentery to the final stage of wasting away'; he died a few days later.

Hamelin, anxious to be home, planned to head directly for France without calling at any port; crossing the equator on 28 March, he set course for Le Havre, but almost within sight of the French coast he was stopped by an English frigate. Unknown to him Britain had declared war the previous week, 'and every captain in King George's navy was alert and eager to get in a blow upon the enemy'.[1] The English captain chose to ignore the French ship's flag of truce and demanded that she strike her colours at once.

> I replied [wrote Hamelin] that I could not, being a cartel ship, and that I would send my papers and passport over to his vessel. Ten times he repeated his demand, and each time I responded with the correct observances. I put the boat into the water and he continued to fire shots aloft. At the fourth round, not wishing to pointlessly get one of my men killed … I gave into [my officers'] very reasonable objections, that it would be futile to expose the topmasts to the frigate's guns. Since they were going to keep on with the same demands, I ordered ensign Moreau to haul down our colours and our pennant.[2]

The frigate captain escorted his 'prize' into Portsmouth harbour, Hamelin protesting furiously against this 'infamous conduct'. If he or Baudin had thus abused their power in their behaviour towards Flinders, he told the port authorities, 'the scaffold in Paris would have been awaiting Commandant Baudin and myself, and rightly so'. Several English officers agreed with him, and it may have been one of them – or perhaps his passenger Surgeon Thomson – who suggested he should appeal to Banks for help. The latter's intervention with the Admiralty brought immediate results; orders for the *Naturaliste*'s release arrived on 6 June, and she entered Le Havre next day 'after an absence of two years, seven months, and 18 days'. On board were 133 cases of specimens, numerous plants and shrubs, and upwards of twenty live animals and birds (including two black swans, four emus or cassowaries, three wombats, and two dingoes).

The timing of their return could scarcely have been worse. French ports and arsenals along the Atlantic and Channel coasts were in turmoil following the renewal of the war with Britain. Within the month, First Consul Bonaparte issued orders to the *Grande Armée* to assemble at the Channel ports in preparation for the seaborne invasion of England; the first troops began arriving in late June. Work on the construction of the armada of landing craft, halted in 1801, now resumed, while Boulogne and other small ports on the straits of Dover were upgraded and extended.

At Le Havre the euphoria of their departure was long forgotten – apart from the families and friends of the crew, their return roused little interest. The expedition, it seemed, was generally held to be a failure, as Hamelin discovered when he visited Paris to present his report. Within the Ministry it was apparent that the personal attacks on Baudin by Lieutenant Gicquel, Bory de St Vincent and others on their return home from Isle de France, added to formal complaints against the commandant's conduct lodged by the colony's administrators, had been taken seriously, and prejudiced official opinion against the voyage.

Though unable to obtain an audience with Admiral Denis Decrès, Minister of Marine and Colonies – unsurprising, perhaps, given the Minister's pressing responsibilities at the time – Hamelin was promoted *capitaine de vaisseau* and posted to the *Grande Expédition* against England. There is no record that he supported the campaign of denigration against Baudin; neither, however, does he seem to have spoken up for his former commander. As a career naval officer his first loyalty was to his service, and he took up his new duties with enthusiasm. 'Charged with the conduct and direction of convoys and divisions of the [invasion] flotilla', between October 1803 and July 1806 he led eighteen voyages along the French coast 'under the fire of superior enemy forces'.

His personal good fortune did not blind Hamelin to the misfortune and neglect that had befallen many of the men who had sailed under him. On learning in 1806 that an official account of the voyage would be published, he wrote to the Minister:

> *Voilà* the reward for my painful labours, and I thank you for it! ... If my voyage has been able to win me your esteem and confidence, I am well enough paid for my efforts, but wages are still owing to many of the seamen [who sailed] on the French expedition. Every day I meet sailors who beg me to intercede with you, requesting payment of their wages for the voyage of discovery ...[3]

The return of the *Géographe*, 1804

Port North-West, Isle de France. Friday, 16 December 1803:

> *We raised anchor on the 16th December. I had aboard a crew of one hundred men, a score of passengers, some 40 tubs or casks containing living plants, as well as several kangaroos, lions, tigers, panthers, monkeys, cassowaries, a large number of parrots, and many other birds from New Holland and the East Indies. As we got under way I ordered a 21-gun*

salute in honour of the colony. As I passed under the stern of the Dutch flagship [of Admiral Dekker] I gave it a 13-gun salute. My salute was returned shot for shot. [Milius, 1987, p. 53]

Milius's orders from Admiral Linois spelt out his personal responsibility for transporting the expedition's live animals and plants safely to France, and for the security of the various charts of New Holland made on the voyage – his 'glory' (more precisely, his future career) rested on their safe delivery. Baudin, he found on investigation, had housed the kangaroos and plants beneath the quarterdeck, where the officers and *savants* also had their quarters. Believing that these arrangements endangered their preservation, Milius had most of the cabins dismantled, apart from a few at the stern:

> I allocated the great cabin and the gun-room as officers' quarters. This was not to every one's liking, especially to those unused to giving up their personal comforts. I had some pens built beneath the gangways for the quadrupeds, and small cages for the birds. The pens were raised about four inches above the deck, which made cleaning easy and also allowed the air to circulate … I dedicated the whole area beneath the quarterdeck to the plants, and had this section enclosed with a safety net to prevent these beautiful Asiatic specimens being damaged.[4]

Ashore, Péron and Lesueur prepared their collections for the voyage, adding to them by 'a most careful and persevering study' of the fish populations around the island's coasts; 'they discovered a multitude of new species', wrote Freycinet, 'which were described and drawn with the exactitude typical of the work of these scientific travellers'. Similarly the hardworking gardener Guichenot added substantially to the collection of live plants, obtaining many new specimens from Céré, director of the Botanical Gardens at Pamplemousses. Finally, six animals (among them a panther and a tiger) – a personal gift from Captain General Decaen to Mme Bonaparte – were taken aboard. The separation of carnivores from herbivores, of Asian, African and Australian species, was time-consuming, and it took several days to settle them all in their new quarters.

Before sailing, Milius 'informed the members of the expedition who had disembarked … at our first visit to Isle de France [in 1801] that I would offer them a return passage to France'. The artists Milbert and Lebrun gratefully accepted the offer, along with several other passengers wishing to return to the homeland, and 'a number of orphans recommended to me by respectable persons in the colony'. He took care 'that none of these people in any way compromised my neutrality'.

Milius left the island with a profound sense of relief. His relations with General Decaen had been far from easy, and at times he feared they would be held indefinitely; at one stage the general, as hard as nails, had proposed to keep the *Géographe* in the colony and use the crew to help man the island's privateers. Fortunately nothing came of the plan, Decaen having agreed to a petition signed by the officers and *savants* that they be allowed to complete their mission. The concession had come at a personal price; voluntarily or otherwise, Milius prepared a detailed report for the captain general on the Port Jackson fortifications, and also embarked Major Barois, Decaen's brother-in-law and aide-de-camp, as the expedition's 'geographer' on the homeward passage. Pierre Faure withdrew in Barois' favour.

The voyage is broken by a three-week stopover at Cape Town, once more under Dutch rule. Admiral Dekker has provided an introduction to the governor of the Cape Colony, General Jenssen, and again the expedition is given a more generous welcome on foreign soil than at Isle de France.

Milius receives the governor's permission to advertise to the colonists that he will purchase all animals that might be offered for sale, hoping by this means to obtain 'a precious collection' for the zoological gardens in Paris. Jenssen offers several 'curious beasts' from the company's garden as a gift for the Institut National and for Mme Bonaparte's menagerie. As a mark of appreciation, Milius presents him with two kangaroos (a male and a female), together with a splendid double-barrelled gun. All told, some thirty animals and birds – including a gnu, a zebra, two ostriches, two apes, two porcupines, a civet and a squirrel – are added to the floating zoo.

With Jenssen's aid, Lesueur and Péron make an expedition into the interior, accompanied by the colonial physician. Péron's objective is to study the physical differences between the various African tribes, and specifically the anatomy of the 'Hottentot apron' – a distinguishing feature of female Bushmen that had generated many travellers' tales. His clinical study of the phenomenon, supported by Lesueur's meticulous sketches, demonstrate that the 'apron' is no more than the labia minora, though greatly enlarged.

The Géographe *sails on 26 January. Conditions on board fast become intolerable for passengers (human and animal) and crew. Water rationing is introduced as they approach the equator. The officers bitterly resent their cramped quarters, and (like Baudin before him) Milius is obliged to relieve them of their duties one by one. Luckily the passage is relatively quick, though at the last moment bad weather in the Bay of Biscay destroys many of the plants; the animals all survive. They drop anchor at the port of Lorient on 23 March 1804.*

On arrival, Milius gave orders for the plants and animals to be landed, then prepared to set out for Paris to present his report to Minister Decrès. His health, always poor since his near-drowning on Wonnerup Beach and again at Swan River, now gave way; admitted to hospital in a state of collapse, he remained there for several weeks. In his memoir he wrote:

> The hardships and pains of all kinds that I endured on this voyage led to the complete breakdown of my health. It is pointless to name here those responsible for all my ills.[5]

M. Thevenard, the maritime prefect at Lorient, sought advice from the sole naturalist on board concerning the care of the living cargo[6] now filling his storehouses. Péron, taking advantage of Milius's illness, had already written to Decrès requesting his authority to disembark the collections and arrange their dispatch to Paris; he now persuaded Thevenard to send him to the capital to formalise these arrangements with the authorities there. Thus, notes Frank Horner, 'Péron was the first man from the *Géographe* to present himself at the Ministry of Marine and the Museum'. Milius was not well enough to travel for some six weeks.

Meanwhile, at Lorient Lesueur organised the dispatch of the animals, plants, and natural-history collections, under the supervision of Geoffroy Saint-Hilaire, a celebrated naturalist, who formally took possession of them for the government and the Muséum d'Histoire Naturelle in Paris. They left in separate road trains, the animals in their cages on nine carriages led by Lesueur and guarded by eight men, and the plants and trees in almost 300 tubs under the care of the gardener Guichenot. Geoffroy described the collections as 'the richest ever received by the Museum'.

Péron in Paris

Paris, 19 April 1804. State Counsellor Fourcroy, director of the Muséum d'Histoire Naturelle, to Citizen Milius, commanding the Géographe at Lorient:

Citizen, – The collection that the Géographe *has brought back to us is the largest we have ever received. Everything from New Holland was until now unknown, and in this respect enormously valuable to the Museum, which seeks to bring together and collate specimens from around the world. If we owe much to those who by their efforts have obtained these items for us, everyone who has assisted them in their work, and who has developed measures for conserving and transporting their collections in good condition, must also share the grateful thanks of those who study natural history. [Milius, 1987, p. 61]*

While Milius remained in hospital at Lorient, Péron arrived in Paris in early April. First calling on Minister Decrès and the administrators of the museum, he set about counteracting the 'bad rumours' about the expedition held in many quarters. Decrès was discouraging, and although approving funds for the naturalist's immediate needs, he refused an application for backdated promotion as 'not the Navy's concern'.

Péron's reception at the museum was another matter. When the collections arrived by road from the port, the professors were dazzled by their scope and magnificence. Péron, as the man on the spot, was singled out for special mention. In a certificate signed by Fourcroy, Jussieu, Geoffroy and others, the work of Péron and Lesueur was described as superior to all previous expeditions of this kind.

Milius did not report to the Ministry until May, then took leave to recover his health. Decrès confirmed his promotion as *capitaine de frégate* in June, and later that year Milius was appointed to the *Patriote* (seventy-four); in 1805 he received his own command, the *Diodon* (forty). The *Géographe*'s remaining officers also returned to active service, while the ship's papers (among them Baudin's journals) were lodged with those of the *Naturaliste* in the Ministry. Though it had always been intended to write an official account of the voyage, national priorities had since undergone a sea change – *la gloire* was now to be found in battle, not in scientific discovery. Decrès would not countenance the release of experienced officers such as Hamelin and Milius just to write the history of what he considered a failed voyage.

In June 1804, at Peron's request, Professor Jussieu submitted a report to the ministers of Marine and the Interior and to Mme. Bonaparte (now empress of the French following her husband's elevation on 18 May), defending 'an expedition which has been judged unfavourably'. Jussieu stressed the great value of the natural-history collections to French science, and called for the earliest possible publication of 'this fine voyage'; yet again the request fell on deaf ears.

Péron, meanwhile, was soliciting support in more exalted circles. Josephine's bird-keeper had accompanied the caged animals and birds from Lorient to the capital, and subsequently Péron had corresponded with her concerning those items (animals, plants, and artefacts) destined for Malmaison. In May, with the museum's agreement, he presented the empress with the expedition's entire collection of native artefacts from New Holland, Van Diemen's Land and the South Pacific,[7] at the same time begging her to intercede with 'her glorious spouse' to support his own work on the zoography of New Holland. The appeals, to the empress and the emperor, continued through 1805, without noticeable effect. Horner quotes a letter from Péron to Josephine at this time, sounding, as Horner says, a new note of desperation:

> More than ever you are indispensable to us … our work is suspended … we are refused the smallest advance. The salaries from our voyage are almost entirely owing … our needs are pressing … we have obligations of honour to meet.[8]

The emperor's prodigious energies were engaged elsewhere. The invasion fleet in the Channel ports grew to more than 2000 craft, and some 150 000 men of the 'Army of England' trained month after month for cross-Channel amphibious operations, awaiting only the arrival of the French fleet to shield the crossing; more than 2000 were drowned in a disastrous exercise in July 1804. The Royal Navy's blockade remained intact, however, and in September 1805 Napoleon marched east to attack the Austrians massing on his eastern borders. The crushing victories at Ulm (October) and Austerlitz (December) forced Austria to sue for peace, but meanwhile, Nelson at Trafalgar had destroyed once and for all Napoleon's dream of invading England.

Throughout this time, Péron and Lesueur, now settled in a crowded apartment near the museum, continued their work on the expedition's collections, cataloguing specimens, and acclimatising living plants and animals into their new environments in the Paris Botanical Gardens and Zoo and at Malmaison; very probably Péron advised the empress personally on her acquisitions. (If Napoleon's alleged remark concerning Baudin[9] – '[He] did well to die; on his return I would have had him hanged' – was indeed made, perhaps it stemmed from his wife passing on the naturalist's denunciations of his former commander.)

Péron also wrote extensively on scientific topics – on naval hygiene, dysentery in the tropics, the Hottentot 'apron' (lavishly illustrated), the temperature of the sea, and others – and began drafting his account of the voyage. He and Lesueur seem to have survived on loans, some back pay, and perhaps occasional *ex gratia* payments by Josephine.

Matters improved in 1806, following a fresh report by the Institut (now Imperial rather than National), signed by its perpetual secretary, Georges Cuvier, and other eminent members, Bougainville and Fleurieu among them. Cuvier and his colleagues again singled out the two friends for extravagant praise – 'Messrs. Péron and Lesueur alone have discovered more new animals than all the naturalist voyagers of our time' – and urged that the young naturalist (just turned thirty) should be appointed to write the official narrative of the voyage. Significantly, their report was addressed not to Vice-Admiral Decrès at the Marine Ministry, but to Jean-Baptiste Champagny, Minister of the Interior. After fobbing off the importunate Péron for two years, Decrès was more than content to pass the ministerial responsibility for overseeing the *Voyage* to his colleague.

Less than two months later, on 4 August, Napoleon issued an imperial decree approving publication of the *Voyage* at government expense. It was to appear in three parts – historical, 'customs and description' of the native peoples visited, and navigation and geography. A fourth part, covering natural history, would be financed later by public subscription. The narrative was to be written by Péron and Lesueur, who were granted pensions of 2000 and 1500 francs respectively. As Horner points out, 'the decree was also an unmistakeable Imperial damning of Baudin – the commandant's name appeared nowhere in the proclamation'.

François Péron's story of the voyage

Paris, 1808. The geographer Conrad Malte-Brun reviews Péron's Voyage de découvertes aux terres australes, *volume I, published the previous year:*

> *If one is to believe the authors of this narrative, the impressive travellers who at the call of glory threw themselves into a dangerous career, saw themselves delivered over to the ineptitude of a leader who neglected all his instructions, ran headlong into all the obstacles he had been warned to avoid, knew nothing of how to use the winds or the currents, hindered*

even the investigations he was required to promote, and, to crown all, sac-
rificed to a sordid avarice or to a culpable negligence, the health and life
of all his comrades. [Horner, 1987, p. 3]

'All voyage-narratives are self-serving,' writes Philip Edwards in his study
of sea narratives in 18th-century England,[10]

and to watch the development of a narrative is to see the record being
adjusted, massaged and manipulated ... The writing, in all its devious-
ness, is a continuing involvement with and a continuing attempt to domi-
nate the reality it is claiming to record[11]

in the interests of the writer's self-image and reputation. Bligh's account
of the *Bounty* mutiny is a case in point: although he avoided 'fabrica-
tion and lying', Bligh was nonetheless

adept at suppression ... it was clearly in [his] interest to record a voyage
as trouble-free as possible until the disaster, and this frequently meant
suppressing incidents and comments which indicated friction or could
imply grounds for resentment or revenge.[12]

Viewed from this perspective, Péron's *Voyage de découvertes aux*
terres australes takes on added meaning, revealing much more of the
author's personality than he intended. In this respect, present-day readers have a marked advantage over our 19th-century forebears, because
we can compare the naturalist's account of events with that of Baudin,
now available in English translation.[13] The commander's narrative has
yet to be published in France.

The first volume, accompanied by an *Atlas (Historique)* illustrated
by Lesueur and Petit, and covering the first part of the voyage from Le
Havre to Port Jackson, was the only part of the history to appear in
Péron's lifetime. It is a narrative of happenings, and a richly detailed
account of the natural history and of the appearance and customs of
the inhabitants of the regions visited (including Tenerife, Isle de France,
Shark Bay, Timor, Van Diemen's Land, and Port Jackson).

Péron, like Dampier a century earlier, was a natural Baconian scientist, absorbed by everything he saw; his limitless enthusiasm for his
research, whether collecting data for measuring the temperature of sea
water at various depths, observing the phenomena of luminescence at
sea, or recording the burial rites of Maria Islanders, is reflected in his
writing. His literary style encompasses a scientific detachment, as when

describing the islands of the Forestier Archipelago off the north-west coast, and the effusive chronicling of a land seen for the first time:

> at this moment [Port Cygnet] appeared to our delighted view. The serenity of the atmosphere; the last rays of the setting sun reflected on the waves; the shade of the forests; the darkened tinge of their verdure; the grand appearance of the mountains in the interior, the tops of which appeared above the clouds; the numerous little creeks and small bays to be seen on each shore; the companies of elegant black swans sailing with majestic motion ...[14]

Underlying all else, however, one senses an obsessive hatred for his captain – an emotion that, to be fair, was reciprocated by Baudin. Reading the latter's journal, one is struck by the numerous derogatory references to Péron, often charged with a contemptuous or spiteful sarcasm. Now, as he wrote the *Voyage*, Péron had the captain's journal open in front of him, and gleaned the full extent of Baudin's contempt. It was enough, writes Professor Oskar Spate,[15] to enrage a saint – and Péron 'was assuredly very far from saintliness'. Not once does he mention Baudin by name – instead it is '*notre Chef*' or '*notre Commandant*', the term often prefaced by a demeaning phrase or epithet, such as 'fault of ...', 'obstinacy of ...', even 'this wretched man'.

Not satisfied with turning Baudin into a non-person, Péron 'was also under a compulsive obsession to paint a full-length portrait – and what a portrait! Unbalanced, tyrannical, ever willing to abandon his people without a show of reason' (here Spate comments that even in his own version Péron appears eminently abandonable), 'an oaf and a fool; and moreover, as a sailor utterly incompetent'. It was this incompetence, according to Péron, that enabled Flinders to claim prior discovery of the Unknown Coast, thus ceding to Britain the prestige rightfully due to France.

Specifically, Péron charged that Baudin's plan to follow the West African coast on the passage from Le Havre to Isle de France cost the expedition precious weeks at sea (some forty days in all); that his squabbles with the colonial authorities caused another month's delay; and that on arriving at Cape Leeuwin at the end of May 1801, ten weeks behind schedule, through cowardice or obstinacy he ignored his instructions and sailed north for Timor instead of south-east to the wintry seas off Van Diemen's Land. For more than a century, most historians uncritically accepted the charges, and they became part of the canon.

In recent years Professor Spate and Dr Horner, among others, have convincingly refuted them, although the evidence was always there, in Baudin's journal and other records of the voyage.

Put simply, the *Géographe* was in the position of an ocean racehorse yoked to a carthorse. Ships in convoy must sail at the speed of the slowest, and the *Naturaliste* was notoriously slow; day after day we find Baudin lamenting the hours lost as he shortened sail to wait for his consort to catch up. As for sailing too close to the African coast, for most of the passage south from Tenerife he was further west than Flinders, and despite being encumbered by the *Naturaliste* he took only seven sailing days longer to reach the Cape of Good Hope.[16] From there to Isle de France he was delayed by unseasonable adverse winds.

The delay at Isle de France was also mostly beyond his control. The island was under blockade, provisions were short, the authorities had not been warned of the expedition's coming, and in any case were wary of the Republican government and its attitude towards slavery (the mainstay of the island's economy). Their refusal, or inability, to reprovision the ships owed as much to these factors as to any hostility towards Baudin personally.

Finally, Baudin's decision to sail north from the Leeuwin does not seem to have been questioned by his officers at the time, and makes good sense. He was short of basic supplies, winter was closing in, and he was sailing into largely unknown waters. As it was, he discovered, and charted, lengthy stretches of the west Australian coast, including Géographe Bay and the Bonaparte Archipelago, before reaching Timor in September 1801.

Péron's 'authorised' narrative of the voyage is, as might be expected, invariably favourable not only to the writer, but also to his friends and colleagues in their frequent disputes with Baudin. Only since the publication (in translation) of Baudin's journal in 1974 have readers been able to judge the extent of the naturalist's exaggeration and 'massage' of the facts. There is little point in detailing this process incident by incident, but the following two examples will serve to illustrate the technique – remembering that whereas Baudin was writing at the time, Péron wrote after his return to France, and with the captain's account before him.

During the *Géographe*'s stay in Shark Bay in the winter of 1801, Péron was twice 'lost' while exploring on Bernier Island at the entrance to the bay. On the first occasion, according to Péron, Sublieutenant Picquet, in charge of the boat's crew, was ordered to wait for him until moonrise, and then, if he had not returned, was to abandon him and return to the ship. Picquet disobeyed the order, camped ashore with his crew overnight, and set off at dawn to find the naturalist. Baudin's journal merely records that Picquet was instructed to return to the ship by moonrise because 'the sky did not look too fine' – he did not want to be caught on a lee shore with a threatening wind; there is no mention of abandoning Péron.

A week later Péron became lost once more. According to Baudin, he was 'as usual' alone, and had gone to the beach to collect shells:

> … wanting to climb on some rocks, where the sea appeared not to break too roughly, he was knocked head over heels by a wave which carried off most of the beautiful shells he claims to have found. The fall caused him several wounds in different parts of the body, and these decided him to return. But instead of orienting himself to go from West to East, he took a North-South course, so that … he became completely lost. It was nine o'clock at night before he rejoined us, worn out with weariness and exhaustion, having had to abandon on the way his tin box, his shirt and his crabs.[17]

In his account, Péron reworked and amplified the incident; while searching for shells on a 'dangerous reef', a sudden surge

> broke with such force … that I was driven against the neighbouring rocks, and over these frightful reefs; all my clothes were in a moment torn to pieces, and I was in an instant covered with wounds and weltering in blood. I recovered myself, however, and exerting all my strength to escape from the surge, which as it retreated would have carried me back against the reefs, I clung to the point of a rock, and thus succeeded in avoiding this last misfortune, which doubtless would have been my destruction.[18]

Such graphic descriptions won Péron a reputation for intrepid bravery with his readers: 'All these observations, with the collections which I have just described, are the fruit of many labours, and many dangers, which twice nearly cost me my life.'[19]

The *Voyage* sold well in Paris, and English and German translations soon followed.

The English version of 1809 – Horner calls it 'a cheap job for a London bookseller, unfortunately a careless and in parts tendentious translation, with many passages omitted' – brought an immediate reaction from the Admiralty. The *Quarterly Review* of 1810 carried a scathing critique of the French original, while including lengthy extracts from the translation. The anonymous reviewer (probably John Barrow, the Second Secretary at the Admiralty) vehemently attacked the French claim to have discovered the extensive coastline of Terre Napoléon, seeing it as a brazen attempt to 'defraud [Flinders] of his well-earned reputation, by adopting the name of the usurper'. Indeed,

> a strong suspicion arises that the whole has the premeditated design to snatch the merit of the discovery from its rightful possessor, for the purpose of setting up a claim, at some future day, to this part of New Holland. The circumstances of Captain Flinders' unjust detention, as a prisoner, on the Isle of France, was an admirable incident to favour this design ...

Barrow exempted both Baudin and Péron of personal complicity in this attempted fraud. 'We know not much of Captain Baudin's character,' he wrote, 'but we cannot think so ill of him as to suppose that he would lend his authority, in so wanton and unjustifiable a manner, "to pluck the laurels from a brother's brow".' As for Péron,

> The perusal of his book has certainly afforded us considerable pleasure, although in the course of our examination of it we shall feel ourselves called upon to reprobate, in the strongest manner, the mean and illiberal conduct into which he must have been betrayed by superior influence. Of M. Péron, as a man of general science, we are disposed to think highly; but we repeat, that in the publication of the work before us, we do not and cannot consider him as a free agent.[20]

Flinders was disposed to agree:

> How came M. Péron to advance what was so contrary to truth? Was he a man destitute of all principle? My answer is, that I believe his candour to have been equal to his acknowledged abilities; and that what he wrote was from over-ruling authority, and smote him to the heart: he did not live to finish the second volume.[21]

Death of a scientist

Paris, December 1810. The Chief Naval Medical Officer, Dr Keraudren, suggests an epitaph for a proposed memorial to François Péron in his native village of Cérilly:

He had great talent yet many friends. He wasted away like a tree laden with the finest fruit which succumbed to the excess of its fertility. [Wallace, 1984, p. 163]

'It is a great misfortune for me not to have had a single day of good health since my return to France,' Péron wrote shortly before his death. Though he had shaken off bouts of fever, dysentery and scurvy during the voyage, on the homeward passage from the Cape of Good Hope he developed an annoying cough that left him pale and sweating. It persisted after his move to Paris, and (with Baudin's drawn-out agony fresh in his mind) needed no diagnosis – he too had tuberculosis. The outcome, as he knew, was inevitable – only the number of years remaining to him to complete his work was in doubt.

At first Péron intended to write all volumes of the *Voyage*. His is the only name on the title page of the first volume of the history, although in chapter I he goes out of his way to acknowledge 'my worthy assistant, my estimable friend, M. Lesueur, the dear companion of all my dangers, of my privations, and of my zeal'. Even before the volume's publication, however, Louis de Freycinet had joined the two naturalists; invalided ashore through illness, Freycinet requested secondment to the project, and the Ministry agreed. For most of 1806 he worked with Lesueur preparing Part I of the *Atlas* for publication; despite the name, it included the latter's natural-history drawings and Petit's sketches of New Holland natives. In the event, Petit's contribution was cruelly cut short by a fatal road accident in Paris, while lack of funds delayed the appearance of Freycinet's charts until Part II was published in 1811.

While Freycinet began work on the third volume of the *Voyage* ('Navigation and Geography'), Péron continued with the second part of the history, covering the return voyage from Port Jackson to France. Despite flagging energies as his health worsened, he still managed to produce several specialist papers, among them a lengthy memoir for the emperor on the economy, political organisation and strategic importance of Port Jackson, concluding with a detailed proposal for the colony's military overthrow (an extension of his 1803 report for General Decaen). He also planned a major anthropological study, a comparison of the races of mankind, which he envisaged would require three lengthy sea-voyages and take fifteen years to complete.

By mid-1808 Péron must have realised the impossibility of finishing even the second volume of the history. He spent summer recuperating at the estate of Count Mollien, Napoleon's Treasurer, but the enforced idleness brought on a depression as enervating as the heat of Paris; in autumn he returned to the capital and resumed his writing. The following year, at the insistence of his doctors (among them the Navy's chief medical officer, Dr P.M. Keraudren), he journeyed south with Lesueur to winter at Nice, on the Riviera. In the benign climate, Péron's symptoms entered partial remission, and as his spirits improved his appetite for field work returned. Instead of resting, he and Lesueur renewed their interests in marine zoology; hiring a small boat, the friends spent much of their time at sea, netting small marine creatures in the coastal waters off Nice and measuring water temperatures in all weathers.

Back in Paris in the spring of 1810, Péron suffered an alarming deterioration of his health. Pale, emaciated, and coughing blood, suffering recurrent fevers and insomnia, enduring agonies from an infected throat, he laboured desperately to complete the manuscript of the return voyage. Louis de Freycinet, filled with admiration for his colleague's courage, helped with the narrative, but by October it was clear that the task was hopeless: 'Nothing can stop or even delay the rapid progress of my decline,' Péron wrote to a friend. 'What does console me and sustains my courage is that I will not die without some honour or without leaving behind me people who will miss me.'

On 19 October – ten years to the day since the guns at Le Havre signalled the expedition's departure – François Péron returned to his birthplace, the village of Cérilly in the Bourbonnais, to die among his family and childhood friends. He was too weak to travel on his own, so Lesueur accompanied him. According to another friend and fellow scientist, J.P.F. Deleuze, Péron

> wanted to finish his days in the place of his birth, close to his sisters who had been the principal objects of his affection. He said to me and to his friends in Paris an eternal 'adieu', and this separation was cruel.[22]

For some time Péron had supported his two sisters financially from his own meagre resources. They found him lodgings with a local dairy farmer, M. Bonnet, who – in keeping with the customary treatment for tuberculosis – made up a comfortable bed for the dying man in his cowshed. Here the scientist spent his last days, with Lesueur and his sisters providing loving care and affection, and feeding him fresh warm milk a spoonful at a time – the only food he was able to swallow. He died, aged thirty-five, on 14 December 1810.

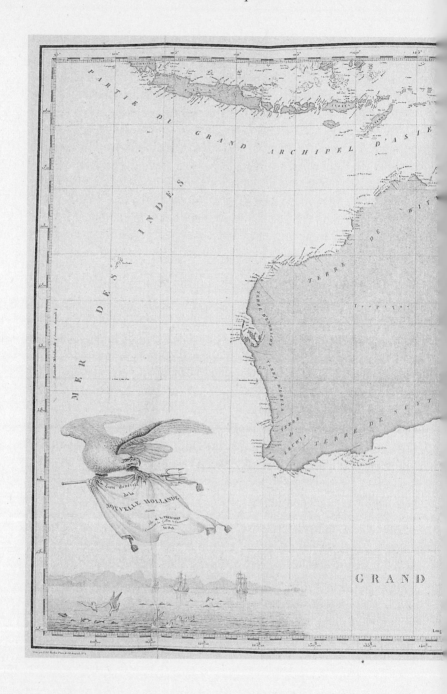

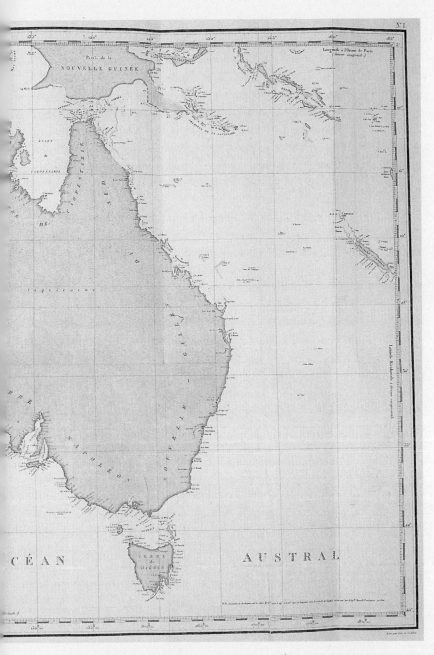

General chart of Australia, drawn and engraved by Freycinet (1808).
[Courtesy of the Royal Geographical Society of South Australia]

A few days before Péron's death, Lesueur drew a final portrait of his friend. Seated in an armchair beside the fireplace in Bonnet's farmhouse, Péron – warmly clad in a fur-lined coat and cap as protection against the December cold – reads a page of manuscript from his work. A bound copy of his *Voyage* lies on a table at his side, next to more manuscript pages and a map of New Holland, clearly displaying the words *Terre Napoléon* on the south coast. Three bound volumes of notes stand next to an inkstand and two quill pens. Measuring instruments and preserved specimens from the voyage complete the drawing. For the artist and his subject, the voyage to the southern lands was clearly the most memorable event of their lives.

Péron bequeathed his scientific papers to his friend. After the funeral, Lesueur returned to Paris, where he and Freycinet prepared the second part of the *Atlas* for publication the following year. However, the naturalist's death ended hopes for early publication of the second volume of the history. Although much of it was already written, funds were lacking for its completion, and Lesueur, it seems, may not have been considered a suitable editor. Freycinet was occupied with the navigation and geography volume and its accompanying atlas and charts (published in 1815 and 1812 respectively). Further work on the historical narrative was deferred indefinitely.

Meanwhile, Napoleon's Götterdämmerung drew closer – the retreat from Moscow in 1812, defeat in the Battle of the Nations at Leipzig in 1813, the invasion of France and his abdication in 1814, his final overthrow at Waterloo in 1815. When work on the history at length resumed after the restoration of the monarchy in 1815, it was Louis de Freycinet who was commissioned to complete it, with Joseph Ransonnet (his former sublieutenant in the *Casuarina*) as his assistant. Volume II eventually appeared the following year. Lesueur, bitterly disappointed, sailed for America, to build a career as artist and naturalist across the Atlantic.

THE BRITISH EXPEDITION

A sword for Lieutenant Fowler

London, 11 August 1804. The Times *reports a naval victory off Pulo Auro, at the approaches to the Malacca Straits:*

> *The signal defeat of [Admiral] Linois by a fleet of loaded merchant ships,*
> *without one ship of war in company, is, perhaps, the most complete triumph*

that British sailors have ever enjoyed over the enemies of their country.
The victory appears to be more complete and decisive when we consider
that these merchantmen engaged the enemy with an equal number of
ships, five to five, and under every disadvantage ...

We hope ... that the Committee at Lloyd's, whose liberality is equal to
their wealth and munificence, will not fail to confer some honourable mark
of distinction, a sword at least, on the India Captains who bore down on
the French line as well as some substantial mark of gratitude on the crews
of those vessels who conducted themselves with such gallantry, as to prove
that they are not unworthy of the 'Tars of Old England'. [Gillespie, 1935,
vol. 21, p. 183]

After farewelling Flinders and the *Cumberland* at Wreck Reef, Lieutenant Fowler and most remaining members of the *Porpoise*'s crew
(including many former 'Investigators') sailed in the merchantman
Rolla for Canton, where the annual China fleet was being assembled
for the voyage back to England. The 1804 convoy was the richest to
date – carrying cargo valued at some £8 million sterling – and included sixteen sail of tea-ships (East Indiamen averaging more
than 1200 tons) and a dozen 'country ships' bound for India. Captain
Nathaniel Dance, master of the *Earl Camden*, was elected commodore
by his fellow captains and the East India Company's representatives
at Canton.

Without a naval escort, and fearing attack by the French squadron
under Admiral Linois, known to be raiding in the region, Dance welcomed the arrival of Fowler and his Navy-trained seamen. Fowler and
a party of his best men joined Dance in the *Earl Camden*, Samuel Flinders and another party went aboard the *Royal George*, and a third group
under a petty officer boarded the *Ganges*. They set to work putting
the three Indiamen on the same footing as sixty-four-gun ships of the
line. Their normal armament was thirty to thirty-six guns (although not
all were mounted), and perhaps deceptive measures, such as painting
false gunports on the ships' sides, were employed. The convoy sailed
on 21 January 1804.

They sighted the island of Pulo Auro, at the eastern approaches to
the Malacca Straits, at daybreak on 14 February. Shortly afterwards the
leading ship, *Royal George*, signalled four strange sail to the south-west.
Fowler boarded a fast-sailing brig to inspect them, returning soon to
confirm that they were indeed Linois's squadron of five ships – a

battleship (the *Marengo*, eighty-four guns), two heavy frigates, a corvette, and a small Dutch brig. Dance recalled his lookout ships, and with Fowler's assistance 'formed the line of battle in close order'.

Neither side made any immediate move to attack. Linois had learned from a Portuguese vessel that 'the India convoy had fitted out three of their ships as 64's'; the news was unsettling, for it meant that the convoy's armament almost matched his. For his part, Dance's priority had to be the safety of his heavily laden and slow-moving merchantmen and 'the immense property at stake'. He continued under easy sail, with the French in the rear: 'I was in momentary expectation of an attack there,' he wrote, 'but at the close of the day we perceived them haul to windward.' Fowler was sent off in the brig to station the 'country ships' on the lee bow, where they were protected against enemy fire by the larger Indiamen; having done so, he returned with some volunteers with war experience to help man the guns.

Both fleets lay to all night, the men at their quarters, the ships cleared and ready for action. Next morning Dance hoisted his colours, offering the enemy battle 'if he chose to come down'. Linois's ships raised French colours, the battleship flying a rear-admiral's flag. At about 1 p.m., as Dance reported to the Honourable Court of the East India Company on his return,

finding they proposed to attack and endeavour to cut off our rear, I made the signal to tack and bear down on him, and engage in succession – the *Royal George* being the leading ship, and *Ganges* next, and then the *Lord Camden* [sic]. This manoeuvre was correctly performed, and we stood towards him under a press of sail. The enemy then formed in a very close line and opened their fire on the headmost ships, which was not returned by us till we approached him nearer.

The *Royal George* bore the brunt of the action, and got as near the enemy as he would permit him. The *Ganges* and *Earl Camden* opened their fire as soon as their guns could have effect; but before any other ship could get into action, the enemy hauled their wind and stood away to the Eastward under all the sail they could set. At 2 p.m. I made the signal for a general chase, and we pursued them till 4 p.m., when fearing a longer pursuit would carry us too far from the Mouth of the Straits ... I made the signal to tack, and at eight p.m. we anchored.

The *Royal George* had one man killed and another wounded, many shot in her hull and more in her sails; but few shot touched either the

Camden or *Ganges*, and the fire of the enemy seemed to be ill-directed, his shot either falling short or passing over us.[23]

Dance paid generous tribute to his 'brother Commanders':

I found them unanimous in the determined resolution to defend the valuable property entrusted to their charge ... and this spirit was fully seconded by the gallant ardour of all our officers and ships' companies.

Fowler was singled out for commendation:

I received great assistance from the advice and exertion of Lieutenant Fowler, whose meritorious conduct in this instance I hope the Honourable Court will communicate to the Lords of the Admiralty.[24]

Two warships of Admiral Peter Rainier's East Indies Fleet, belatedly sent to provide protection through the straits, escorted the convoy to St Helena in the South Atlantic, where they joined other England-bound ships. All arrived safely at English ports in early August.

For Commodore Dance, there was a knighthood, a ceremonial sword worth £100, and a silver vase. The Honourable East India Company, conscious of what might have been lost, also rose to the occasion, presenting Dance with 2000 guineas and plate worth 200 guineas, and Captain Timmins of the *Royal George* with 1000 guineas, a sword, and a piece of plate for his 'bravery and good conduct ... on that occasion in which the bravery and Undaunted intrepidity of British seamen was eminently conspicuous'. The other captains received 500 guineas each, a sword, and plate, while junior officers and crew were rewarded proportionately, down to £6 each for ordinary seamen.

Lieutenant Fowler was not overlooked, receiving a ceremonial sword from the Patriotic Fund Committee, and 300 guineas from the company's Court of Directors; the court also wrote to the Admiralty praising his 'very material assistance' to Captain Dance during the battle. Fowler was promoted commander in 1806, but could rise no further in the service until his former commanding officer, Matthew Flinders, was released by the enemy and received *his* promotion to post captain. It is not clear whether Second Lieutenant Samuel Flinders and the other officers and men from Wreck Reef Bank who served on the Indiamen shared in the company's generosity. Seaman Smith of the *Investigator*, for one, was impressed in The Downs before setting foot on English soil, and was 'retained in the Naval Service until the year 1815'.

Return of the *Investigator*, 1805

Sydney, 21 July 1805. Governor King writes to Sir Joseph Banks:

Although [Flinders] has in a great measure been unsuccessful, a more correct and zealous officer I believe His Majesty's service has not. I hope no carping cur will cast any censure on him respecting the Investigator's *bottom getting Home after the entire ship was condemned here. Should it be so it will be an act of great injustice, for there is not a doubt but her upper works would have separated from her bands had she gone again to sea in her then state; and any person must be convinced how ineligible she was to prosecute the survey in her reduced state if he had waited till she could have undergone the alterations that have been made to her.* [Historical Records of New South Wales, *vol. V, 1803-1805, p. 671]*

Condemned in June 1803 as unseaworthy and 'not worth repairing in any country', the *Investigator* had been laid up as a hulk in Sydney Harbour for less than a year when Governor King had second thoughts about her potential usefulness to the colony. Following a close examination, a new board of surveyors concluded in May 1804 that with suitable repairs she would prove a serviceable vessel for several years yet. By December, King could report to the Admiralty that

the great Repairs given to the *Investigator* have made her in the General Opinion as good as a new Ship altho' considerably reduced in her Tonnage – She is now ready for Sea and will sail immediately for Norfolk Island to execute the Service of removing a part of the Establishment from that Island.[25]

Her former commander would scarcely have recognised his old ship. Her decayed upper works had been removed and the sides cut down to the level of the gunports, transforming the former lower deck into an upper (and only) deck; the guns were removed and stowed, leaving her defenceless. She was now rerigged as a brig (having two masts instead of three), and the original masts and spars were correspondingly reduced in length. Her estimated tonnage (formerly 334 tons) was halved. The *Investigator's* old crew long dispersed, the *Sydney Gazette* carried advertisements during November for 'healthy, active Seamen … able and willing to serve on board the said ship'. Two months' advance would be paid when the ship was ready for sea.

HM Brig *Investigator* sailed for Norfolk Island on 8 January, arriving at her destination after a rough passage of five weeks that badly damaged her newly rigged masts and yards. There she embarked her passengers – mostly settlers bound for the Derwent River and members of the convict settlement's civil and military establishment, but including also the artist Ferdinand Bauer, returning to Port Jackson after an involuntary six-month stay on the island.[26] The voyage left her captain, Lieutenant Houston, disillusioned with his new command and highly critical of her sailing abilities.

Following Flinders's departure in the *Porpoise* in August 1803, Robert Brown and Bauer had been permitted to remain in the colony for a further twelve months to complete their botanical observations. They shared a house in Sydney to the end of November, when Brown left for Van Diemen's Land in the *Lady Nelson* while Bauer remained at Port Jackson. The naturalist's expedition to the southern settlements was planned to take ten weeks, but in the event stretched to nine months – he later called it 'uncomfortable' – and took him to the Kent Group in Bass Strait, Port Phillip Bay, Port Dalyrymple (Launceston) and the Derwent River. It was August 1804 before he arrived back in Sydney, to find that Bauer had just left for Norfolk Island. The two men did not meet again until the *Investigator*'s return in early March.

Their studies completed, Brown and Bauer waited impatiently for a passage to England. They were dismayed when the governor offered them a place on the *Investigator*, which was returning home with urgent dispatches. She was, he said, 'the best conveyance for the collection that was likely to offer'. Brown's concerns were partly put to rest when King appointed Captain William Kent (Flinders's former superior officer and long-time friend) to replace Houston as commander for the homeward voyage. Not only was Kent's seamanship highly respected, but he also listened attentively to Brown's requests regarding secure accommodation for the more perishable part of the collections – 'they were', Brown wrote to Banks on his return, 'provided for as well as the Reduc'd state of the vessel and her crazy condition would admit of'. The ship departed Port Jackson on 24 May and sailed direct for England, by way of Cape Horn.

Liverpool, Sunday, 13 October 1805. An eyewitness observing the arrival of the Investigator *in the port is struck by*

> *The extraordinary appearance of this wonderful old ship, her sides being covered with barnacles and seaweed, and her sails, masts, and rigging*

presenting the usual signs of a vessel that had been abandoned ... [on board] a sight was presented still more astonishing – plants we had never before beheld, black swans and other curious birds and animals surrounded us on every side.[27]

The following day, Captain Kent writes to Mr Marsden at the Admiralty:

Although the sufferings of the men have been considerable in rounding Cape Horn in the dead of winter in this low cut-down ship, whose single deck was almost constantly under water, yet we have had the good fortune not to lose any of them. From the line, which we crossed on the 20th of August, easterly winds have been nearly constant, which, inclining somewhat to the southward, obliged us to press between the Mull of Cantire [Kintyre] and Ireland. Yesterday ... we were overtaken by a great storm from the south-west, and having but three days water, the men without clothing, and the masts, sails, and rigging in a shatter'd condition, I bore up for this port.[28]

A month later Kent reports to Marsden again, this time from Falmouth:

At the time of our departure [from Liverpool] sickness was pretty general through the ship; the two lieutenants, carpenter, and midshipman (all the officers I had) were confined below incapable of doing duty, and a third part of the men. The long voyage we had performed without stopping at any place for refreshment, the old and bad provisions with which we had been supplied, and the short time we had remained in port for the reinstatement of health, all tended to create debility and thereby induce disease. In the passage to this port the carpenter, who had been with me eleven years ... died of fever – also one of the seamen. This valuable officer in his line died miserably, without medical assistance.[29]

Robert Brown wastes no time in reassuring Sir Joseph of his concern for the botanical collection, writing from Liverpool the day of their arrival:

... since it had been on board I have receiv'd from Cap'n Kent every assistance in his power to preserve it. While within the tropics the plants were carefully examin'd and those that most requir'd it were chang'd into dry paper, but such has been the wet state of the ship that they must again be suffering, and that I fear considerably. I am therefore most anxious to have them removed on shore, and I earnestly beg they may not be again put on board the Investigator *for the purpose of being brought round to Portsmouth or the river, as she is not only a crazy but absolutely a defenceless ship. The expense of land carriage cannot, I suppose, be very great.*[30]

Brown and Bauer left the ship at Liverpool and travelled overland to London by coach. On Tuesday, 5 November, they presented themselves at the Admiralty to request an interview with Secretary Marsden – less than twelve hours earlier Lieutenant Lapenotière of the schooner *Pickle* had arrived from Gibraltar with dispatches from Admiral Lord Collingwood. Although it was then after midnight, Lapenotière was taken straight to the First Secretary's apartments. 'In accosting me [Marsden recorded] the officer used these impressive words: "Sir, we have gained a great victory; but we have lost Lord Nelson!".'[31]

Marsden had no time that morning for two men of science. After waiting vainly for three hours, Brown penned a short note for the Secretary:

Robert Brown Botanist and Ferdinand Bauer Painter of Natural History belonging to His Majesty's Ship *Investigator* have the honor of acquainting Mr. Marsden, for the information of their Lordships of their arrival in London.[32]

Even the safe arrival at Banks's mansion in Soho Square of the collection – specimens of nearly 4000 species of dried plants, a large number of them (1700) new to science – received a rather muted welcome. At the time, writes Marlene Norst, 'Sir Joseph happened to be suffering from gout in the right arm, and so could give it neither his undivided attention nor the enthusiastic admiration it deserved'.

In January 1806 Banks proposed to the Admiralty that Brown and Bauer should continue their work on the botanical collection at government expense. He offered his services to

overlook and direct the progress of these gentlemen, to quicken them if they are dilatory, to assist them when it is in my power, and to report to their Lordships the progress made by each.[33]

The work was still proceeding when Flinders returned to London in 1810.

Also in January, Kent received his deserved promotion to post captain, and two months later was summoned to an interview with Lord St Vincent (now commanding the Channel Fleet). Congratulating Captain Kent on his seamanship in bringing the crank *Investigator* home, St Vincent asked bluntly if he would complete the survey of New Holland, so unfortunately left unfinished. Kent, to his great credit,

did not wish to interfere with the future prospect of my esteem'd friend, Cap'n Flinders, whose views I would at all times be happy to forward, were it in my power, instead of impeding.[34]

The Refuge, Wilhems Plains, November 1806. News of peace negotiations at Paris is the principal topic of conversation on Isle de France, but for General Decaen's prisoner of state 'the hope of peace [seems] too feeble to admit of indulging in the anticipation'. Matthew's dejection of spirits, though less oppressive than before, leaves him with no pleasure in life:

> *there is a weight of sadness at the bottom of my heart that presses down and enfeebles my mind. Everything with respect to myself is viewed on the darkest side. The little knowledge I have is not reckoned or is unappreciated; that which I have not is exaggerated, the errors or faults I have committed are exaggerated whilst those of my actions which might bear the name of good are depreciated ... Miserable state! the energy of my mind is I fear lost for ever.[35]*

Sleep, 'that sweet calmer of human woes', is his best resource.

At the beginning of November he receives from his friend Charles Baudin a copy of Steel's Navy List for the previous year, recovered from a British merchantman taken as a prize by the Piémontaise. *Leafing through it, Matthew is startled to come across a familiar name –* Investigator. *Surely it cannot be his ship – she is shown as a brig, lying at Plymouth. Brief as it is, the information makes no sense – his letters from home have not mentioned a new* Investigator. *Worried and uncertain, knowing he cannot expect an answer for nine months or more, he dashes off a letter to Ann:*

> *What is the meaning of it? Is she coming here to me to finish my voyage of discovery? Or is it some other officer to reap the harvest of my labours whilst I am let to remain in prison. I know not what to think; but if the first takes place, particularly under the command of my brother, I request thou will apply to the Admiralty by way of Sir J. B. for permission to take passage to Port Jackson, where, my love, in that case I would place thee ... whilst the examination of Australia should be completed.[36]*

Four months later, visiting Port North-West on parole, Flinders meets an American seaman whose ship had lately been at Port Jackson. From him he learns of the alterations made to the Investigator *and of her homeward voyage. His unease increasing, he writes to Samuel:*

> *I learned from [him] that the* Investigator *had, after all, been repaired, and that Captain Kent had taken her to England. I then supposed, that the* Investigator *in Steel's List must be the same and had arrived safe; but*

*you say not a word of it, though you frequently mention Captain Kent,
and tell me of his promotion. There is something in this affair, Samuel,
which I cannot comprehend. If Captain Kent received his promotion for
taking home the* Investigator, *it follows that I am censured for quitting
her! Is it possible that the old ship has arrived in this manner in England,
and neither you nor my wife made any mention of the circumstance! I
begin to think that I was wrong informed, but in what ship did Captain
Kent return?*[37]

A sailor home from sea

Spithead, Wednesday, 24 October 1810. The Olympia *cutter anchors
in the port, after an uneventful run of eight weeks from Cape Town.
Matthew Flinders takes the night stage to the capital:*

*Thursday 25 October. Arrived in town at 7 o'clock [a.m.]. Went to Mr.
Bonner from whom I learned that Mrs. Flinders was in town. Took a lodging at the Norfolk Hotel, and went thence to the Admiralty, where I saw
Messrs. Croker and Barrow, the two secretaries, and was treated with flat-
tering attention. I learn my promotion took place on Sep. 24 last, previously to the late extensive promotion of post-captains, and from the day it
was known I had arrived at the Cape of Good Hope. At noon, my Mrs. F.
came to me with Mrs. Proctor. I was obliged to leave them in order to send
up my card to Mr. Yorke the first lord ... During the time of waiting, I sat
with my friend Pearce who let me a little into circumstances and characters. When sent up for, Mr. Yorke received me with urbanity and appeared
to appreciate my sufferings in the Isle of France. He told me that my commission should be dated back to the time of my embarking in the cartel,
and a further conversation allowed me to present a memorial for its being antidated [sic] considerably. [Flinders, Private Journal, 1803-1814,
Mitchell Library]*

Flinders remained at the Cape for seven weeks before finding a passage home in the *Olympia* cutter – 'an indifferent sailing vessel, very
leaky, and excessively ill-found', which must have brought back unwanted memories of the *Cumberland*. Despite her drawbacks, she
made St Helena in only fifteen days, on 11 September, and from thence
made a fast passage to Spithead. He had been away for nine years and
three months.

For Flinders the overnight journey to London was far from restful, the emotions kindled by his return banishing all thought of sleep. Following a thorough scrub and a hearty English breakfast at the Norfolk Hotel, he called on his friend Charles Bonner – from whom he learned that Ann, alerted to his return by Sir Joseph, was already in town – before hastening to present himself at the Admiralty. Assured by secretaries Croker and Barrow that his promotion had been gazetted, he made his way back to the Hotel, where Ann awaited him. Their long-anticipated reunion was interrupted briefly by Lieutenant John Franklin, who withdrew in some confusion: 'I felt so sensibly the affecting scene of your meeting Mrs. Flinders,' he apologised later, 'that I could not have remained any longer in the room under any consideration; nor could I be persuaded to call a second time that day.'

The afternoon meeting with the First Lord, the Right Honourable Charles Yorke, somewhat dampened his exhilaration. Yorke was affable enough, sympathising with his hardships as a prisoner of the French, but on the crucial matter of his promotion could only offer to backdate it to 7 May, the date of his own accession to office. Matthew argued vainly that before his departure Earl Spencer had promised him promotion on his return, which assuredly would have been in 1804 (as witnessed by the safe arrival of his fellow-officers) but for the action of the French in seizing him and his ship against the spirit of his passport. His goodwill slipping away, Yorke repeated there was nothing he could do – Flinders's only remedy was an appeal to the King in Council.

The pall of acrid, yellowish smoke congealed above the city matched Flinders's mood as he walked to his hotel. In his absence London had grown still dirtier and more chaotic than he remembered, the strains of seventeen years of war showing clearly. The crowds thronging streets fouled and blocked by a melee of carriages, coaches, coal-wagons and cursing draymen seemed more ill-natured than before, with more soldiers, sailors and beggars evident among them.

Before reaching the hotel, Flinders recognised a familiar figure across the street – Ferdinand Bauer, on his way from Sir Joseph's house in Soho Square. They greeted each other warmly, Flinders anxious to hear the full story of the *Investigator*'s homeward voyage and Bauer congratulating him on his safe return. The artist added that he and Robert Brown were still working on the botanical collections; Sir Joseph, however, was away in Lincolnshire.

Matthew's reunion with Ann was again interrupted that evening when Robert Brown called, having learnt from Bauer of Flinders's arrival. Ann listened absorbed as the two men recounted their adventures since parting at Port Jackson in 1803. When Brown mentioned before leaving that Sir Joseph was expected back within a week or so, following the sudden death of his librarian and friend Jonas Dryander, Flinders took the opportunity to write to his patron: 'I have the happiness to inform you of my arrival', he began, but went on to raise the issue of his promotion, again arguing that it should be backdated to 1804, 'when it may be supposed I might have arrived, had not my unjust detention taken place'. Depressed by his friend's death, Banks was not disposed to agree, writing to Barrow that he thought Yorke's choice of 7 May 'liberal in the extreme'.

After twenty-one years of service in the Royal Navy, Flinders remained a relatively poor man. Instead of service on a ship of war – with its hopes of lucrative prize money – he had chosen the career of an explorer. He had returned from six and a half years of detention with substantial debts and credits to his name. He was still owed his pay from the time the *Investigator* left England in 1801 (approximately £1750, or about £140 000 in modern currency), along with compensation pay for the servants allowed him on the ship. In his absence all bills drawn at Port Jackson and Isle de France had been paid by the Navy Office, but had yet to be debited against his account. His first concern, therefore, was to bring some balance into his finances; his efforts to do so all too soon embroiled him with the Admiralty's bureaucracy.

First and foremost was the request to back-date his post-captaincy – an important issue for Flinders, because it represented not just a significant increase in back pay due to him, but also a substantial rise in seniority on the Post Captains' List. The First Lord (perhaps caught unawares by what he no doubt considered an importunate request) had suggested presenting a memorial to the King in Council, but could scarcely have expected the applicant to pursue it so doggedly. Flinders solicited in turn Secretary Barrow, Admiral Sir Joseph Yorke, the First Lord's brother and also a member of the Board of Admiralty ('who was good enough to allow, that my claims upon the service exceeded what the Admiralty could do for me without superior authority'), the former First Lord Earl Spencer, and his patron Sir Joseph Banks.

Sir Joseph, torn between sympathy for Flinders's situation and doubts about the merits of the case, went so far as to obtain 'the skeleton of a Memorial' from the Privy Council office in Whitehall, advised his protégé on how best to compose it, and suggested amendments to successive drafts. Finally, Sir Joseph's patience seems to have worn thin. 'Called on Sir Jos. Banks,' wrote Flinders on 20 November,

> and found that he had not seen Mr. Yorke yesterday; but it appeared some difficulties were likely to be made about the memorial. Sir Jos. has sent it by Mr. Barrow to Mr. Yorke, stating that both he and I had no intention of proceeding in it, without his approbation. After learning this, I went down to the Admiralty, and sent up my card ... Mr. Yorke ... mentioned the subject of the Memorial, and said he certainly look[ed] upon it favourably; but that he had not yet made up his mind upon it, and wished me not to flatter myself too much.[38]

Flinders could sense how the wind was blowing: 'If the whole time I prayed for was thought too much', he replied, then perhaps the date of the French marine minister's order to set him at liberty might be considered. But in any case, he and Sir Joseph wished to have the First Lord's 'full and entire approbation of the measure, before proceeding further; and I should receive with gratitude any accession of rank, which might accord with his ideas'. Yorke had more positive news on the subject of his compensation pay; although their Lordships had decided against paying compensation as such, because of the precedent it would set, they had granted £500 in lieu, 'for my services after leaving the *Investigator* and on account of my expenses'.

Eventually, faced with the tacit disapproval of both the First Lord and Banks, Flinders allowed the memorial to lapse. He had better luck with the East India Company's Court of Directors, who without hesitation granted £600 in table money retrospectively for the final part of the voyage; of this he received half, the remainder being shared between the officers and scientists of the *Investigator*. He had to wait a further six months, until May 1811, before the Navy Board made up his back pay, from 16 February 1801 to 25 October 1810 – after all expenses were deducted he was left with a credit balance of £104 11s 4d (some £8400 today), barely enough for a year's rent in London. At his request he was placed on the half-pay list as a post captain – thus removing himself from the active-service list.

Amid this time-consuming haggling with the bureaucracy, Flinders did not forget his debt of gratitude to his many friends on Isle de France, several of whom had relatives who were prisoners of war in England.

Top: *Mme Kerivel's house, where 'M. Baudin ceased to exist'. Engraving by Jacques Gerard Milbert, 1812.* [*Courtesy of M. Ly-Tio-Fane*]

Above: *Baie du Cap, Mauritius (formerly Isle de France).* [*T. Bradshaw,* Views in the Mauritius, *1832; RGSSA Library*]

Above: *Port Louis, Mauritius, from the eastern side. [Bradshaw, 1832; RGSSA Library]*

Left: *Monument to Matthew Flinders on Mauritius, on the road to Le Mare aux Vacoas, south of Curepipe, at the junction with La Brasserie road, looking from the direction of Mahebourg. [Photo courtesy of Mr T. Young, Hon. Consul for the Republic of Mauritius]*

On 11 November he and Samuel, who had arrived in London a week before, took the stage to Odiham, in Hampshire, to pay a visit to four of them, for whom he had letters and money from home. They supped with the prisoners, and the following morning

> I had the same four French gentlemen to breakfast with me; they could scarcely enough express their thanks for what I had done for them, [n]or did they appear fully convinced that I should have gone down to Odiham solely on their account.[39]

Happy to have had the chance to repay the kindness he had received as a prisoner, Flinders returned to the capital. To the end of his life he did what he could to assist these and other prisoners of war, writing to them regularly and, when the occasion offered, supporting their claims for release.

Writing the *Voyage*

London, Sunday, 13 January 1811. Captain Flinders leaves his lodgings at King Street, Soho, for Soho Square:

> *Went to Sir Jos. Banks at one to meet Mr. Barrow, to make arrangements for my writing the* Investigator's *voyage. It was generally settled that the admiralty should pay the expenses of reducing and engraving the charts, landscapes, figures, and the parts of natural history, and that the produce of the voyage should pay the printing and paper and all other expenses. I ask if, during the time I was employed writing the voyage, I could not be put on full pay, which would make up the difference of expense I should be at betwixt living in the town and privately in the country, but Mr. B. thought the Admiralty would object from want of a precedent. Thus, in all appearance, my time and labour must be given in – it will cost me £500 or £600, and I cannot be employed during the time in any way that might be advantageous to my fortune. [Flinders, Private Journal, 1803-1814 Mitchell Library]*

Matthew and Ann left London on 23 November to visit relatives and friends in Cambridge, Lincolnshire, and south Yorkshire. Ann at last had the opportunity to show off her 'lion' to her 'old friends, nurses, servants, etc.', while Flinders was eager to see his stepmother, sisters, and nephews and nieces after so long an absence. Former friends and neighbours flocked to meet him, some requesting that a brother or son might sail with him on his next voyage. Among the visitors was John Allen, his shipmate in the *Investigator*, to whom he gave £20 as his share of the East India Company's *batta*, or table money.

Mixing business with relaxation, Flinders settled his father's will, under which he inherited £500, received his share (£195) of an uncle's will, and negotiated the sale of the former family home for £550. Christmas – his first at home for ten years – was spent amid his family, and he and Ann returned to London 'about 10 at night of Thursday the 3rd [January], after an absence of six weeks, and having been 34 hours in the coach'.

Next day he called on Sir Joseph to discuss the writing of the voyage, only to find that nothing had been decided during his absence. At the Admiralty he saw Barrow, 'but found the First Lord too much occupied' with other matters to receive him. More than a week passed before he could meet Yorke, who seemed to 'have much cooled' on the subject of his memorial; as for the narrative of the voyage, he had 'referred it to a committee of Sir J. Banks, Mr. Barrow, and myself; and between the two first I got one o'clock tomorrow fixed for the discussion'. Flinders spent the evening at home reading Péron's account of Baudin's voyage, borrowed from Sir Joseph's library.

Banks, meanwhile, had offered to their Lordships

the same Superintendence in the management of Engravers, Draughtsmen, &c. &c. as I had the honour to execute under the direction of the then Board of Admiralty, in the Publication of the third Voyage of Capt. Cook.[40]

Thanking him on their behalf, Barrow wrote that 'my Lords are too sensible [of its Value] not to accept of it with pleasure and satisfaction'. There could be little doubt where the influence lay in the committee of three, which held its first meeting at Banks's mansion on 13 January, a Sunday.

Chagrined by his less than happy experience with the memorial, Flinders did not press his case for adequate compensation for his labour – which by his estimate left him £500 or £600 out of pocket. Probably regretting his impulsive application[41] in November to be placed on half-pay, he now had no option but to 'consider with Mrs. F. upon the various modes of living in London, proportionate to the expense of living even in our present way'.

Pushing aside his resentment, Flinders threw himself into the twin tasks facing him – preparing his charts of the coasts of New Holland for publication, and writing the narrative for *A Voyage to Terra Australis*. Together they would consume, in a very real sense, the remainder of his life. The documentary materials needed for the narrative history were mostly ready to hand: accounts of previous voyages to New Holland (some in his collection and others in Sir Joseph's library); the first two volumes of his journals with the record of the *Investigator*'s voyage (though not the third, with its irreplaceable data from the voyage of the *Cumberland*, still in General Decaen's hands); his logbooks; and his memoirs, reports and letters to the Admiralty and to Banks. Moreover, he wrote in the preface, even before his return to England 'much had been done to forward the account'. The charts posed a different problem, and will be considered separately.

Inevitably the book's structure, as it emerged from the committee, was an unwieldy compromise; the title itself became a matter for argument. Flinders had first proposed the name *Australia*, or *Terra Australis*, to denote 'the whole body of what has generally been called New Holland', in a letter to Banks in November 1804. He favoured *A Voyage to Australia* for the title, with *Terra Australis* as his second choice. Banks objected to both, preferring New Holland as the name in general use, and did not finally withdraw his opposition to 'the propriety of calling New Holland and New South Wales by the collective name of *Terra Australis*' until August 1813 – henceforth Matthew could use the term for his charts and his book. Officially, at least, *Australia* was ruled out of contention.

The published work, in two large volumes, came to nearly 1100 pages, some 400 000 words in all. The Introduction alone, an historical survey of prior discoveries in Terra Australis by Dutch, French and British navigators (including those by George Bass and Flinders) exceeded 200 pages. The voyage proper occupied almost 750 pages, arranged in three books (covering the voyage out; the circumnavigation; and the return), although most of Book III dealt with his captivity on Isle de France. As well, there were four appendices, two on 'latitudes, longitudes, and bearings', one on compass variations due to magnetism, and the fourth an eighty-page monograph by Robert Brown (now Banks's librarian) on the botany of Terra Australis. It was a complicated format, and the lack of an index further added to readers' difficulties. Perhaps it was Sir Joseph who complained about the latter:

I heard it declared [Flinders wrote in the preface] that a man who published a quarto volume without an index ought to be set in the pillory, and being unwilling to incur the full rigour of this sentence, a running title has been affixed to all the pages ... this, it is hoped, will answer the main purpose of an index, without swelling the volumes.[42]

London, 1814. Extract from the preface to A Voyage to Terra Australis*:*

Having no pretension to authorship, the writing of the narrative, though by much the most troublesome part of my labour, was not that upon which any hope of reputation was founded; a polished style was therefore not attempted, but some pains have been taken to render it clearly intelligible ... Matter, rather than manner, was the object of my anxiety; and if the reader shall be satisfied with the selection and arrangement, and not think the information destitute of such interest as might be expected from the subject, the utmost of my hopes will be accomplished.[43]

Flinders could point to numerous precedents in disavowing any pretensions to authorship. Cook (in his preface to *A Voyage Towards the South Pole and Round the World*, 1777) had begged his readers to excuse his lack of polish, hoping that 'candour and fidelity' would 'balance the want of ornament'. When Flinders came to write his *Voyage* in the early months of 1811, however, he was unusually well-equipped for the task.

His education at Horbling Grammar School in Lincolnshire had given him an early grounding in classics and mathematics. The impetus to go to sea had come from reading Defoe's *Robinson Crusoe*. Though Decaen may have circumscribed his physical surroundings at Wilhems Plains, the friendship of Thomi Pitot and his literary, artistic and scientific circle expanded Flinders's mental horizons in ways that would never have been possible at sea.

Throughout the years of his detention, Flinders in one way or another was practising the writer's craft – apart from his official correspondence and letters to family and friends, there were appeals to anyone with influence who might help secure his release, in England, France, India and Australia; he prepared major scientific papers (some subsequently included in the *Voyage*); rewrote his journals from the *Investigator*; maintained from the first day a private journal recording his imprisonment;[44] and, towards the end of his detention, wrote *A Biographical Tribute to the Memory of Trim* (a 'contrived little essay' clearly influenced by Laurence Sterne's *Tristram Shandy* in style and content).[45] At the same time he was reading widely in English and French, in books borrowed from Pitot and other friends.

Flinders's mature style reflects a range of literary influences – classical and contemporary, sea narratives from Dampier to d'Entrecasteaux, novels from Defoe to Sterne. He prided himself on writing calmly and accurately, taking pains 'to render [the narrative] clearly intelligible'. Thus, describing his arrival at Cape Town in 1801,

> I had the satisfaction to see my people orderly and full of zeal for the service in which we were engaged; and in such a state of health, that no delay at the Cape was required beyond the necessary refitment of the ship.[46]

Elsewhere he brings to life scenes such as the shipwreck of the *Porpoise* in vivid and forceful prose, verbs, nouns and adjectives combining effortlessly to drive the narrative forward:

> On going up, I found the sails shaking in the wind, and the ship in the act of paying off; at the same time there were very high breakers at not a quarter of a cable's length to leeward ... the ship was carried amongst the breakers; and striking upon a coral reef, took a fearful heel over on her larbord beam ends ... A gun was attempted to be fired, to warn the other vessels of the danger; but owing to the violent motion and the heavy surfs flying over, this could not be done immediately; and before lights were brought up, the *Bridgewater* and *Cato* had hauled to the wind across each other.[47]

Nevertheless Flinders, no less than Péron, found himself caught in what Philip Edwards calls the 'schizophrenic dithering' between the demands of science and the general reader's claim for entertainment.[48] While the latter's numbers alone made voyage narratives profitable, it was the voyage sponsors (governments and/or national scientific institutions) who demanded that the account should serve patriotic ends – by enhancing national prestige through detailed observations of geographic and scientific discoveries, promoting future trade and settlement, and so on. No 18th-century writer, says Edwards, achieved a satisfactory balance between the two, and Flinders proved no more successful in the 19th.

Also, like most voyage narrators, Flinders had a personal agenda. For more than six years he had been obsessed with vindicating his voyage and himself – writing to Banks in 1809 that 'they search at Paris to deprive me of the little honour with the scientific world which my labours might have produced me'. The *Voyage* provided Flinders not only the means to reclaim his rightful position in that world, but also the opportunity to argue his case against General Decaen. The temptation was understandable, and irresistible – nothing could exorcise the bitterness of his captivity – but it served only to further distract the reader's attention from the book's scientific content.

Throughout 1812 and 1813 Flinders worked painstakingly on his narrative and his charts. Sometimes, after a day spent on the latter, he continued late into the night correcting his rough account of the *Voyage*. By October 1813 the introduction (204 pages) had been typeset and corrected; from now on his waking hours were occupied preparing fair copy, reading proofs, and crosschecking the narrative against the charts. As late as May 1814 he was still 'examining for Errata and correcting proofs of the preface; but it was very little and ill done'. His chronic 'gravelly' complaint had returned in November 1813, and grew steadily worse. By May, with Flinders no longer able to leave the house, Arrowsmith and Nicol (printers of the charts and the *Voyage*) visited him at home, bringing samples of their final prints for his inspection.

Death of a navigator

London, 1814. An obituary for Captain Matthew Flinders, RN, appears in the Naval Chronicle*:*

> *Nor among the celebrated names that in the various paths of life shed lustre upon the annals of this manly island, must the patriot navigator neglect a grateful tribute to that of FLINDERS. At the precise period when this hardy and persevering mariner had just made good his title to be enrolled among those worthies of his profession, who have merited the gratitude of all 'they that go down to the sea in ships, that do business in great waters', by the final delivery of the elaborate account of his labours to the public has he been cut off from amongst us in the prime of life, by maladies, originating from sufferings that consecrate his memory as that of a martyr in the cause of his country and of science. [Naval Chronicle, vol. XXXII, 1814, pp. 177 et seq.]*

On his return to England in 1810, Flinders anticipated few problems in completing his charts for publication – all those sent back from Isle de France 'were nearly prepared for the engraver'. To ensure their accuracy, however, he first checked his calculations of longitude during the voyage (based on the astronomical predictions published in the Nautical Almanacks for 1801 to 1803) against the solar and lunar observations made at the Greenwich Observatory for the same period. To his consternation, the Almanack's estimates differed so much from the Greenwich observations as to render 'a reconstruction of all the charts ... indispensable to accuracy'. So daunting was the prospect that the Board of Longitude agreed to employ John Crosley, the *Investigator*'s former astronomer, to recalculate the data, assisted by Samuel Flinders.

The unexpected delay at least allowed an opportunity to correct another anomaly that had worried Flinders for some time:

> A variety of observations with the compass had shown the magnetic needle to differ from itself sometimes as much as six, and even seven degrees, in or very near the same place, and the differences appeared to be subject to regular laws; but it was so extraordinary in the present advanced state of navigation, that they should not have been before discovered and a mode of preventing or correcting them ascertained, that my deductions, and almost the facts were distrusted; and in the first construction of the charts I had feared to deviate much from the usual practice.[49]

In April 1812 the Admiralty gave Flinders permission to conduct experiments with the compass aboard HM ships lying at Sheerness and Portsmouth. Spread over two months, these experiments verified Flinders's belief that compass errors were the result of specific identifiable factors – such as changes in the direction of a ship's head, the presence of iron objects in the vicinity, and so on – that 'acted upon the needle in the nature of a compound force', and could be neutralised or corrected when taking observations. Flinders spent a further month writing up his findings in a forty-five-page report,[50] and more time passed while he awaited an official response. Calling at Whitehall a fortnight later he learnt the report had yet to be considered: 'No person at the Admiralty,' he wrote in frustration, 'feels himself competent to form an opinion upon the subject; so that nobody will interfere, and the discovery runs the risk of being neglected.'

It was Banks who came to the rescue. He sought expert comment on Flinders's paper, was impressed by the latter's arguments in its defence, and duly recommended that it should be published: 'There seems to be no question but that Capt. Flinders has made out his case.' Their Lordships compromised, desiring Flinders to summarise the report in

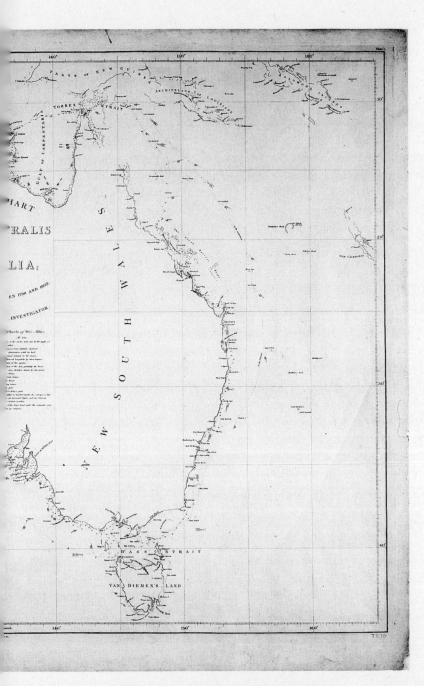

General chart of Australia, by Flinders (1814).
[By permission of the National Library of Australia]

a few pages; in this form it was circulated under the title *Magnetism of Ships* to all officers commanding His Majesty's Ships. His discovery of compass deviation and its causes proved of the utmost value to navigation; over the years the compensating bars around a magnetic compass – still known as the Flinders Bar – became standard equipment on all ships. The discoverer, notes Ingleton, claimed £17 16s 1d in expenses (about £1400 in today's money) for his investigations, and waited six months for settlement; he remained on half-pay throughout.

Family matters also preoccupied Flinders at this time. On 1 April 1812 he recorded in his journal: 'This afternoon Mrs. Flinders was happily delivered of a daughter, to her great joy and mine.' The child, he wrote to his stepmother next day,

> is a little black-eyed girl, without blemish, neither fat nor lean and has a decent appearance enough, so that I hope you will have no occasion to be ashamed of it as a grand-daughter.[51]

By the end of the month, though, baby Anne had been handed over to the care of a wet-nurse, and lived apart from her parents for the next fifteen months. The practice was common enough, and it seems probable that her mother's indifferent health, allied to her father's irregular working hours, were responsible for the decision.

Flinders's compass experiments and the birth of his daughter further delayed the reconstruction of his charts. However, by the end of 1812 Crosley's computations had been completed and checked, and work on the final fair copies could proceed. The task required infinite patience. 'I am at my voyage,' he wrote to a friend, 'but it does by no means advance according to my wishes. Morning, noon and night I sit close at writing, and at my charts, and can hardly find time for anything else.'

Without doubt the charts were closest to Flinders's heart, their detail and accuracy the touchstone by which he would be judged by his peers. His passion for both is well-documented – during his last illness in 1814 he returned the proof of one of Arrowsmith's engravings with ninety-two corrections marked on it. The end result was his *General Chart of Terra Australis or Australia,* defining the continent's newly established boundaries (apart from parts of the west and north-west coasts he had not seen), and a series of detailed charts of his discoveries, which filled in most existing gaps and served as archetypes for his successors.

Writing to Banks in 1801, Flinders had vowed to 'accomplish the important purpose of the voyage ... in a way that shall preclude the necessity of any one following after me to explore'. In large measure he succeeded – I know of yachtsmen sailing the gulf waters of South Australia today who carry his charts in their cabins. Robert Brown, who had more opportunity than most to observe Flinders practising his craft, placed him 'next to Cook among modern navigators [for] the amount of discovery and remarkable accuracy of survey'. Even the French agreed, the geographer Conrad Malte-Brun lamenting that by his death 'the geographical and nautical sciences have lost in the person of Flinders one of their most brilliant ornaments'.

During 1813 the long hours, unremitting labour, and lack of exercise began to take their toll. Towards the end of the year, as noted above, Flinders again fell ill with the urinary complaint that had first afflicted him during his tour of duty at Port Jackson in the 1790s. In the damp and fog of late November he caught a cold, accompanied by a painful inflammation of the kidneys and bladder. The trouble persisted, and in February he 'communicated to Mr. Hayes our surgeon all the symptoms of the complaint which alone ever troubles me, which appears to be either stone or gravel in the bladder'. The treatment brought little relief, and he 'continued to pass gravel, consisting of oblong small crystals ... more or less enveloped in mucus forming a pulpy mass'.

Flinders's chronic illness has been variously blamed on stone in the kidneys or bladder (Austin), a lack of vitamin A or a malignant bladder tumour (Mack), and ill-treatment at the hands of the French (the common 19th-century view).

Ingleton attributes its origin to a venereal infection contracted at Tahiti in 1792, based on evidence from the pay-book of the *Providence* that Flinders twice received the surgeon's mercury treatment (at 15 shillings a consultation) on the ship; if so, in the longer term the 'cure' may have proved as deadly as the disease, causing irreversible damage to his urinary system.[52]

By April 1814 it was clear there would be no remission. On the 1st (his daughter's second birthday), Flinders wrote to Robert Brown:

Since the day I saw you, my complaint has so much increased as to confine me at home, and during some days caused excessive pain, and it still gives pain though in a less degree and with intermissions.[53]

He was 'obliged to move very snail-like', and passed much time 'on the sopha'. Even so, he spent hours at his desk, correcting final proofs of his charts and the *Voyage*. Towards the end of June he had the great satisfaction of receiving from Arrowsmith a presentation copy of the *Atlas*, containing a set of all his charts – his life's achievement condensed into tangible form.

In a letter to Thomi Pitot, Ann described Matthew in the last weeks of his life as 'so dreadfully ... altered, he looked full 70 years of age, & worn to a skeleton'. The last entry in Flinders's journal, dated 10 July, notes: 'Did not rise before two being, I think, weaker than before'. On the 18th he lapsed into a coma; hours later an advance copy of *A Voyage to Terra Australis* was delivered to his lodgings, and placed on his bed – thirteen years to the day since the *Investigator* 'made the signal and weighed' to begin her great voyage. Flinders died early the following morning without regaining consciousness.

In a memoir of her husband's career, Ann Flinders later wrote:

> Thus in the prime of life, fell an officer unexampled in his devotions to the service of his country – from the age of 16 to 40, a period of 24 years, he served it faithfully and unremittingly – never for twenty years, or until his arrival in England from imprisonment having been unemployed, or on half pay, not even a single day & never allowing himself in his visits to his family & friends more than three weeks absence from his duty – if the six years & a half passed in imprisonment & four years constantly occupied in laying his labours before the World, by orders from the Admiralty, be not looked upon in that light.[54]

Notes

1. Scott, 1910, p. 241.
2. Faivre, 1962, p. 53. (Passage translated by J. Treloar.)
3. Ibid., p. 56.
4. Milius, 1987, p. 53. (Passage translated by J. Treloar.)
5. Ibid., p. 60.
6. Milius lists the animals and birds on board as follows: lions, tigers, panther, hyena, gnu, zebra, dog (jackal?), kangaroos (large), kangaroos (small), monkeys (large), lemurs, squirrels, deer (male and female), porcupines, civet-cats, cassowary (Java), cassowaries (New Holland; that is, emus), ostrich, numerous parrots and parakeets, white cockatoo (yellow-crested),

aquatic and other birds. Since leaving Timor several of the Australian animals – including the wombats and some kangaroos – must have died.

7. These artefacts included a collection of items from the South Pacific Islands donated by George Bass to the expedition at Sydney for the proposed Museum of Mankind (Musée d l'Homme) in Paris. The Societé des Observateurs de l'Homme, which had promoted the museum, was now defunct.

8. Horner, op. cit., p. 330.

9. According to Péron's biographer Louis Audiat, writing in 1855 (quoted in Horner, 1987, p. 335).

10. Edwards, 1994.

11. Ibid., p. 10.

12. Ibid., p. 133.

13. Baudin, 1974, translated by C. Cornell.

14. Péron, 1809, p. 184.

15. Spate, 1974, pp. 52-57.

16. Baudin and Hamelin took ninety-four sailing days to the Cape, Cook in 1772 took 105 in the *Resolution* and the *Adventure* – not bad, says Alan Villiers, for two ships in company.

17. Baudin, 1974, p. 215.

18. Péron, op. cit., p. 100.

19. Ibid., p. 98.

20. *Quarterly Review*, August 1810.

21. Flinders, 1814, vol. I, p. 193.

22. Wallace, 1984, p. 159.

23. Gillespie, 1935, vol. 21, pp. 174-175.

24. Ibid.

25. *Historical Records of Australia*, vol. IV, p. 633.

26. Bauer had planned a stay of about two months, but no ship from Port Jackson had called at the island in the meantime. The delay allowed him time for an exhaustive study of its flora and fauna.

27. Mabberley, 1985, p. 124.

28. Ingleton, 1986, p. 328.

29. Ibid., p. 329.

30. Ibid.

31. Hibbert, 1994, p. 381.

32. Norst, 1989, p. 68.

33. Ibid.

34. Ingleton, op. cit., p. 329.

35. Flinders, Private Journal, 20 December 1806. Ms, Mitchell Library.

36. Ingleton, op. cit., p. 330.
37. Ibid.
38. Flinders, Private Journal, op. cit., 20 November 1810.
39. Ibid., 12 December 1810.
40. Mack, 1966, p. 220.
41. Flinders's motives for this request are unclear, as the Admiralty does not appear to have pressed him to go on half-pay. Banks may have suggested it – Flinders had met him earlier on the day he submitted the application to Secretary Croker.
42. Flinders, 1814, vol. I, p. iv.
43. Ibid., p. ix.
44. See appendix C: 'A Day in the Life of Matthew Flinders, prisoner'.
45. See also Austin, 1966, pp. 39-46.
46. Flinders, 1814, vol. I, p. 37.
47. Ibid., vol. II, p. 289.
48. Edwards, 1994, p. 7.
49. Flinders, op. cit., vol. I, pp. iii-iv.
50. *An Account of some experiments to ascertain the effects produced on the Compass, by the attractive power in ships; with the modes by which they may be obviated* (National Maritime Museum, London).
51. Ingleton, op. cit., p. 403.
52. Relics of the wreck of HMS *Pandora* (1791), displayed in the Museum of Tropical Queensland in Townsville, include several medical implements assumed to have belonged to the ship's surgeon, George Hamilton. Among them is an ivory syringe with a wooden plunger, used *inter alia* to inject anti-venereal treatments such as mercury compounds into the male urethra. The effect on the urinary system can only be imagined.
53. Mack, op. cit., p. 233.
54. Retter and Sinclair, 1999, p. 144.

CHAPTER TWELVE

⚓

Epilogue:
Baudin and Flinders – Reputations

The lost reputation of Nicolas Baudin

'At ev'ry word a reputation dies,' wrote the poet Alexander Pope in 1714. Character assassination by the written word is common enough in history (witness Richard III of England), but can rarely have been applied to greater effect than in Baudin's case. His must be the only major voyage of discovery that by and large achieved its objectives, returned safely home without losing a vessel, brought back an unrivalled collection of living animals and plants and of botanical specimens, and yet whose commander was so vilified that his name appeared just once in the official history – and then only to record his death.

Baudin has been a shadowy figure in Australian history, known if at all for the fact that he met Flinders at Encounter Bay, South Australia, in April 1802.

No statues honour him in his own country, no placenames acknowledge his explorations on our coasts, save for a tiny island in Shark Bay, Western Australia (so insignificant it was left unnamed for decades), and Baudin Rocks off Robe, South Australia (named by Flinders). The reader is compelled to ask: Why has Baudin's name slipped from sight? Why was he denied the recognition and honours his achievements merited? Why is his name virtually missing from the map of Australia, when those of his subordinates are scattered liberally around the coasts of south-eastern Tasmania and Western Australia?

Most obviously, Baudin made the ultimate mistake of dying on the voyage, thus leaving its history in the hands of his enemies – notably François Péron. Well before the latter's *Voyage de découvertes* appeared in 1807, however, the commandant's reputation with the authorities in Paris lay in tatters, thanks to successive attacks from other alienated members of his staff. The first reports to cause concern came from

459

officers and scientists who had left the expedition at Isle de France in 1801, and sought to justify their desertion by portraying the voyage as a doomed enterprise, bound to fail under an incompetent, venal, and cowardly commander.

Lieutenant Gicquel, for example, reported to friends in the Ministry that 'the disgusted sailors of the two ships had all deserted', that Baudin had antagonised everyone of note in the colony, and that, 'M. Baudin has no qualities, either moral or social. He is neither naturalist nor mariner. His hair stands on end in the least squall.' For good measure he attacked Hamelin as well: 'he is no more a mariner than Baudin – he is a coward!'

The young naturalist Bory de St Vincent (also a serving officer, on leave of absence from the Army) had joined Governor Magallon's staff after the expedition sailed for New Holland, and visited Réunion and St Helena before returning home in 1803; he published an account of his travels the following year. Like Gicquel, he set out deliberately to disparage Baudin, but unlike his colleague he had sailed on Hamelin's ship, and his snide criticisms were mostly attributed to anonymous 'enemies of the captain' in the *Géographe* – suggesting that even at this early stage there was a coterie of junior officers and scientists on both ships bitterly opposed to the commandant. The reports from Bory de St Vincent and Gicquel undoubtedly prejudiced the Ministry against the expedition – all the more so with the change in Minister from the supportive Forfait (who knew Baudin well) to the authoritarian Decrès.

The *Naturaliste*'s return in 1803 further reinforced the Ministry's view of the expedition as a failure. Hamelin's reception was cool – national priorities had changed, and were now clearly geopolitical and military rather than scientific; by disregarding his instructions it was apparent that Baudin had allowed the British to discover the unknown southern coast of New Holland. Anxious to resurrect his naval career, Hamelin returned to active service at the first opportunity, leaving young Hyacinthe de Bougainville to provide his father the admiral with a jaundiced account of the commandant's failings.

In 1804, following Milius's breakdown at Lorient, it was left to François Péron to present the Ministry with the first detailed report on the expedition's achievements – made, he was at pains to point out, in the face of Baudin's obstinacy, incompetence, ill will and malice towards his subordinates. Péron's *Voyage*, published with the authority of an imperial decree, capped the process of annihilation with an official

NOUVELLE-HOLLANDE : Ile King.

LE WOMBAT. (Phascolomis Wombat S.)

De l'Imprimerie de Langlois.

Top: *Kangaroos, by*
C.-A. Lesueur. [Collection
Lesueur, Muséum d'Histoire
Naturelle, Le Havre]

Above: *Wombats, King*
Island, by Lesueur. [Collection
Lesueur, Muséum d'Histoire
Naturelle, Le Havre]

Right: *The snake caught by*
Thistle on Thistle Island, south-
western Spencer Gulf, SA.
Painted by William Westall.
[By permission of the National
Library of Australia]

Top: *Beautiful Firetail
(formerly* Emblema bella, *now*
Stagonopleura bella), *Maria
Island, by C.-A. Lesueur.
[Collection Lesueur, Muséum
d'Histoire Naturelle, Le Havre]*

Above: *Emus, by Lesueur.
[Collection Lesueur, Muséum
d'Histoire Naturelle, Le Havre]*

seal of approval. Little wonder that a later French writer could comment in 1842:

> This oblivion [into which Baudin has fallen] is vengeance for the bad be-
> haviour and incompetence of this officer, so far inferior to the great mis-
> sion entrusted to him by some inept bureaucrat.[1]

By then it mattered little that France's leading scientists, two of her most eminent naval officers (Bougainville and Fleurieu), the Minister of Marine, and Napoleon himself could be numbered among the 'in-ept bureaucrats' who had selected Baudin for the command.

Many years after his death, the disgraced commandant was to find one supporter among his younger contemporaries – the navigator and explorer Dumont d'Urville, little inferior to Cook in Antarctic explora-tion in sailing ships. Calling at Mauritius (at the time a British colony) in 1828, during the first of his two great voyages, he talked with an old seaman who had served under Baudin:

> I delighted in getting him to relate his adventures ... [to] enable me to
> make up my mind on the character and abilities of Baudin. The informa-
> tion I obtained ... confirmed the opinion I have always had of this mari-
> ner. He was a man of resolution and character; his courageous naviga-
> tion along the coasts of New Holland proves as much, especially when
> one compares it with the hydrographic work of his two successors;[2] but
> he was badly supported.
>
> The scientists' ways of thinking [this is the voice of experience speak-
> ing] were inevitably incompatible with those of a mariner accustomed to
> the despotism of shipboard life. These men allied themselves with the
> officers in order to resist with all their might the requirements of the
> commandant. He for his part, moved by a mistaken pride ... believed he
> could bring them to reason by misplaced acts of authority which were
> quite arbitrary. From this arose the hullabaloo of imprecations and re-
> criminations which have covered his memory with a kind of dishonour.
>
> Still, one must agree that if he had lived things might have turned out
> differently; on his return Baudin might have got the advancement and
> credit due to him, while those who made such a clamour against him
> would have been silenced, and would even have hurried to ingratiate
> themselves with him ...[3]

What are present-day readers to make of Nicolas Baudin? We have the benefit of Christine Cornell's splendid translation of his journal,[4] an invaluable primary source not available to or ignored by earlier historians, but an essential corrective to Péron's venomous portrait. More recently Frank Horner has published *The French Reconnaissance* (1987), an authoritative and well-documented account of Baudin's expedition, while Oxford University Press, in association with the Australian Academy of the Humanities, produced a beautifully illustrated volume *Baudin in Australian Waters : The Artwork of the French Voyage of Discovery ... 1800-1804* for Australia's bicentenary in 1988.

Down the years, Australasian historians (rather more so than French) have queried Péron's version, starting with Ernest Scott in 1914, who could not accept his portrayal of a captain to all appearances incapable of steering even a canal barge. In the 1930s, Arthur W. Jose contrasted Baudin – a seaman 'experienced in maritime hardships and always looking ahead for the benefit of his crews, [who] did his best to act on Drake's great principle, that in emergencies the gentlemen should pull and haul with the mariners' – with his younger officers, who thwarted him at every turn, calling him 'a miser, no naval man, merely an old merchant skipper with his head turned, unfit to command real fighting officers'.[5]

John Dunmore in 1969 commented that

> his achievements deserve the highest praise; he rejected all the excuses – unchallengeable most of them – that offered themselves to him; and he continued his voyage in spite of everything, driven by his inexorable will.[6]

For Horner, 'his achievements and qualities, especially in large matters, make all the more inexplicable those actions, especially in small matters, for which we find it hard to respect him'.

Baudin remains an enigma, a creature of contradictions. His early life is poorly documented, as is his career as a merchant captain. From Baudin's writings – and from Péron's – the reader can discern a strange, complex temperament, a man seemingly imbued with an infinite sense of frustration. Lacking in formal education, he possessed a keen natural intelligence, an intense curiosity and a zest for learning, with qualities that won the enthusiastic support of the professors of the museum and stood up to Napoleon's scrutiny.

Given his family background and his barren years in the merchant service, it is scarcely surprising that he was grasping on occasion – he could equally be generous on others. He was clearly a stickler for etiquette, impatient and irritable with those he considered fools, standing too much and too often on the dignity of his rank, and possessed of what seems at times a tragic, prickly, childish vanity. 'It is not hard to imagine', notes Horner, how such a man, in daily contact with Péron's prodigious energy, 'could find so much zeal too much'.

As a commander, at least on the last voyage, Baudin was authoritarian, secretive, distrustful of his subordinates, stubborn, and inflexible. He drove his men hard, determined at all costs to complete the task given him. In his journal he accuses his officers of disobedience, incompetence, and selfish exploitation of stores intended for the crew, but rarely criticises the men themselves, whose qualities of endurance he admires. In comparison to Flinders, though, he lacked the qualities that make a leader of men – in the words of Professor Spate, 'Baudin could drive but not lead, Flinders was a leader born'.

Towards the end, Baudin drove himself hardest of all, knowing as he must have that the tuberculosis bacilli in his lungs would soon kill him. At the very least the disease would have influenced his mood, feeling and behaviour for years before his death. More than likely it was this that affected his judgement, just as Cook's deteriorating health on his final voyage occasioned a character change that brought about the tragedy at Kealakekua Bay. Cook was fifty when he died, Baudin forty-nine – both had been at sea almost continuously for more than thirty years, both by then must have been suffering the cumulative stress effects of 'campaign fatigue'.

Baudin's friendships with Ledru, Riedlé, and Maugé show him in a light far removed from that so vindictively sketched by Péron. With these congenial, hardworking and like-minded companions he felt himself on safe ground, could relax and dispense with his habitual reserve, and pursue his genuine delight in the natural sciences; there can be no doubting his affection for them. With Governor King, too, he enjoyed a warm friendship, his letters revealing him as 'courteous, honourable, tactful and generous' (Ingleton), grateful for King's unstinting help, and anxious to avoid any cause for dispute that might threaten their relationship.

For Baudin's epitaph, Arthur Jose's summing up can hardly be bettered:

He had done good work; he hoped to do better; when the hope vanished, he died. For the present we may content ourselves with that portrait – the experienced mariner, desperate and disabled, facing his one irretrievable failure. Avid of friendships, he had outlived his friends, and died alone among enemies, who took no trouble to disguise their enmity. Yet he could console himself with the remembrance that he had in his time deserved well of the Republic, and had seen his trophies borne in triumph through Paris along with those of France's greatest general and coming Emperor.[7]

Ambition denied – Matthew Flinders's lost dream

In July 1804, from his prison on Isle de France, Matthew Flinders wrote a long and revealing letter to Sir Joseph Banks – the first of several appealing for his patron's aid in securing his release and in supporting his promotion to post captain. It would, he said, 'soften the dark shade with which my reflections in this confinement cannot but be overspread', to be promoted to the all-important list where his rank would thence become progressive. He continued:

> I have too much ambition to rest in the unnoticed middle order of mankind. Since neither birth nor fortune have favoured me, my actions shall speak to the world. In the regular service of the Navy there are too many competitors for fame. I have therefore chosen a branch which, though less rewarding by rank and fortune, is yet little less in celebrity. In this the candidates are fewer, and in this, if adverse fortune does not oppose me, I will succeed; and although I cannot rival the immortalized name of Cook, yet if persevering industry, joined to what ability I may possess, can accomplish it, then I will secure the second place, if you, Sir Joseph, as my guardian genius will but conduct me to the place of probation. The hitherto obscure name of Flinders may thus become a light by which even the illustrious character of Sir Joseph Banks may one day receive an additional ray of glory … [*Historical Records of New South Wales*, vol. 5, 1803-1805, p. 397]

Second only to Cook! Throughout Flinders's career as a navigator, Cook was the benchmark, the standard against which he measured his skills and achievements. As David Mackay (1990) has pointed out,

> Cook is the silent presence, the revered model who offered the key to success. His example was scrupulously followed in terms of fitting out the ship, sailing directions, surveying technique and shipboard routine … In personal terms, the ultimate measure of success for Flinders was how close he could come to the master in terms of achievement and reputation.

Even before the *Investigator* sailed, Flinders had written to Banks
(29 April 1801): 'My greatest ambition is, to make such a minute investigation of this extensive and very interesting country, that no person shall
have occasion to come after me to make further discoveries' – perhaps
unconsciously echoing Cook's words at the close of his first voyage: 'I
have exploard more of the Great South Sea than any that have gone
before me, so much that little remains now to be done ...'. Banks, remembering the *Endeavour*'s brush with disaster, was less sanguine:

> I confess I feel much apprehension of the *Investigator* tracing a Reef on
> the ocean side where there is no anchorage & where in a calm the long
> waves of the ocean will have you upon it ...[9]

Several writers have sought to give Flinders's story the stature of classic tragedy. Thus Sidney Baker, in his 1962 biography *My Own Destroyer,* portrays him as

> a man beleaguered by destiny and not merely subjected to various dis-
> tresses as a result, but slowly and relentlessly ground into dust by forces
> which he himself had invoked, but from then on was utterly powerless
> to check.[10]

While it may be doubted whether Flinders consciously accepted this
tragic vision of his life, in retrospect his single-minded determination
to stand alongside Cook can be seen to have driven him down a particular path; if ambition first lifted him to a pinnacle of achievement, it
then destroyed him. The compulsion to succeed – 'to make so accurate an investigation ... that no further voyage to this country should
be necessary'[11] – occasioned successive errors of judgement that in the
end engulfed him.

The first misjudgement occurred as early as 1801 – prompted in this
case by an interior conflict between head and heart, ambition and love.
Flinders's reckless and futile attempt to conceal his marriage from the
Admiralty and Banks, and to take his bride to Port Jackson on board
the *Investigator* in defiance of orders, almost cost him his command
and his career. Although Sir Joseph generously came to his rescue, the
breach of trust unquestionably damaged Banks's confidence in his protégé and had a lasting effect on their relationship. Ann, too, felt be-
trayed, and full forgiveness took several years.

The turning point in the voyage, and in the course of Flinders's
life, came in early March 1803, off the continent's north coast, with the

forced abandonment of the survey. The decision, though inevitable, shattered him:

> ... with the blessing of God, nothing of importance should have been left for future discoverers, upon any part of these extensive coasts; but with a ship incapable of encountering bad weather, – which could not be repaired if sustaining injury from any of the numerous shoals or rocks upon the coast, – which, if constant fine weather could be ensured and all accidents avoided, could not run more than six months, – with such a ship, I knew not how to accomplish the task.[12]

The return to Port Jackson brought further reverses. Dysentery and fever, legacy of the stop at Timor, claimed nine lives – one in ten of the ship's company. Flinders was crippled with scurvy. Within a week of their arrival at Sydney the *Investigator* was surveyed and found to be 'not worth repairing in any country'. With his ship condemned, Flinders had little option but to return to England, report his discoveries to the Admiralty, and solicit a new vessel in which to complete his investigations. Governor King agreed, and offered HM Armed Vessel *Porpoise* for the homeward voyage. 'I am going home,' Flinders wrote joyfully to his friend Mrs Elizabeth Macarthur at Parramatta, 'with the promise of being attended by fortune's smiles, and with the delightful prospect of enfolding one to whom my return will be a return of happiness.'

Judging the passage he had discovered through Torres Strait to be 'the smoothest as well as the most expeditious at this season', Flinders suggested that the *Porpoise* should return by this route; 'it will [also] give me a second opportunity of seeing whether the strait may safely become a common passage for ships from the Pacific into the Indian Ocean', he told King. When she sailed, however, she was joined by two merchant ships, the East Indiaman *Bridgewater* (750 tons) and the *Cato* (450 tons), whose captains 'desired leave to accompany us through the Strait'. Their company, he wrote,

> gave me pleasure; for if we should be able to make a safe and expeditious passage through the strait with them, of which I had but little doubt, it would be a manifest proof of the advantage of the route discovered in the *Investigator*, and tend to bring it into general use.[13]

Flinders's decision to sail through partly uncharted seas and shallows in company with two much larger vessels (the *Bridgewater*'s tonnage was more than twice that of the *Porpoise*), made to test the safety of an untried route he had but recently discovered, was to say the least unusual. Captains Park and Palmer were experienced and careful seamen (Palmer had been on the board of survey that condemned the *Investigator*), and would hardly have 'desired leave' to accompany Flinders through the strait without his prior assurance that their ships and cargoes would not be at risk.

A week after leaving Port Jackson the three ships were running north under full sail, with *Porpoise* in the lead. Flinders travelled as a passenger, perhaps to allow him the opportunity to work on his charts and journals. Lieutenant Fowler was in command, but was under orders to accept his former captain's directions concerning the navigation and survey of the strait. On the afternoon of 17 August the appearance of an uncharted reef on the port side caused some apprehension, but when no others were observed before nightfall the decision was made to continue under reduced sail during the night. Fowler apparently felt misgivings and considered heaving to, but compromised after talking to Flinders. It was, Ingleton says, 'a fatal error of judgement' on the latter's part – made, one suspects, with a view to demonstrating the value of his new route in cutting sailing times to Batavia.

There is no need to repeat here the story of the shipwreck.[14] From his rescue mission to Port Jackson, Flinders returned to Wreck Reef with the Indiaman *Rolla* and two colonial schooners, the *Cumberland* and the *Francis*, to take off the castaways. At Sydney, Governor King had offered him the *Cumberland* 'to proceed direct to England through Torres's Strait with Officers and Men belonging to the *Investigator* requisite to work her', so that he might 'add to his Survey and arrive with his Charts &c. before any other accounts reach England'. Despite reservations about her small size and inconvenience, the temptation proved too strong to resist:

> ... the advantage of again passing through, and collecting more information of Torres' Strait, and of arriving in England three or four months sooner to commence the outfit of another ship were important considerations; and joined to some ambition of being the first to undertake so long a voyage in such a small vessel ...[15]

Despite her size, leakiness, and substandard sailing qualities – all too obvious on the passage to the reef – Flinders persisted with the plan to continue the voyage in the *Cumberland*. Ingleton, himself a career naval officer and cartographer, considers this 'the biggest blunder in his career

– perhaps a vainglorious attempt to show that he could not be wrong!' Apart from the risks involved, it meant ignoring 'the rule of the captain remaining with his men … with separation, he was powerless to help with deserving promotion, or taking many to a new command'.[16]

James Mack, in contrast, believes the decision was dictated by Flinders's 'unshakeable resolve to reach England, find a ship, and return to the business at hand' with the least possible delay. He accepted the *Cumberland* 'as the least evil, knowing that at every convenient opportunity along the way' he would need to call in for food, water, and repairs – including at Isle de France.[17]

Both assessments, however, take in ambition as a deciding factor, consistent with Flinders's justification for his choice.

And so to the final, calamitous misjudgement – the confrontation with Captain General Decaen at Isle de France. During the 19th century the general was widely denounced by Anglo-Australian writers as 'a malignant tyrant', 'vindictive, cruel and unscrupulous'. Ernest Scott, writing in 1914, tells of 'one admirer of Flinders who had a portrait of Decaen framed and hung with its face to the wall'. Scott, however, in common with many later biographers of Flinders, came to view the general as a patriot, efficient, incorruptible (unlike many of Napoleon's generals and marshals, he did not abuse his office to enrich himself), ambitious, and honourable – but also as brusque, irritable, 'too free with his tongue' (the general's words), strong-willed and inflexible. It is part of Flinders's tragedy that the description applies no less accurately to him.

Flinders's reaction to Decaen's accusation of espionage at their first meeting is understandable – he was exhausted, sick, conscious of his own integrity, and fearful lest this latest disaster further delay his plans to complete the survey of Terra Australis. Yet thereafter it seems that his judgement totally deserted him. Overcome by frustration, thwarted ambition, arrogance – even, one senses, a touch of paranoia – every step seemed calculated to challenge not just Decaen's authority but fate itself. Towards the end of his detention, though, after long and bitter experience, he brought himself to acknowledge that his adversary was said to possess 'the character of having a good heart, though too hasty and violent'. By then it was too late.

Without question Flinders was a fine seaman and outstanding navigator, a worthy successor to Cook. His charts, compiled with meticulous attention to detail and always based on his own observations, doubly checked for accuracy, served as a prototype for those who followed him; some remained in use until World War II. He appears to have been held in genuine respect and affection by his crew, and only one of his select 'volunteers' – his clerk, Olive – refused to join him in the ill-fated *Cumberland*. His relations with his officers were mostly harmonious, with occasional exceptions – notably his younger brother Samuel, from whom he may have expected too much. His capacity for friendship is evident throughout his life – George Bass, his fellow officers in the *Providence* and the *Reliance*, John Thistle and young John Franklin, Governor King; it is most clearly shown at Isle de France, where many French colonists were prepared to risk Decaen's disfavour by openly supporting his prisoner's cause.

Matthew had written to Ann from his prison on New Year's Eve, 1804: 'I shall learn patience in this island, which will perhaps counteract the insolence acquired by having had unlimited command over my fellow men.' It was a lesson Ann had already learnt, forced to eke out a meagre existence on a Navy wife's pension and dependent on the charity of her mother and stepfather. Now that she and Matthew were both deprived of the normal expectations of life, however, she felt drawn far closer to him emotionally than before. In the end she paid as high a price for his lost dream as did Flinders, and it is only just that after his death she should be allowed the final say on his life's work:

So strong was his Inclination for this dangerous service that amongst his friends, he has been frequently heard to declare his belief 'That if the plan of a Discovery Expedition were to be read over his grave, he would rise up awakened from the dead' – For the successful pursuit of his favourite employment, his voyage in the *Investigator* will offer the best proof, & forever remain an example of his undaunted spirit & irresistible perseverance. In him, prudence, ability, & the most unswerved industry were united in an uncommon manner: no difficulty could stop his career, no danger dismay him: hunger, thirst, labour, rest, sickness, shipwreck, imprisonment, Death itself, were equally to him matters of indifference if they interfered with his darling Discovery ...

From close imprisonment in the Isle of France [his illness] became confirmed & at times was excruciatingly painful, & altho he afterwards had intervals of ease, yet the last four years of sedentary life in sitting to prepare his labours for the public eye was too much for his, or perhaps any constitution to resist, & he died, if any Man did, a martyr to his zeal for his country's service – That Country it is hoped will properly appreciate & long retain the memory of his services & the learned & scientific of other Nations will deplore the loss of a Man whose life has been spent in useful research, & whose last days were devoted to laying the result of his labours before the World ...

It may however afford some consolation to his family & friends to believe that the name of *Flinders* will henceforth stand equal in the annals of discovery, with a Dampier, a Wallis or a Byron; equal perhaps with the later & more celebrated names of a Bougainville, a Bligh, a Peyrouse, or even with that of the immortal Cook – His private character was as admirable as his public one, his integrity, uprightness of intention, disdain of deception & liberality of sentiment, wise not to be surpassed, & he possessed all the social virtues & affections in an eminent degree.[18]

Hostages to fortune

Fortune's a right whore,
If she give aught, she deals it in small parcels,
That she may take away all at one swoop.
[John Webster, The White Devil, *1612, act I, scene I]*

In hindsight, the lives of both Flinders and Baudin do seem to unfold with the inevitability of Greek tragedy (though perhaps the Icelandic saga is a better analogy), with neither man able to control the unseen forces that shaped his life and would order the manner of his death. They were brought together in an unknown bay on the far side of the world in the autumn of 1802 by events as distant in time and space as the marooning of a Scottish sailor, Alexander Selkirk, on a lonely island and the subsequent retelling of his story by Daniel Defoe, and an Austrian Emperor's passion for botany.

Together they completed the mapping of Terra Australis, the Great South Land, filling in the last great gap, the unknown southern coast. Both met their nemesis on a tropical paradise in the Indian Ocean – Baudin worn out by disease and failure, dying in disgrace among his enemies, and Flinders dragging out six years of exile before returning to a London too preoccupied with the war against Napoleon to heed his discoveries. So strangely linked in their lives, it is not unfitting that the graves of both are unknown.[19]

Two centuries later we can observe, though not condone, the personal jealousies and national rivalries – inflamed by a long and bitter war – that saw both captains denied due recognition in their own countries. In the event, Baudin escaped the ultimate humiliation – he did not live to see his name erased from the history of his voyage, and his enemies take for themselves the credit for his geographic and scientific achievements. Nevertheless, it is their names, not his, that bear witness to his discoveries on the map of Australia.

Flinders for his part has been compensated by posterity, with a great mountain range, islands, a peninsula – and more recently a prestigious university – all named in his honour. Following the creation of the Commonwealth of Australia in 1901, the new nation cast around for suitable heroes who might wipe away the stain of the convict years. Flinders appeared a prime candidate – young, fearless, a loving husband and a faithful friend, his virtues were exemplary. To him more than anyone was due the very name Australia, 'as being more agreeable to the ear, and an assimilation to the other great portions of the earth'. Professor Ernest Scott of Melbourne University produced the first full-scale biography in 1914; although inevitably written from the standpoint of the victor of the Napoleonic Wars, his work was not markedly unfair to Baudin, or even to General Decaen.

It was left to the Australian writer Ernestine Hill to elevate Matthew Flinders to the status of national hero *sans peur et sans reproche* in her historical novel *My Love Must Wait* (1942). Writing during World War II, when 'home' to many Australians still meant England's green and pleasant land, she 'rescued Flinders from the pages of history and navigation and [gave] him and his age life, passions, and fine romance' (*The Age*, Melbourne). It was a story, she wrote,

> of a life's unending labour for Discovery, of the persecutions of a cruel fate that gave all only to take away, and sent a broken cripple to an unknown grave, and crowned the ruin with undying fame when a century had gone.

Yet even today Flinders holds a relatively minor place in the ranks of British explorers.[20] Not one of the four significant biographies published in the 20th century was written by an Englishman – three are by Australians (Scott, Baker, and Ingleton), and one by an American (Mack). Before I close this chapter, it is pertinent to look at some possible factors behind this neglect.

One clue may be found in the previously quoted comment by Governor King (a good friend to Flinders) in a letter to Banks: 'he has in a great measure been unsuccessful'. Failure is not a word associated with Flinders's name in Australia, but one gets the distinct impression that, as his misfortunes mounted, so his reputation diminished among the authorities at home. Banks, Flinders's 'guardian genius', seems increasingly to have lost confidence in his protégé as the voyage progressed. Apart from Flinders's deception about his marriage, there were his misadventures on the passage to Portsmouth with his wife on board, his delay in sending back the charts of his south-coast discoveries, his failure to complete the survey of New Holland, and his 'haughtiness' towards Decaen – enough to raise doubts in Banks's mind about Flinders's judgement. Aware of the criticisms, Flinders took pains to address these and other issues in the *Voyage*, in self-defence and to vindicate his actions.

Within the higher levels of the Admiralty, preoccupied with wartime naval operations on the world's oceans, there was little enthusiasm for a remote voyage of discovery. Moreover, Banks's influence, paramount in winning approval for the expedition, had waned after his friend Earl Spencer was replaced as First Lord in 1801. During Flinders's detention at Isle de France, his stubborn complaints about his treatment by the French may well have aroused as much irritation as sympathy – it was never stated, but conceivably some senior officials considered it a naval officer's duty to attempt to escape, not to live comfortably on the enemy's charity on parole.[21] Evidently bureaucratic concerns at creating a precedent were the principal reasons for the refusal to back-date his promotion, but there may have been something more. His old friend Captain Kent apparently thought so, writing in disgust of 'our devoted Service being in the hands of a set of Borough-mongers'.

More telling, perhaps, is that Flinders was never elected a Fellow of the Royal Society – the ultimate accolade for a naval officer in the navigational branch. As Mack points out, Cook was elected FRS on his return, and awarded the society's Copley Medal. Bligh, another of Banks's protégés, had become a Fellow in 1801, 'in consideration of his distinguished services in navigation, botany, etc.'. Flinders was well qualified, as a navigator and cartographer, and for his observations on the magnetic needle and the marine barometer – papers sent to Banks from Flinders's prison and read before the society on his behalf – and must surely have expected the honour. A fellowship was in the gift of the society's president, yet Banks did not put his name forward – why not?

Flinders had other reasons to feel disgruntled with his patron, as he makes clear in a letter to his long-time friend, the botanist James Wiles, in July 1811:

> I am obliged to remain here [London] about two years until my Voyage
> is finished; and I trust that the portion of the profits which the Admiralty
> may award me, will at least pay the expense I am put to in writing it. My
> trouble, which altogether is immense, will I fear be given away. This,
> between ourselves, I attribute to the non-exertion of my friend Sir J. B.,
> who, if he pleased, might procure me something that would at least pay
> me extra expenses. He seems to have adopted the usage 'keep them poor
> and they will serve you the better'; fearing perhaps, that if I was comfort-
> able in England, I should not be eager to go another voyage of discovery
> when wanted.[22]

It is the most explicit, though not the only criticism of Banks in Flin-ders's correspondence, and signals a growing discontent with their re-lationship. Like many protégés, he had endured his patron's rebukes in silence; possibly his badly timed request to be placed on half-pay[23] had been Banks's idea; certainly he had cause to resent the latter's nig-gardly financial terms for writing the *Voyage* (especially when he com-pared these with the salaried posts on Banks's staff enjoyed by Brown and Bauer, both unmarried). Later there was the dispute over the naming of the continent whose true form Flinders had been the first to estab-lish, which saw his choice – *Australia* – discarded in favour of Banks's unwieldy compromise, *Terra Australis*.

Fortune dealt Flinders yet another blow with the timing of the book's publication. Not only did it coincide with his death, but it appeared amid the tumultuous events of 1814: the invasion of France, the downfall and exile of Napoleon, the occupation of Paris by the Allied armies, the seaborne invasion of the United States in August by a British army, the sack of Washington and the burning of the White House and other public buildings by British troops – in such times the account of a voy-age of discovery to the distant Terra Australis attracted little interest. Of the 1150 copies printed, more than half were remaindered to the book trade. Nicol the printer failed to cover his costs, and there were no profits at all to benefit Flinders's widow and young daughter.

Left impoverished at her husband's death, Ann Flinders petitioned Their Lordships for an increase in her pension, and sought Banks's support. 'I am well aware', she wrote to him, that the Lords of the Admiralty, are at all times mindful of establishing presidents [*sic*], but as Mrs. Cook received this favor, I would willingly hope the same kind consideration might be extended to me. The claims of Captain Flinders were certainly far from being equal to those of Captain Cook, but his misfortunes and sufferings were greater, & although he died in his native land, he certainly fell a victim to the services of his country, his extreme anxiety to complete the work in which he had engaged, led him for the last two years of his existence, to deprive himself of both proper rest, and proper exercise, & thus a most painful disorder … gained so powerfully upon his constitution that it baffled all the efforts of medicine.[24]

Sir Joseph (now in his seventies) sympathised, declaring that he would not die happy unless something were done for the widow and child of poor Flinders. In the event nothing was done, no-one in government being prepared to risk establishing a precedent – 'the stinginess of a rich nation', wrote Scott a century later, 'is a depressing subject to reflect upon in a case of this kind'. It was left to the new self-governing colonies of New South Wales and Victoria to provide some belated redress. Learning at mid-century that Mrs Flinders was still alive and living in genteel poverty, in 1853 the two colonial governments each voted her an annual pension of £100, with reversion to her married daughter, Mrs Petrie. Though the news reached England too late to assist Ann (she died in 1852, aged seventy-nine), it gave great satisfaction to Anne Petrie, mother of a newly born son:

> Could my beloved mother have lived to receive this announcement, it would indeed have cheered her last days to know that my father's long-neglected services were at length appreciated. But my gratification arising from the grant is extreme, especially as it comes from a quarter in which I had not solicited consideration; and the handsome amount of the pension granted will enable me to educate my young son in a manner worthy of the name he bears, Matthew Flinders.[25]

Her son grew up to become Professor Sir William Matthew Flinders Petrie, the noted Egyptologist (1853-1942).

Flinders and Baudin both died in their beds, fretted out by disease and failure, the full extent of their achievements unrecognised by all but a few of their contemporaries. They had sailed in wartime, and in the end each fell victim to the national rivalries and passions aroused by the war. Their principal object – the advancement of science – was largely forgotten as the struggle for imperial dominance between France and Britain intensified after the renewal of hostilities in 1803.

Lacking influence and wealth, each man had fashioned his own progression – Baudin as a 'botanical voyager', Flinders as a navigator. By 1800 both had won recognition in their chosen field. In that year, in a remarkable display of synchronicity, they each drew up a detailed proposal for a voyage of discovery to New Holland, and submitted it, not to their national government, but to the leadership of the scientific community – Jussieu and the Institut National in Paris, Sir Joseph Banks in London. On both sides of the Channel it was the scientists who promoted the voyage and secured the necessary government backing.

Four years later everything had changed. With national survival at stake, neither government had time or resources to spare for purely scientific discoveries – the strategic implications were all-important. Officers of both expeditions were returned to active service – except for Flinders, detained for six years by the French, then retired on half-pay at his own request, his achievements as a discoverer largely unrecognised by the service to which he had devoted his life. The dead Baudin fared worse – posthumously accused by his enemies at Isle de France of misappropriating funds, the alleged deficiencies in his accounts were charged against his estate.

For the two captains, Fortune had changed seemingly beyond recall, from smiling goddess to 'right whore'. Nicolas Baudin, that 'excellent mariner' of 1798, became a 'non-person', his name erased from the history of his own voyage. Matthew Flinders, neglected in death as in life by his countrymen, had his grave destroyed and the contents 'carted away as rubbish' within forty years of his burial. There seemed little doubt that his achievements too would soon be forgotten.

History's judgement, though, is always multifaceted, and varies over time according to the interests, background, national viewpoint, personality and bias of the observer. Australians gave the historical kaleidoscope a shake and a different picture formed. Flinders's resurrection began early, in the mid-19th century; Baudin remained all but forgotten until the second half of the 20th. Now, two centuries after their voyages, both captains can be honoured for their unique contributions to Australian and scientific history.

Notes

1. Guillonneau, 1988, pp. 258-276.

2. Louis de Freycinet (1817-1820), and Louis Duperrey (1822-1825).
3. Horner, 1987, pp. 372-373.
4. Libraries Board of SA, 1974. The first French edition of the journal is in preparation.
5. Jose, 1934, p. 346.
6. Dunmore, 1969, vol. II, p. 37.
7. Jose, op. cit., pp. 368-369.
8. Mackay, 1990, p. 108.
9. Ingleton, 1986, p. 114, n. 59.
10. Baker, 1962, p. xi.
11. Flinders, 1814, vol. II, p. 143.
12. Ibid.
13. Ibid., p. 297.
14. See chapter 8. Three accounts of the shipwreck survive, by Flinders, Fowler, and Seaman Samuel Smith. The latter's plain-spoken narrative appears at appendix B.
15. Flinders, op. cit., p. 323.
16. Ingleton, op. cit., p. 247.
17. Mack, 1966, p. 168.
18. 'Memoir by Ann Flinders', reprinted in Retter and Sinclair, 1999.
19. It has recently been suggested that Baudin's last resting place was in the Cimitière de l'Ouest at Port Louis (E. Duyker, *NLA News*, September 1999). Flinders was buried in the churchyard of St James Chapel, Hampstead Road, London, but the tomb was destroyed less than forty years after his death. His daughter Anne wrote: 'my aunt Tyler went to look for his grave, but found the churchyard remodelled, and quantities of tombstones and graves with their contents had been carted away as rubbish, among them that of my unfortunate father, thus pursued by disaster after death as in life' (Scott, 1914, p. 397).
20. *The Oxford Companion to Ships and the Sea* (1976), for example, devotes eight columns to Cook, three columns to Dampier, just over one to Vancouver, and less than one to Flinders – overall the ratio indicates his standing fairly accurately.
21. Flinders was not, strictly speaking, a prisoner of war, which led to procedural difficulties for the French and British governments. His value to General Decaen as a hostage, never fully appreciated in London, further complicated his position.
22. Ingleton, op. cit., p. 400.
23. Flinders' request, made on 8 November 1810, followed his promotion to

post captain and his appointment to HMS *Ramillies* (seventy-four); it may well have been viewed as a refusal of service by some Admiralty officials.

24. Ibid., p. 423.
25. Scott, op. cit., p. 401.

APPENDIX A

⚓

Passports

Passport delivered to Captain Baudin by the British government

Passport for the French Ship *Géographe*
                                              Naturaliste
By the Commissioners for executing the Office of
Lord High Admiral of Great Britain and Ireland.

Whereas upon our transmitting to the Right honble. Henry Dundas, one of his Majesty's principal Secretaries of State, for his Majesty's information, an application which had been made to the Commisioners for the care and custody of prisoners of war, by Monsr. Otto on the part of the french Government, for protection against his Majesty's cruisers for the Ships *Géographe* and *Naturaliste*, as described at the foot hereof, intended to proceed from Havre under the direction and command of Captain Nicolas Baudin on a voyage round the world, and that the said Ships might have permission to put into any of his Majesty's Ports in case of stress of weather, or for the purpose of obtaining assistance, if necessary, to enable them to prosecute their voyage, Mr. Dundas hath by his letter of the 18th instant, signified to us his Majesty's pleasure, that the necessary Passports and protection should be granted to the Ships above mentioned;

You are, in pursuance of his Majesty's pleasure, as above signified, hereby required and directed, to suffer the said Ships *Géographe* and *Naturaliste*, their officers, people and effects, to pass free and unmolested during their present intended voyage accordingly, and to permit them to put into any of his Majesty's Ports in foreign parts in case of stress of weather, or for the purpose of obtaining assistance if necessary, to enable them to pursue their said voyage, provided nevertheless that there shall not be any reason to believe from the circumstance of time or place that they, or either of them, shall have willfully deviated from the regular course of their voyage, or that they, or either of them, shall have committed, or be about to commit any hostilities against his Majesty or his Allies, or have afforded or be about to afford any assistance to his Majesty's enemies, or have carried on or be about to carry on, any contraband Trade.

Given under our hands and the Seal of the office of Admiralty the 25th June 1800

(Signed) Spencer, S.H. Stephens, Hambur, H. Young

To the respective flag officers, Captains and commanders of his Majesty's Ships and Vessels; the commanders of ships and vessels, having letters of marque, and to all others whom it may concern.

By command of their Lordships
Evan Nepean

Description of the French Ships Géographe and Naturaliste

Each Ship measures about 350 tons, is armed with 8 carriage guns 4-pdrs. and 8 swivels, and is to have on board a complement of one hundred men – exclusive of ten artists and men of science.

[Archives Nationales, Ministry of Marine Records, Paris; SLSA microfilm]

Passport delivered to Captain Flinders by the French government

LE PREMIER CONSUL DE LA RÉPUBLIQUE FRANCAISE, sur le compte qui lui a été rendu de la demande faite par le LORD HAWKESBURY au *Citoyen Otto*, commissaire du gouvernement Francais à Londres, d'un Passeport pour la corvette *Investigator*, dont le signalement est ci-après, expédiée par le gouvernement Anglais, sous le commandement du capitaine *Matthew Flinders*, pour un voyage de découvertes dans la Mer Pacifique, ayant décidé que ce passeport seroit accordé, et que cette expédition, dont l'objet est d'étendre les connaissances humaines, et d'assurer davantage les progrès de la science nautique et de la géographie, trouveroit de la part du gouvernement Francais la sureté et la protection nécessaires.

LE MINISTRE DE LA MARINE ET DES COLONIES ordonne en conséquence à tous les commandants des batiments de guerre de la République, à ses agens dans toutes les colonies Francaises, aux commandants des batiments porteurs de lettres de marques, et à tous autres qu'il appartiendra, de laisser passer librement et sans empechement, ladite corvette *Investigator*, ses officiers, équipage, et effets, pendant la durée de leur voyage; de leur permettre d'aborder dans les différents ports de la République, tant en Europe que dans les autres parties du monde, soit qu'ils soient forcés par le mauvais tems d'y chercher un refuge, soit qu'ils viennent y reclamer les secours et les moyens de réparation nécessaires pour continuer leur voyage. Il est bien entendu, cepandant, qu'ils ne trouveront ainsi protection et assistance, que dans le cas ou ils ne se seront pas volontairement détournés de la route qu'ils doivent suivre, qu'ils n'auront commis, ou qu'ils n'annonceront l'intention de commettre aucune hostilité contre la République Francaise et ses alliés, qu'ils n'auront procuré, ou cherché à procurer aucun secours à ses ennemis, et qu'ils ne s'occuperont d'aucune espéce de commerce, ni de contrebande.

Fait à Paris le quatre Prairial an neuf de la République Francaise,
Le Ministre de la Marine et des Colonies
(Signed) *FORFAIT*
Par le Ministre de la Marine et des Colonies
(Signed) *Chzs. M. JURIEN*

Signalement de la corvette

La corvette *l'Investigator* est du port de 334 tonneaux. Son équipage est composé de 83 hommes, outre cinq hommes de lettres.
Son artillerie est de 6 carronades de 12
2 ditto de 18
2 canons de 6
2 pierriers
Le soussigné, commissaire du gouvernement Francais à Londres, certifie le signalement ci-dessus conforme à la note qui lui a été communiquée par le ministre de Sa Majesté Britannique

Londres le 4 Messidor An IX
(Signed) *OTTO*

[Flinders, 1814, vol. I, pp. 12-14]

⚓

Shipwreck – Samuel Smith's Account of the Disaster at Wreck Reef

... the 10th August we stood out of [Sydney] Cove, in Comp^y with the *Bridgewater* and *Cato* Indiamen: in order to conduct them through Torris Straights, we proceeded on our passage, with all the care possible, & nothing perticulour hapned untill the 17th about 2 O C. in the Afternoon the *Cato* & *Bridgewater* in Comp^y – the *Cato* made a Sign'l that they saw land or A reef, Accordingly we haul'd our Wind & stood for the same the before Mention'd Ships kept their course; upon Examination it proov'd part of A reef, thus being Satisfied, we made Sail, & came up with the *Cato* & *Brigewater*, made A sign'l to Shorten Sail, lower'd our Topsails, & took in A reef & run with Square Yards, the Watch call'd at 8 O C & all well, having A Steady Breeze & thought ourselves out of Danger

about 10 O C. at Night the look out on the Forecastle Call'd out Breakers A Head; Mr Aken the Master had the Watch on Deck, desired the Hands to be turn'd up, & before the Yards cou'd be Braced up she struck upon the Reef A Dreadfull Shock, & Creend [careened?] down upon her Larboard side, by this time the *Cato* & *Bridgewater* was coming down upon the reef as fast as possible notwithstanding the Shouts & Cries to them to be aware of the Danger into whitch we where thrown, the Bridgewater got round off the reef but the *Cato* shared in our Fate: our first care was to get our Boats out, which we did with great Dificulty, one being nearly swamp'd along side. Capt'n Flinders went away in one Boat & Mr Ollive & Mr Allen in the Other, & lay to Leward of the Ship all Night, in case the Ship shou'd part they might have an Opportunity of saving some lives, as there was no other appearance but of Death, there being such heavy Seas striking her Momentary & Flying all over us:

we Cutt our Foremast away to ease the Ship it was likewise proposed to take the Hatches off, to get A Boat out we had stow'd between Decks;

but this was Countermanded by reason it wou'd Weaken the Ship so that she wou'd go to pieces directly; the Ship by this time was bilg'd in the Bottom, our fear was of her drifting into deep Water then she wou'd go down Instantaniously, we then Cut our Weather Main rigging & the Mast, which then fell over the Side, in this mizarable situation we spent the Night, every breast filld with Horror, continual Seas dashing over us with great Violence, at length Day light Arriv'd, we saw the *Cato* had parted in the Night, & her ships Comp^y standing in A Cluster & hanging in A Cluster, by each other on part of the Wreck, having nothing to take to, but the Unmerciful Waves, which at this time bore A dreadful Aspect, they having lost all their Boats, provissions, & Water; but as providence Ordaind it there appeard upon our Lee about A Mile Distant, A Dry Sand Bank; Every heart before, that Expected nothing but Death had now a Glimmer of hope left, between the thoughts of our Ship holding together, our 2 Boats being safe, and the smal patch of Sand so nigh to us, & that the *Bridgewater* wou'd send her Boats to our Assistance; she Appeard the next Day but offerd no Assistance.

The Catoes Ships Comp^y threw themselves into the Surf & made best of their Way to our Ship assisted by our Boats & Men, & Arriv'd with the loss of one Man & 2 Prentices: our Ship still remained together & we Work'd with all the Expedition possible we cou'd, to get out our Provissions & Water, attended by the Catoes Ships Comp^y's Assistance; this being our only support, & was attended with great hardship & Dificulty, having to take every thing through the Surf, to the Sand Bank which is About half a Mile in Circumferrence, & about 150 yds Wide, & have every reason to think that at high Spring tides & bad Weather it is over Flow'd by the Sea; by our diligence we got our Stores upon our small portion of Sand by the 24th, I will leave the reader to Guess our Situation between the Catoes Ships Comp^y & ours Consisting of upwards of 90 Men, upon a Small Uncertainty 150 Miles from the Nearest land & upwds of 900 from the Nearest Port:

having all our Stores on Shore we erected Store houses, & habitations for ourselves Consisting of Tents made of the Sails made also of Spars &c that we recoverd from the Wreck, our Boat was imploy'd each Day at convinient Seasons, in getting every thing from the Wreck, that Wil be of any use, on the 29th Mr Fowler took the Boat to Try to Catch some Fish as at this time it wou'd be A rarety, in the Evening return'd with A Quantum of Snappers each Fish Weighing 3 lbs, at other times the Men Was allow'd the Same privaledge being supplied with Hooks & lines from the Store;

Capt'n Flinders proposed to take the largest Boat with Men accordingly, to go to Port Jackson, to bring a Ship to take us out of this perilous situation; accordingly the Carpenters rep'd & alterd her & she was fitted out for the Voyage, the Mean While the Carpenters was to Build 2 large Boats to take us away if he Miscarried in his Attempts. Accordingly all being ready Capt'n Flinders, & Capt'n Parks of the *Cato*, and his second mate & our Boatswain, with Ten seamen Sail'd on the 26th of Sept'r – all hands gave them 3 Chears, which was returnd by the Boats Crew, it continues A fair Wind & Moderate Weather –

from this time our hands are Imploy'd, some about our new Boat, whose Keel is laid down 32 Feet, others Imploy'd in getting anything servisible from the Wreck; our Gunns & Carriadges we got from the Wreck & placed them in A Half Moon Form, close to our Flag Staf, our Ensign being Dayly hoisted Union downward, our Boats sometimes is Imploy'd in going to An Island about 10 Miles Distant; & sometimes caught Turtle & Fish, this Island was in general Sand Except on the Highest part it produced Sea Spinage, Very plentifully stock'd with Birds & Egs : in this Manner the hands are Imploy'd, & the month of October is set in, still no Acct of our Capt'ns success –

our Boat likewise ready for Launching the rigging also Fitted over her Mast head, & had the Appearance of A rakish schooner; on the 4th of Oct'r we Launch'd her & gave her the name of the *Hope*, on the 7th we loaded her with Wood in order to take it over to the Island before Mention'd, to make Charcoal, for our Smith to make the Iron Work for the next Boat, which we Intend to build Directly, she Accordingly sail'd;

About 2 O C. PM we espied A Sail, afterwards 2 More, our Schooner & small Boat in sight at the same time. Every Heart was Overjoy'd at this unexpected Delivery; accordingly they came close & brought up to Leward of our Bank, we Fired a salute of 13 Gunns, & upon Capt'n Flinders stepping on shore we Cheer'd him, the Ship he brought with him was the *Rolo* Indiaman & 2 Schooners belonging to Sidney. Capt'n Flinders Inform'd us those that Chose might go to Port Jackson in the Schooners under the Governers protection, others go to China in the *Rolo* – in this manner we took our Departure leaving the *Porpoise* Totally lost in Latt. 22- 20 S. Long. 155-52 E.

[Seaman Samuel Smith's journal, MS, Mitchell Library, State Library of NSW]

APPENDIX C

⚓

A Day in the Life of Matthew Flinders, Prisoner

Port North-West, Maison Despeaux. Sunday, 18 August 1805.

Rose at half past six. Slipped on my shoes and morning gown, and went down to walk in the garden. Met the sergeant and bid him bonjour. Think the old man looks a little melancholy at the prospect of his last prisoner leaving the house, for he will lose his situation. The dogs came running after me. Meditated during my walk upon the extreme folly of General Decaen keeping me prisoner here, for it can answer no good purpose either to him or the French government; and some expense, and probably odium, must be incurred by it. This injury that is to me is almost incalculable – but this will not bear to be dwelt upon, it leads almost to madness.

Got up into the tall almond tree to see if there was any ship off; none to be seen. Could have seen much further round from the top of the house if the General had not shut it up, and taken away my spyglass. Got down from the tree and continued my walk. General Decaen's conduct must have originated in unjust suspicion – been prosecuted in revenge, his dignity being injured at my refusing to dine with him, and continued from obstinacy and pride; but seeing a shower coming on hurried into the house …

Half past seven, went up into my room, stripped myself naked, and washed from head to foot in the little tub. Put on my clothes. Called Smith to bring me hot water to shave with. Think he walks better upon his broken leg every day, and that it will soon be as well as ever. Elder not being returned from the Bazar [*sic*], read five pages in La Condamine's voyage down the River Amazones from Quito to Para. Condamine writes in easy perspicuous French, and it seems to be a good language.

Half past eight. Elder not returned from the Bazar yet. Can't think what keeps him. Laid the cloth and breakfast things myself and ordered Smith to bring me the tea kettle. Used plenty of milk in my tea, and made a good breakfast. Took three pinches of snuff whilst I sat thinking of my wife and good friends in England. Memo. Must not take so much snuff when I return, for it makes me spit about the rooms.

Find myself better this morning than usual, and less head-ache. Took up my flute and played the 1st and 5th Duo of Playel's opera 9. Note. The first commences in a great stile [*sic*], and is sweetly plaintive in some parts of it. Hope Mrs. F. [his wife Ann] will have got the better of the inflammation in her eyes; it is now fine weather in England and she will be able to ride out. Must take a house in the country when I return, and enjoy myself two or three months before I engage in any service; but, God knows, it is now three years since I heard from anybody at home; and what may have happened it is impossible to say ...

Ten o'clock. Sat down to my writing. Transcribed four pages of the chapter in my log book upon the state of the barometer upon the different coasts of Australia, into a letter to Sir Joseph Banks for the Royal Society. Found a giddiness and aching in my head; left off writing, and walked backwards and forwards in my room. Think I have advanced rather too much of my own opinions in this letter, and wish I had confined myself more closely to facts; but think Sir Joseph will strike out what he thinks is incorrect. Hope he will be alive and well when I return; but is now advancing in years. If any accident should have happened to him, there is great doubt whether my voyage will ever be completed; or much notice taken of what I have already done ...

My head still aching; went out into the garden, and walked under the shade of the trees. Consider that Mr. Aken will be arrived in America before this with my charts and log book. There is some hopes the Admiralty may put me upon the post captains list on his arrival in England but it is much to be feared they will not before I can return. Determine to have a court-martial upon the loss of the schooner,[1] as it will give me an opportunity of making known my treatment in this island in an official manner.

Returned up stairs, washed, and sat down to dinner at two o'clock. The French beans are very good in this island. Made a tolerably good dinner and drank three glasses of Madeira. Am determined to persevere in the plan of eating more of puddings and vegetables, and less meat. Find my head-ache better after dinner. Think there must certainly be some river or large opening upon the north-west coast of Australia. Hope the Admiralty will not give any more passports to French ships to go out on discovery, whilst I am kept prisoner here. Cannot conceive how it is that there should be no copy of Tasman's chart of that coast remaining, spoken of by Dampier.

If there is an opening near the Rosemary Isles, a settlement there would be advantageous for the East India Company, on account of the high tides, the proximity of the position to the Spice Islands, as a place for their ships to touch at and take in Spices for China, as a naval station for the eastern cruizers, and to counteract the armaments of the French at this island. Determine to propose it to the company on my return. Would I go out as governor of a settlement there, should it be proposed to me? I can't tell, it would depend on many circumstances. Wish to finish the examination of the whole coasts of Australia before I do anything else ...

Half past three. Find myself a little sleepy. Don't know whether to go down and play a game of billiards with the old sergeant, to drive it off, or take a nap. Determined on the latter and laid down on my bed.

Five. Waked. Had a head-ache and looked very pale. Went down to the gate and sat down in the sentry box, looking at the people who passed by. The mulatto creoles are very thin and tall, but their legs have but little calf, and have something of the negro curve in them. They have pleasant countenances. The sentry talking with his comrade said Bonaparte was only 38 years old when he made himself Emperor. The soldiers don't understand the comparative rank between the army and navy officers. Note. French soldiers talk much more than English soldiers do, and seem happy enough upon an allowance that would scarcely keep an English soldier from starving.

At sunset, returned upstairs and walked till Elder got tea ready. Do not wish my friend Pitot to give me introductions to more than two or three families when I go into the country. Hope Captain Bergeret will be able to procure me the remainder of my books, papers, &c., from the general ... After tea, walked a few turns. Necessary to read over the article Meteorology in the Encyclopedia Britannica. Read over also, the artices Weather and Wind. Agree with the writer of the article that the moon has little, if anything, to do with the weather.

Went to bed at half past nine. Lay considering for some time upon the causes of the trade and of the westerly winds, especially upon the earth's revolution round its axis. Think they are certainly owing in some parts to this cause, as well as the rarefaction of the air under the vertical sun. Must have some kind of trap set for that rat, which comes disturbing me every night; but as I am soon to leave the house it does not signify.

Dropped asleep soon after ten. Waked about one by the noise of the soldiers in the guard house, who are playing about and running after each other like children. Wish that loud-voiced fellow had taken a dose of opium. Fell asleep again. Dreamed that general De Caen was setting a lion upon me to devour me, and that he eat me up. Was surprised to find devouring so easy to be borne, and that after death I had consciousness of existence. Got up soon after six, much agitated, with a more violent head-ache than usual, and with bilious sensations in my stomach.

[Flinders, Private Journal, 1803-1814, Mitchell Library, State Library of NSW][2]

Notes

1. The *Cumberland.*
2. The author acknowledges his use of the facsimile edition, published by Genesis Publications Ltd, in association with Hedley Australia, 1986.

APPENDIX D

⚓

Roll Call

The French expedition – biographical notes

Naval officers

BAUDIN, Charles (1784-1854). Naval officer (admiral). Born at Sedan, July 1784, son of elected member of National Convention (1792-1795). Embarked as *aspirant* (midshipman) in *Géographe* 1800. Promoted *enseigne* (sublieutenant) 1804. Lost arm in combat with English frigate in Indian Ocean 1808. Lieutenant 1807, *capitaine de vaisseau* (post captain) 1814. Retired at restoration of monarchy and established import business at Le Havre. Rejoined Navy 1838 and appointed rear admiral. Commanded expedition of reprisal against government of Mexico, captured 'impregnable' fortress of Saint-Jean d'Ulloa at Vera Cruz. Promoted vice-admiral 1839. Commander-in-chief Mediterranean fleet 1848, awarded *Grand Cordon de la Légion d'Honneur*. Admiral 1854, died same year in Paris.

BOUGAINVILLE, Hyacinthe-Yves-Philippe Potentien de (1781-1846). Naval officer (rear admiral) and navigator. Born at Brest, December 1781, son of circumnavigator Louis-Antoine de Bougainville. Entered Navy 1800, joined Baudin expedition as *aspirant*. Returned to France in *Naturaliste* 1803. Served with North Sea and Channel fleets, promoted lieutenant 1808, commander 1811. Taken prisoner 1814. Post captain in Royalist Navy 1821, took part in expeditions to Far East and West Indies 1819-1820. Commanded voyage of circumnavigation 1824-1826 (including three months at Sydney 1825). Returned to France with extensive natural-history collections. Promoted rear admiral 1838, served as naval commander at Algiers. Died in Paris, December 1846.

FREYCINET, Henri-Louis de Saulses de (1777-1840). Naval officer (rear admiral) and colonial governor. Born at Montélimar (Drome), December

1777. Older brother of Louis (q.v.). Joined Navy with Louis, 1794. Saw action against British vessels 1795-1800. Embarked in *Géographe* as *enseigne* (sublieutenant) autumn 1800. Promoted acting lieutenant by Baudin, and served as senior deck-officer in *Géographe* from November 1801. On return to France in 1804, promoted lieutenant, and resumed active service. Wounded and taken prisoner in West Indies. After restoration of monarchy appointed successively governor of Réunion, Guiana and Martinique. Later Maritime Prefect of Rochefort. Promoted rear admiral before his death at Rochefort, 1840.

FREYCINET, Louis-Claude de Saulses de (1779-1842). Naval officer and navigator. Born at Montélimar, August 1779. Entered Navy 1794 with older brother Henri. Embarked as *enseigne* in *Naturaliste* 1800. Promoted acting lieutenant by Baudin 1801, and appointed to command schooner *Casuarina* at Port Jackson, November 1802. Returned to France in *Géographe* 1804, and from 1806 charged with preparing charts and atlas of Baudin's voyage. Following Péron's death in 1810, responsible for completing latter's history of voyage. Remained in Navy after restoration of monarchy, and promoted *capitaine de frégate* 1817. Commanded corvette *Uranie* on scientific voyage around world 1817-1820 (against orders, carried wife Rose on board). Shipwrecked on Falklands on homeward voyage, but saved scientific collections. Received by King Louis XVIII on return, and promoted *capitaine de vaisseau* (post captain) on spot. Devoted remaining years to preparing his *Voyage autour du Monde* (seventeen volumes, 1824-1844), numerous scientific papers, and revised edition of his and Péron's *Voyage de découvertes aux Terres Australes* (1824). Foundation member of Paris Société de Géographie. Elected to Academy of Sciences 1825. Created baron. Died 1842.

HAMELIN, Jacques-Félix-Emmanuel (1768-1839). Naval officer (rear admiral) and administrator. Born at Honfleur, 1768. Served in merchant marine before joining Navy 1792. Appointed lieutenant 1795, sailed on expedition to Ireland 1796 as *capitaine de frégate* (commander). Second-in-command to Baudin, and captain of *Naturaliste* 1800-1803. Promoted *capitaine de vaisseau* (post captain) on return, commanded division of Boulogne flotilla preparing for invasion of England. Engaged in several actions against British vessels blockading French ports. Commander of 'Hamelin Division' in Indian Ocean and East Indies 1808-1810. Took part in battle of Grand Port (Mauritius), in which four British frigates were destroyed, before surrendering to superior forces. After return to France in 1811, promoted rear admiral and created baron. Following restoration of monarchy, appointed major general at Toulon 1818-1822; commanded blockade of Cadiz 1823. Later director-general of Navy's Office of Maps and Plans. *Grand-Officier de la Legion d'Honneur.* Died in Paris, 1839.

MILIUS, Pierre-Bernard (1773-1829). Naval officer (rear admiral) and colonial governor. Born at Bordeaux, January 1773. Went to sea as boy on father's merchant ships, 1786-1792. Joined Navy 1793, served on several campaigns. As *lieutenant de vaisseau* (lieutenant commander), appointed Hamelin's second-in-command on *Naturaliste* 1800-1802; promoted *capitaine de frégate* at Timor, 1801. Left expedition at Port Jackson due to ill health, and returned to Isle de France (Mauritius) via China. Appointed commander of *Géographe* after Baudin's death, September 1803. Returned to active service 1804, but captured by British in following year; paroled 1806. Appointed director of Port of Venice, then became post captain in Italian Navy (under French control). Returned to France 1814. After restoration of monarchy, appointed director of Port of Brest 1815; governor of Bourbon (Réunion) 1818-1821, and of Guiana 1823-1825. Commanded *Scipion* at Battle of Navarino, 1827; promoted rear admiral on return. Created baron 1819, *Chevalier de la Légion d'Honneur* 1821. Died August 1829.

Scientific staff

BORY DE ST VINCENT, Jean-Baptiste-Georges-Marie (1780-1846). Naturalist and army officer. Born 1780. Volunteered for Baudin expedition while serving in Army of West, sailed in *Naturaliste* as zoologist. Defected at Isle de France 1801, attacked Baudin's character after returning to France following year. Resumed army career, serving as staff officer with marshals Davout, Ney and Soult during Napoleonic Wars; reached rank of colonel. Exiled after restoration of monarchy, 1815, remained abroad until 1820. Published *Dictionnaire Classique d'Histoire Naturelle* (seventeen volumes, 1822-1831). Elected to Academy of Sciences. Appointed *chef du bureau historique* in War Office, and promoted *maréchal de camp* (brigadier) in Corps of Engineers. Died in Paris, 1846.

LESCHENAULT DE LA TOUR, Jean-Baptiste (1773-1826). Naturalist. Born November 1773 into lesser nobility. Imprisoned with family during the

Terror, 1793-1794. Volunteered for Baudin's expedition 1800. Became interested in breeding of merino sheep at Port Jackson. Left at Timor, June 1803, on homeward voyage, too ill to continue. Made his way to Dutch headquarters at Batavia (Jakarta), botanised extensively on Java 1803-1806. Returned to France with rich collections 1807. Appointed *inspecteur des brebis mérinos* (merino sheep) 1811. After restoration of monarchy, travelled widely in India and Ceylon 1816-1822 under patronage of Sir J. Banks. Pioneered acclimitisation of useful plants (sugarcane, spices, coffee, cotton, and so on) and animals in French colonies. Awarded Cross of the Legion of Honour, 1822. Scientific visits to Brazil and Guiana 1823-1824. Died in Paris, March 1826.

LESUEUR, Charles-Alexandre (1778-1846). Naturalist and artist. Born at Le Havre, January 1778, son of naval official. Enlisted in National Guard 1797-1799. Recruited by Baudin as 'assistant gunner' on *Géographe*, in reality to illustrate, with N.-M. Petit, the commandant's journal of voyage. Close friend of F. Péron. Returned to France 1804, worked with Péron on history of voyage until latter's death 1810. Emigrated to USA, 1815. Elected member of American Philosophical Society and Academy of Natural Sciences. Travelled widely in North America, lived for time at Robert Owen's experimental community at New Harmony, Indiana. Returned to France 1837. *Chevalier de la Légion d'Honneur* 1845. Appointed first director of Muséum d'Histoire Naturelle at Le Havre, March 1846. Died at Le Havre, December 1846.

Colonial military and administration

DECAEN, Charles-Mathieu-Isidore (1769-1832). Army officer (general) and colonial governor. Born at Caen, 1769. Served in marine artillery 1787-1790, then studied law. Re-enlisted at outbreak of war 1792, promoted captain, fought in Vendée against royalist rebels. Promoted *général de brigade* in Rhine Army, took part in battle of Hohenlinden against Austrians, 1800. Captain general of French territories east of Cape of Good Hope 1803-1810. On surrender of Isle de France returned to France. Commanded French army in Catalonia, 1811, and in the Netherlands, 1813. In 1814 appointed to command 'Gironde Corps' to retake Bordeaux from Wellington's troops. At restoration of monarchy, given a divisional command, but rallied to Napoleon during the Hundred Days. Imprisoned in 1815, amnestied after fifteen months. Reserve general until 1830. Wrote his memoirs in retirement. Died near Paris, 1832.

LINOIS, Charles-Alexandre-Léon Durand (1761-1848). Naval officer (admiral). Born at Brest, 1761. Entered Marine 1776, served as volunteer in American War of Independence. Captain 1795. Lost eye and captured in battle off French coast 1795, exchanged some months later. Took part in expedition to Ireland, 1796. Rear admiral 1799. Defeated English fleet at Algeciras, 1801. Commanded East Indies Division 1803-1806. Met British squadron on homeward voyage, 1806, and surrendered after fierce battle. Remained prisoner until 1814. Appointed governor of Guadeloupe by Louis XVIII, but rallied to Napoleon; after latter's defeat, surrendered colony to British forces. Court-martialled for revolt and insubordination, but acquitted. Died at Versailles, 1848.

The British expedition – biographical notes

Naval officers

FLINDERS, Samuel Ward (1782-1834). Naval officer (lieutenant). Born at Donington, Lincolnshire, November 1782, younger brother of Matthew Flinders. Enlisted as volunteer in HMS *Reliance* 1784 (at age twelve) to accompany Matthew (senior master's mate) on voyage to New South Wales. Returned to England 1800. Promoted lieutenant March 1801, appointed to *Investigator* at Matthew's request. After shipwreck returned to England with Commodore Dance, awarded ceremonial sword for his part in action off Pulo Auro. Served in Channel Fleet, appointed to command HMS *Bloodhound* 1806; court-martialled for disobeying orders 1808, dismissed ship and docked three years' seniority. Retired on half-pay. Died 1834, buried at Donington.

FOWLER, Robert Merrick (1778-?). Naval officer (rear admiral). Born 1778 at Horncastle, Lincolnshire. Entered Navy May 1793 as volunteer. Served as midshipman in *Royal William* (100), *Hector* (seventy-four), *Cumberland* (seventy-four), and *Royal George* (100), flagship of Admiral Lord Bridport. Promoted lieutenant February 1800, joined *Xenophon* (later *Investigator*) as first lieutenant; Flinders's second-in-command 1801-1803. Appointed to command *Porpoise* on homeward voyage, wrecked August 1803; exonerated of responsibility for shipwreck at court-martial, 1804. Reached Canton, embarked as passenger on board East Indiaman *Earl Camden*, Captain Nathaniel Dance. Joined Dance in driving off French squadron under Admiral Linois at Pulo Auro, February 1804. Promoted commander 1806. On active service in home waters and West Indies Station, 1805-1811. Post captain 1811, retired on half-pay. Rear admiral 1846.

FRANKLIN, Sir John (1786-1847). Naval officer (post captain), governor of Van Diemen's Land, and Arctic explorer. Born at Spilsby, Lincoln-shire, 1786; related to Flinders family. Entered Navy as midshipman, present at battle of Copenhagen, 1801. Midshipman in *Investigator* 1801-1803. Signals officer in *Bellerophon* at Trafalgar, 1805. First voyage to Arctic 1818-1822. Promoted post captain on return and elected Fellow of Royal Society. Second voyage to Arctic 1825-1827. Knighted 1829, served as governor of Van Diemen's Land 1836-1843. Left on third Arctic voyage May 1845; his ships *Erebus* and *Terror* sighted in Baffin Bay on 26 July, then disappeared. Cairn found in 1859 revealed ships had been trapped in ice and abandoned after eighteen months. Crews had left to march overland, but none survived. (Remains of some bodies have been excavated from permafrost in recent years.)

MURRAY, John (1775- ?). Naval officer (lieutenant). Following Lieutenant Grant's resignation, Governor King appointed Murray acting lieutenant to command surveying brig *Lady Nelson*, September 1801; sailed mid-November to survey north-west coast of Bass Strait. Explored Port Phil-lip Bay February-March 1802. After return to Port Jackson, *Lady Nelson* sailed in company with *Investigator* on survey of east and north coasts, but Flinders sent Murray back to port following damage to brig's keel. Murray's application for promotion to lieutenant supported by King, but refused by Admiralty because he had not served six years at sea as re-quired by regulations. Returned to England 1803, undertook hydro-graphic surveys of south coast 1804-1808. Promoted lieutenant Octo-ber 1807. Superseded in his command 1809, subsequent career unknown.

Scientific staff

BAUER, Ferdinand Lucas (1760-1826). Natural-history artist and naturalist. Born in Feldsberg, Austria, January 1760, son of court painter and gal-lery curator. Following studies as botanical draughtsman at Vienna, visited Italy, Greece, and Asia Minor 1784-1787 as artist with Professor Sibthorp, Professor of Botany at Oxford; returned to England with Sibthorp. Appointed to *Investigator* expedition by Sir Joseph Banks, and worked closely with Robert Brown throughout voyage; spent eight months on Norfolk Island 1804-1805. Returned with Brown to England,

1805, with more than 2000 natural-history sketches. Employed by Admiralty to make finished drawings and paintings under Banks's supervision; completed around 300 splendid watercolours, but unable to find publisher in wartime. Returned to Austria at war's end, 1814, died March 1826. Now considered 'the Leonardo of natural history painting' (Bernard Smith).

BROWN, Robert (1773-1858). Botanist. Born at Montrose, Scotland, December 1773. Studied medicine at Edinburgh University. Offered position of naturalist in *Investigator* by Banks, 1800. After completing circumnavigation with Flinders, he and Bauer remained in colony to continue botanical investigations; Brown spent ten months in Van Diemen's Land. Returned to England in 1805 with nearly 4000 plant species. Appointed Bank's librarian and curator 1810; on latter's death in 1820 was bequeathed life tenancy of Soho Square and custody of its collections. Transferred collections to British Museum in 1827, and appointed first Keeper of Botany – held position until his death in 1858. First observer of Brownian movement. Fellow of Royal Society 1810, president of Linnean Society 1849-1853.

WESTALL, William (1781-1850). Artist and topographical illustrator. Born at Hertford, October 1781. Probationary student at Royal Academy when appointed landscape artist on voyage at age nineteen. Shipwrecked at Wreck Reef, 1803. Boarded *Rolla* for Canton, remained there to paint 'picturesque' subjects; later visited Ceylon and India before returning to England 1805. Following Flinders's return in 1810, Westall prepared drawings made on voyage for publication; also completed oil paintings of selected views for Admiralty – several of these exhibited at Royal Academy, 1812. Elected associate of Royal Academy the same year. In later years specialised as topographical and architectural illustrator. Died in London, 1850.

Colonial military and administration

KING, Philip Gidley (1758-1808). Naval officer (post captain), governor of New South Wales 1800-1806. Born at Launceston, Cornwall, April 1758. Joined Navy as captain's servant, December 1770. Served in East Indies 1771-1775, in American waters 1775-1778. Promoted lieutenant 1780; served under Captain Arthur Phillip 1782-1783. Accompanied

Phillip, first governor of New South Wales, as second lieutenant in HMS *Sirius*, flagship of First Fleet; established settlement on Norfolk Island, February 1788. Lieutenant governor of Norfolk Island 1791-1796. Promoted post-captain 1798, and appointed third governor of New South Wales 1800. King encouraged exploration, increased colony's trading links, and sought to control trafficking in liquor; in 1803 he established first settlement in Van Diemen's Land. Failing health and opposition to his policies in London induced him to offer his resignation 1803, but he was not replaced until 1806 by Governor Bligh. Returned to England 1807, died in London, September 1808.

PATERSON, William (1755-1810). Soldier (lieutenant colonel), explorer, and lieutenant governor. Born August 1755 in Scotland. Visited South Africa 1777-1780; made four journeys into interior collecting botanical and other specimens. Joined Army 1781, served in India to 1785. Appointed captain in New South Wales Corps 1789, commanded troops on Norfolk Island 1791-1793. Promoted major 1795, returned to England following year. Elected Fellow of Royal Society 1798. Paterson returned to colony in 1799 as officer commanding NSW Corps, appointed lieutenant governor of NSW 1800. Lieutenant governor of new settlement at Port Dalrymple (Launceston) 1804-1808. Returned to Sydney 1809, assumed office as acting governor following Rum Rebellion against Bligh. Recalled after Governor Macquarie's arrival 1810; died at sea on homeward voyage in June 1810.

Casualties of history

BECKWITH, Mary (1786? –). Convict. At age fourteen found guilty of theft of goods valued at £5 by Court of King's Bench, Old Bailey, London, in July 1800. Death sentence commuted to transportation for life. Arrived at Port Jackson on transport *Nile* December 1801. Met Nicolas Baudin during French visit to Sydney, left with him on *Géographe* November 1802. Disembarked at Port Louis, Isle de France, August 1803. Probably left colony with Captain Augustin Baudin for Danish settlement of Tranquebar, India, in 1804. Nothing is known of her later life.

JORGENSON, Jorgen (1780-1841). Seaman, adventurer, convict, writer. Born at Copenhagen, Denmark, 1780, second son of royal watchmaker. Joined British merchant service; by his own account was press-ganged into Royal Navy. Arrived at Port Jackson on brig *Harbinger* 1801, enlisted (as John Johnson) as second mate in *Lady Nelson*; accompanied Flinders on first stage of his circumnavigation. Returned to Denmark via London 1806. Commanded Danish privateer in Anglo-Danish war, taken prisoner 1808. In 1809 visited Iceland on English merchant ship, organised coup and overthrew Danish governor; proclaimed Iceland independent and himself 'Protector'. After nine weeks surrendered to British warship and voluntarily returned to London. British agent on continent 1815-1816, present at Waterloo. Imprisoned for theft 1820, sentenced to transportation to Van Diemen's Land 1825. Granted conditional pardon 1830, worked as constable, farmer, explorer and writer. Died in Hobart, January 1841.

SMITH, Samuel (1771? -1821). Seaman. Obituary from the *Manchester Gazette*, Saturday, 10 February 1821:

[Died] in Thornton's Court, on the 29th ultimo, Samuel Smith, aged 50. In 1801, he sailed a volunteer on board the *Investigator*, Capt. Flinders, on a voyage of discovery; which vessel after keeping the sea for rather more than two years, was wrecked in the South Sea, upon a sand-bank, about half a mile in circumference – being 150 miles from the nearest land, and nine hundred miles from any port. In this wretched situation, the crew were apprehensive of being washed away by the high spring tides, or by the rolling of the sea in bad weather. After remaining on the bank upwards of eight weeks, they were released by the *Rolla*, Indiaman, bound for China. On his return to Europe, he was impressed in the Downs, and retained in the Naval Service until the year 1815. As an instance of extraordinary application, we should state, that although the deceased was but a private seaman, he kept a regular journal of occurrences during the whole of the time he remained at sea.

Bibliography

Matthew Flinders and the British voyage of discovery, 1801 to 1810

Primary sources

Brown, Robert. In preparation. *Nature's investigator: The diary of the naturalist Robert Brown in Australia, 1801-1805.* Edited by T.G. Vallance, E.W. Groves and D.T. Moore.

Flinders, Matthew. 1801. *Observations on the coasts of Van Diemen's Land, on Bass's Strait and its islands, and on part of the coasts of New South Wales ...* London: Nicol. (Reprinted 1946, Australian Historical Monographs vol. XXIX (new series), Sydney. Introduction by G. Mackaness.)

———. 1814. *A voyage to Terra Australis: Undertaken for the purpose of completing the discovery of that vast country, and prosecuted in the years 1801, 1802, and 1803 in His Majesty's Ship the In-*vestigator, *and subsequently in the Armed Vessel* Porpoise *and* Cumberland *schooner, with an account of the ship-wreck of the* Cumberland ... 2 vols and atlas. London: G. & W. Nicol.

———. 1985. *A biographical tribute to the memory of Trim, Isle de France 1809.* With an introduction by S. Murray Smith and a learned discourse on the subject by T.M. Perry. Sydney: Halstead Press.

———. 1985. *Narrative of Tom Thumb's cruize to Canoe Rivulet (from 25 March 1796 to 2 April 1796).* Edited by K. Bowden. South-Eastern Historical Association, Victoria.

———. n.d. Private journal, 17 December 1803 - 8 July 1814. MS, Mitchell Library, Sydney. (Facsimile edition published 1986 by Genesis Publications in association with Hedley Australia.)

[Flinders, Samuel.] 1910-11. Historical sketch of the life of the late Captain Flinders. *Victorian Geographical Journal.* Vol. 28. (Original dated 1817, attributed to Samuel Flinders by Ingleton (1986).)

Good, Peter. 1981. The journal of Peter Good, gardener on Matthew Flinders' voyage to Terra Australis 1801-1803. Edited by Phillis I. Edwards. *Bulletin of the British Museum (Natural History), Historical Series.* Vol. 9.

La Trobe Library Journal. 1974. 4(3): March. Matthew Flinders Issue.

National Maritime Museum (Greenwich, UK). n.d. Flinders Papers. FLI/25. (Letters received by Mrs Flinders from

Matthew Flinders, together with four letters to her family; also available on microfilm in selected Australian libraries.)

Smith, Samuel. n.d. Journal written by Samuel Smith, Seaman, who served on board the *Investigator*, Cap'n Flinders, on a voyage of discovery in the South Seas. MS, Mitchell Library, Sydney.

Secondary sources

Austin, K.A. 1964. *The voyage of the* Investigator *1801-1803: Commander Matthew Flinders RN*. Adelaide: Rigby.

———. 1965-66. Matthew Flinders as an author. *Proceedings*. Royal Geographical Society of Australasia (SA Branch). Vol. 67.

Baker, Sidney J. 1962. *My own destroyer: A biography of Matthew Flinders, explorer and navigator*. Sydney: Currawong.

Baldwin, B.S. 1964-65. Flinders and the French. *Proceedings*. Royal Geographical Society of Australasia (SA Branch). Vol. 66.

Brown, Anthony J. 1998. Flinders on Mauritius: Ulysses bound. Paper presented to the Royal Geographical Society of South Australia, Adelaide, 19 February.

Colwell, Max. 1970. *The voyages of Matthew Flinders*. Sydney: Hamlyn.

Cooper, H.M. 1953. *The Unknown Coast: Being the explorations of Captain Matthew Flinders RN, along the shores of South Australia 1802*. Adelaide. (Also 1955, *Supplement*.)

Findlay, Elisabeth. 1998. *Arcadian quest: William Westall's Australian sketches*. Canberra: Australian National Library.

Giblin, R.W. 1929. Flinders, Baudin and Brown at Encounter Bay. *Papers*. Royal Society of Tasmania.

Hill, Ernestine. 1941. *My love must wait*. Sydney: Angus & Robertson. (Biographical novel about Matthew Flinders.)

Ingleton, Geoffrey C. 1986. *Matthew Flinders: Navigator and chartmaker*. Genesis Publications in association with Hedley Australia.

Ly-Tio-Fane Pineo, Huguette. 1988. *In the grips of the eagle: Matthew Flinders at Ile de France 1803-1810*. Moka, Mauritius: Mahatma Ghandi Institute.

Mack, James D. 1966. *Matthew Flinders 1774-1814*. Melbourne: Nelson.

Mackay, D. 1990. 'In the shadow of Cook: the ambition of Matthew Flinders' in *European voyaging towards Australia*. Edited by J. Hardy and A. Frost. Canberra: AAH.

Norst, Marlene J. 1989. *Ferdinand Bauer: The Australian natural history drawings*. Melbourne: Lothian.

Perry, T.M., and D.H. Simpson (eds). 1962. *Drawings by William Westall, landscape artist on board HMS* Investigator *during the circumnavigation of Australia by Captain Matthew Flinders, RN, in 1801-1803*. London: Royal Commonwealth Society.

Retter, Catharine, and Shirley Sinclair. 1999. *Letters to Ann: The love story of Matthew Flinders and Ann Chappelle*. Sydney: Angus & Robertson.

Royal Geographical Society of Australasia (SA Branch). 1910-11. General Decaen's report on the detention of Flinders. *Proceedings*. Vol. 12.

Russell, R.W. (ed.). 1979. *Matthew Flinders:*

The ifs of history. Adelaide: Flinders University.

Smith, Keith Vincent. 1992. *King Bungaree: A Sydney Aborigine meets the great South Pacific explorers 1799-1830*. Sydney: Kangaroo Press.

Stubbs, B.J., and P. Saenger. 1996. The Investigator Tree: Eighteenth century inscriptions, or twentieth century misinterpretations? *Journal*. Royal Historical Society of Queensland. 16(3).

Sutton, Ann, and Myron Sutton. 1966. *The endless quest: The life of John Franklin, explorer*. London: Constable.

Watts, Peter, JoAnne Pomfrett and David Mabberley. 1997. *An exquisite eye: The Australian flora and fauna drawings 1801-1820 of Ferdinand Bauer*. Sydney: Historic Houses Trust of NSW.

Nicolas Baudin and the French voyage of discovery, 1801 to 1804

Primary sources

Baudin, Nicolas. 1974. *The journal of Post-Captain Nicolas Baudin, commander-in-chief of the corvettes* Géographe *and* Naturaliste – *assigned by order of the government to a voyage of discovery*. Translated by Christine Cornell. Adelaide: Libraries Board of SA.

———. In press. *Mon Voyage aux Terres Australes: Journal personnel du commandant Baudin*. Transcribed and edited by Jacqueline Bonnemains. Paris: Imprimerie Nationale.

Bory de St Vincent, J.-B.G.M. 1805. *Voyage to, and travels through the four principal islands of the African seas, performed by order of the French government, during the years 1801 and 1802, with a narrative of the passage of Captain Baudin to Port Louis in the Mauritius*. Translated. London.

Hélouis transcripts. Copies held by Mitchell Library, Sydney; LaTrobe Library, Melbourne; National Library of Australia: Canberra. (Extensive selection of documents in the French archives made by Prof. Ernest Scott of Melbourne prior to 1914, and transcribed by Mme Robert Hélouis; includes copies of Baudin's correspondence and reports, and extracts from journals of Hamelin, the Freycinet brothers, and other officers.)

Historical records of New South Wales. 'Appendix: Baudin papers.' IV:941-1009.

Milius, Pierre-Bernard. 1987. *Récit du Voyage aux Terres Australes par P.-B. Milius, Second sur le* Naturaliste *dans l'expédition Baudin (1800-1804)*. Transcribed by J. Bonnemains and P. Haughel. Le Havre: Muséum d'Histoire Naturelle. (Original MS in Kerry Stokes Collection, Perth.)

Péron, F. 1809. *A voyage of discovery to the Southern Hemisphere, performed by order of the Emperor Napoleon, during the years 1801, 1802, 1803, and 1804*. Translated (vol. I only). London: R. Phillips. (Republished 1975, Marsh Walsh Publishing, Melbourne.)

Péron, F., and L. Freycinet. 1807-1816. *Voyage de découvertes aux terres australes, exécuté par ordre de Sa Majesté l'Empereur et Roi, sur les corvettes le* Géographe, *le* Naturaliste, *et la goelette le* Casuarina, *pendant les années 1800, 1801, 1802, 1803 et 1804*. 3 vols and 3 atlases. Paris: De l'Imprimerie impériale. (2nd ed., 4 vols and 1 atlas, 1824.)

————. n.d. A voyage of discovery to the Southern Hemisphere … Vol. II. Unpublished MS, translated by Christine Cornell. SA Museum.

Secondary sources

Ageorges, Roger. 1994. *Ile de Ré/terres australes: les voyages du capitaine Baudin marin et naturaliste.* Ste-Marie-de-Ré: Groupement d'Etudes Rétaises.

Baldwin, B.S. 1962-63. French sources of South Australian history. *Proceedings.* Royal Geographical Society of Australasia (SA Branch). Vol. 64.

Blackman, Maurice (ed.). 1990. *Australian Aborigines and the French: Papers from a symposium held at the University of New South Wales 21-23 July 1988.* Occasional Monographs No. 3. French-Australian Research Centre.

Bonnemains, Jacqueline. 1989. *Les Artistes du 'Voyage de découvertes aux terres australes' (1800-1804): Charles-Alexandre Lesueur et Nicolas-Martin Petit.* Le Havre: Muséum d'Histoire Naturelle.

———— (ed.). 1996. *Les velins de Charles-Alexandre Lesueur: exposition du 4 mai au 2 juin 1996.* Le Havre: Muséum d'Histoire Naturelle.

———— (ed.). 1997. *Oeuvres de Nicolas-Martin Petit, Artiste du Voyage aux Terres Australes (1800-1804): exposition du 1er juin au 31 décembre 1997.* Le Havre: Muséum d'Histoire Naturelle.

Bonnemains, J., E. Forsyth and B. Smith. 1988. *Baudin in Australian waters: The artwork of the French voyage of discovery to the Southern Lands 1800-1804.* Melbourne: Oxford University Press/ Australian Academy of Humanities.

Bouvier, René, and E. Maynial. 1947. *Une Aventure dans les mers australes: l'expédition du commandant Baudin 1800-1803.* Paris: Mercure de France.

Brown, Anthony J. 1995. The shadow of a captain on a ghost ship: The lost reputation of Nicolas Baudin. Paper presented to the Royal Geographical Society of South Australia, Adelaide, 26 October.

————. 1998. The captain and the convict maid: A chapter in the life of Nicolas Baudin. *South Australian Geographical Journal.* 97:20-32

Cooper, H.M. 1952. *French explorations in South Australia: With especial reference to Encounter Bay, Kangaroo Island, the two gulfs and Murat Bay, 1802-1803.* Adelaide.

Cornell, Christine. 1965. *Questions relating to Nicolas Baudin's Australian expedition 1800-1804.* Adelaide: Libraries Board of South Australia.

Cullity, Thomas Brendan. 1992. *Vasse: An account of the disappearance of Thomas Timothée Vasse.* Perth.

Degérando, Joseph-Marie. 1969. *The observation of savage peoples.* Translated by F.C.T. Moore. London: Routledge & Kegan Paul.

Duyker, Edward. 1999. In search of Mme Kerivel and Baudin's last resting place. *News.* National Library of Australia. September.

Faivre, Jean-Paul. 1953. *L'Expansion francaise dans le Pacifique de 1800 à 1842.*

Paris: Nouvelles Editions Latines.

———. 1962. *Le Contre-amiral Hamelin et la marine francaise.* Paris: Nouvelles Editions Latines.

———. 1965. La France découvre l'Australie: l'expédition du *Géographe* et du *Naturaliste* (1801-1803). *Australian Journal of French Studies.* 2(1).

———. 1966. Les Idéologues de l'an VIII et le voyage de Nicolas Baudin en Australie (1800-1804). *Australian Journal of French Studies.* 3(1).

Guicheteau, T., and J.-P. Kernéis. 1990. 'Medical aspects of the voyages of exploration, with particular reference to Baudin's expedition 1800-1804' in *European voyaging towards Australia.* Edited by J. Hardy and A. Frost. Canberra: AAH.

Guillonneau, Bernard. 1988. *Les grandes heures de l'Île de Ré.* 3rd ed. La Rochelle: Editions Rupella. ('Le Rétais Nicolas Baudin fut-il un explorateur et un savant dérisoire ou génial?')

Horner, Frank. 1987. *The French reconnaissance: Baudin in Australia 1801-1803.* Melbourne: Melbourne University Press.

Hughes, M. 1988. 'Philosophical travellers at the ends of the earth: Baudin, Péron and the Tasmanians' in *Australian science in the making: Bicentennial essays.* Edited by R.W. Home. Melbourne: Cambridge University Press in association with the Australian Academy of Science.

Hunt, Susan, and Paul Carter. 1999. *Terre Napoléon: Australia through French eyes 1800-1804.* Sydney: Historic Houses Trust of NSW.

Jose, A.W. 1934. Nicolas Baudin (with comments by F.J. Bayldon and W.M. Dixon).

Journal & Proceedings. Royal Australian Historical Society. Vol. 20.

Kelly, Michael. 1965. Francois Péron's hatred of Nicolas Baudin. Address to WA Historical Society, Perth, 27 August.

Ly-Tio-Fane, Madeleine. 1982. Contacts between Schonbrunn and the Jardin du Roi at Isle de France in the 18th century: An episode in the career of Nicolas-Thomas Baudin. *Mitteilungen des Osterreichischen Staatsarchivs.* Vol. 35.

Mander-Jones, P. 1964-65. The artists who sailed with Baudin and Flinders. *Proceedings.* Royal Geographical Society of Australasia (SA Branch). Vol. 66.

Micco, Helen. 1971. *King Island and the sealing trade 1802. A translation of chapters 22 and 23 of the narrative by François Péron published in the official account of the 'Voyage of discovery to the Southern Lands'.* Roebuck Series No. 3. Canberra: Roebuck Society.

Plomley, N.J.B. 1983. *The Baudin expedition and the Tasmanian Aborigines 1802.* Hobart: Blubber Head Press.

Review of F. Péron's *Voyage du decouvertes aux terres australes ...*, Paris 1807. 1810. Attributed to J. Barrow. *Quarterly Review.* Vol. IV. London.

Scott, Ernest. 1910. *Terre Napoléon: A history of French explorations and projects in Australia.* London. Methuen.

Somerville, J.D. 1947. Discovery of Murat Bay. *Port Lincoln Times.* 10 July.

Spate, O.H.K. 1974. Ames damnées: Baudin and Péron. *Overland.* No. 58, Winter.

Triebel, L.A., and J.C. Batt. 1943. *French exploration of Australia.* Sydney: Editions du Courrier Australien.

Turnbull, John. 1813. *A voyage of discovery around the world in 1800, 1801, 1802, 1803, and 1804*. London. (Contains appendix, 'Narrative of the proceedings of *Le Géographe* and *Naturaliste*, sent on a voyage of discovery by the French government in 1800', by Surgeon Thomson of New South Wales.)

Wallace, Colin. 1984. *The lost Australia of Francois Péron*. London: Nottingham Court Press.

Terra Australis/Terres Australes: Exploration and scientific discovery

Secondary Sources

Badger, Geoffrey. 1988. *The explorers of the Pacific*. Sydney: Kangaroo Press.

Beaglehole, J.C. 1974. *The life of Captain James Cook*. London: Hakluyt Society.

Beriot, Agnès. 1962. *Grands Voiliers autour du monde: les voyages scientifiques 1760-1850*. Paris.

Bowden, Keith Macrae. 1952. *George Bass 1771-1803: His discoveries, romantic life and tragic disappearance*. Melbourne: Oxford University Press.

Brosse, Jacques. 1983. *Great voyages of exploration: The golden age of discovery in the Pacific. Translated from the French*. Translated by Stanley Hochman. Sydney: Doubleday.

Cameron, Ian. 1987. *Lost paradise: The exploration of the Pacific*. Topsfield: Salem House.

Clancy, Robert, and Alan Richardson. 1988. *So came they south*. Silverwater: Shakespeare Head Press.

Dampier, William. 1981. *A voyage to New Holland ... in 1699*. Edited and with an introduction by James Spencer. Gloucester: Sutton.

————. 1998. *A new voyage round the world: The journal of an English buccaneer*. Foreword by Giles Milton. (1st published 1697.) London: Hummingbird Press.

de Beer, Gavin. 1960. *The sciences were never at war*. London: Nelson.

Dunmore, John. 1965, 1969. *French explorers in the Pacific*. 2 vols. Oxford: Clarendon Press.

Eisler, William, and Bernard Smith. 1988. *Terra Australis: The furthest shore*. Sydney: Art Gallery of NSW.

Estensen, Miriam. 1998. *Discovery: The quest for the Great South Land*. Sydney: Allen & Unwin.

Finney, C.M. 1984. *To sail beyond the sunset: Natural history in Australia 1699-1829*. Adelaide: Rigby.

Fisher, Robin, and Hugh Johnston (eds). 1979. *Captain James Cook and his times*. Canberra: ANU Press.

Frost, Alan. 1998. *The voyage of the* Endeavour: *Captain Cook and the discovery of the Pacific*. Sydney: Allen & Unwin.

Horner, Frank. 1995. *Looking for La Pérouse: D'Entrecasteaux in Australia and the South Pacific 1792-1793*. Melbourne: Miegunyah Press.

Ingleton, Geoffrey C. 1944. *Charting a continent: A brief memoir on the history of marine exploration and hydrographical surveying in Australian waters from the discoveries of Captain James Cook to the war activities of the Royal Australian Navy Surveying Service*. Sydney: Angus & Robertson.

Jones, Ian, and Joyce Jones. 1992. *Ocea-nograpby in the days of sail.* Sydney: Hale & Iremonger.

Kennedy, Gavin. 1989. *Captain Bligh: The man and his mutinies.* London: Duckworth.

Kenny, John. 1995. *Before the First Fleet: Europeans in Australia 1606-1777.* Sydney: Kangaroo Press.

Lincoln, Margarette (ed.). 1998. *Science and exploration in the Pacific: European voyages to the southern oceans in the 18th century.* Greenwich: National Maritime Museum.

Mabberley, D.J. 1985. *Jupiter botanicus: Robert Brown of the British Museum.* London and Braunschweig: British Museum and Cramer.

Marchant, Leslie A. 1982. *France Australe: a study of French explorations and attempts to found a penal colony and strategic base in south western Australia 1503-1826.* Perth: Artlook Books.

Moyal, Ann. 1986. *A bright & savage land: Scientists in colonial Australia.* Sydney: Collins.

Perry, T.M. 1982. *The discovery of Australia: The charts and maps of the navigators and explorers.* Melbourne: Nelson.

Price, A. Grenfell (ed.). 1971. *The explorations of Captain James Cook in the Pacific: As told by selections of his own journals 1768-1779.* New York: Dover.

Radok, Rainer. 1990. *Capes and captains: A comprehensive study of the Australian coast.* Chipping Norton, NSW: Surrey Beatty & Sons.

Scott, Ernest (ed.). 1929. *Australian discovery.* Vol. I, *By sea.* London: Dent.

Sharp, Andrew. 1968. *The discovery of Australia.* Oxford: Clarendon Press.

Shelton, Russell C. 1987. *From Hudson Bay to Botany Bay: The lost frigates of Lapérouse.* Toronto: NC Press.

Smith, Bernard. 1984. *European vision and the South Pacific.* 2nd ed. Sydney: Harper & Row.

Sobel, Dava. 1996. *Longitude: The true story of a lone genius who solved the greatest scientific problem of his time.* London: Fourth Estate.

Spate, O.H.K. 1988. *The Pacific since Magellan.* Vol. III, *Paradise found and lost.* Canberra: ANU Press.

State Library of NSW. 1998. *Dare to know: The art and science of Pacific voyages.* Sydney: State Library of NSW.

Steven, Margaret. 1988. *First impressions: The British discovery of Australia.* London: British Museum (Natural History).

Taillemite, Étienne. 1995. *Sur des Mers Inconnues: Bougainville, Cook, Lapérouse.* Paris: Gallimard.

Taylor, H.G. 1973. *The discovery of Tasmania.* Hobart: Cat & Fiddle Press.

Villiers, Alan. 1967. *Captain Cook, the seamen's seaman: A study of the great discoverer.* London: Hodder & Stoughton. (Republished 1969, Penguin Books.)

Withey, Lynne. 1987. *Voyages of discovery: Captain Cook and the exploration of the Pacific.* London: Hutchinson.

New South Wales: Foundation and settlement 1788-1810

Primary sources

Bowes Smyth, Arthur. 1979. *The journal of*

Arthur Bowes Smyth: Surgeon, Lady Penrhyn, *1787-1789.* Edited by P.G. Fidlon. Sydney: Australian Documents Library.

Collins, David. 1971. *An account of the English colony in New South Wales.* Facsimile ed., 2 vols. First published 1798, London. Adelaide: Libraries Board of SA.

Historical Records of New South Wales. Vol. I, pt 1, *Cook 1762-80.*

———. Vol. I, pt 2, *Phillip 1783-92.*

———. Vol. II, *Grose and Paterson 1793-95.*

———. Vol. III, *Hunter 1796-99.*

———. Vol. IV, *Hunter and King 1800, 1801, 1802.*

———. Vol. V, *King 1803-05.*

———. Vol. VI, *King and Bligh 1806-07, 1808.*

———. Vol. VII, *Bligh and Macquarie 1809-11.*

Public Library of NSW. 1963-1973. *The Sydney Gazette and New South Wales Advertiser.* Vol. 1 (1803-1804) to vol. 9 (1811). Sydney: Trustees of the Public Library of NSW in association with Angus & Robertson.

Tench, Watkin. 1996. *1788: Comprising 'A narrative of the expedition to Botany Bay', and 'A complete account of the settlement at Port Jackson'.* Edited and introduced by Tim Flannery. Melbourne: Text Publishing.

Secondary sources

Atkinson, Alan. 1997. *The Europeans in Australia: A history.* Vol. 1, *The beginning.* Melbourne: Oxford University Press.

———. 1999. Richard Atkins: The women's judge. *Journal of Australian Colonial History.* 1(1).

Bassett, Marnie. 1956. *The governor's lady: Mrs. Philip Gidley King, an Australian historical narrative.* 2nd ed. Melbourne: Melbourne University Press.

Berzins, Baiba. 1988. *The coming of the strangers: Life in Australia 1788-1822.* Sydney: State Library of NSW.

Caley, George. 1966. *Reflections on the colony of New South Wales.* Edited by J.E.B. Currey. Melbourne: Lansdowne.

Clark, C.M.H. 1962. *A history of Australia.* Vol. I, *From the earliest times to the age of Macquarie.* Melbourne: Melbourne University Press.

Clune, Frank. 1964. *Bound for Botany Bay: Narrative of a voyage in 1798 aboard the death ship* Hillsborough. Sydney: Angus & Robertson.

Clune, Frank, and P.R. Stephensen. 1954. *The Viking of Van Diemen's Land: The stormy life of Jorgen Jorgensen.* Sydney: Angus & Robertson.

Cobley, John (comp.). 1987. *Sydney Cove 1788.* Rev. ed. Sydney: Angus & Robertson.

Crowley, Frank (ed.). 1980. *A documentary history of Australia.* Vol. 1, *Colonial history 1788-1840.* Melbourne: Nelson.

Day, David. 1996. *Claiming a continent: A history of Australia.* Sydney: Harper Collins.

Evans, Susanna. 1983. *Historic Sydney as seen by its early artists.* Sydney: Doubleday.

Flannery, Tim (ed.). 1999. *The birth of Sydney.* Melbourne: Text Publishing.

Frost, Alan. 1980. *Convicts and empire: A naval question 1776-1811*. Melbourne: Oxford University Press.

———. 1987. *Arthur Phillip 1738-1814: His voyaging*. Melbourne: Oxford University Press.

———. 1994. *Botany Bay mirages: Illusions of Australia's convict beginnings*. Melbourne: Melbourne University Press.

Hainsworth, D.R. 1981. *The Sydney traders: Simeon Lord and his contemporaries 1788-1821*. Melbourne: Melbourne University Press.

Holt, Joseph. 1988. *A rum story: The adventures of Joseph Holt, thirteen years in New South Wales (1800-1812)*. Edited by Peter O'Shaughnessy. Sydney: Kangaroo Press.

King, Jonathon, and John King. 1981. *Philip Gidley King: A biography of the third governor of New South Wales*. Sydney: Methuen.

King, Robert J. 1990. *The secret history of the convict colony: Alexandro Malaspina's report on the British settlement of New South Wales (1793)*. Sydney: Allen & Unwin.

Lacour-Gayet, Robert. 1976. *A concise history of Australia*. Translated by J. Grieve. Ringwood, Vic.: Penguin Books.

Maiden, J.H. 1909. *Sir Joseph Banks: The 'Father of Australia'*. Sydney: Government Printer.

Martin, Ged (ed.). 1978. *The founding of Australia: The argument about Australia's origins*. Sydney: Hale & Iremonger.

Moore, John. 1987. *The First Fleet marines 1786-1792*. St Lucia, Qld: University of Queensland Press.

O'Brian, Patrick. 1987. *Joseph Banks: A life*. London: Collins Harvill.

Robinson, Portia. 1988. *The women of Botany Bay: A reinterpretation of the role of women in the origins of Australian society*. Sydney: Macquarie Library. (Republished 1993, Penguin Books.)

Robson, L.L. 1994. *The convict settlers of Australia*. 2nd ed. Melbourne: Melbourne University Press.

Rusden, G.W. 1874. *Curiosities of colonization*. London.

Steven, Margaret. 1983. *Trade, tactics and territory: Britain in the Pacific 1783-1823*. Melbourne: Melbourne University Press.

Ward, Russel. 1987. *Finding Australia: The history of Australia to 1821*. Melbourne: Heinneman.

The wars at sea

Secondary sources

Baynham, Henry. 1969. *From the lower deck: The old navy 1780-1840*. London: Hutchinson.

Crowhurst, Patrick. 1989. *The French war on trade: Privateering 1793-1815*. Aldershot: Scolar Press.

Fregosi, Paul. 1989. *Dreams of empire: Napoleon and the first world war 1792-1815*. London: Hutchinson.

Frere-Cook, Gervis, and Kenneth Macksey. 1975. *The Guinness history of sea warfare*. Enfield, Mx: Guinness.

Gillespie, R. St J. 1935. Sir Nathaniel Dance's battle off Pulo Auro (1804). *Mariner's Mirror*. 21(2).

Hibbert, Christopher. 1994. *Nelson: A personal history*. London: Viking.

James, William. 1902. *The naval history of Great Britain: From the declaration of war by France in 1793 to the accession of George IV*. New ed. 6 vols. London: Macmillan.

Jenkins, E.H. 1973. *A history of the French Navy: From its beginnings to the present day*. London: Macdonald & Jane's.

Lavery, Brian. 1989. *Nelson's navy: The ships, men and organisation 1793-1815*. London: Conway Maritime Press.

Lewis, M.A. 1960. *The social history of the Navy 1793-1815*. London: Allen & Unwin.

Lloyd, Christopher. 1970. *The British seaman 1200-1860: A social survey*. London: Paladin.

McAteer, William. 1991. *Rivals in Eden: A history of the French settlement and British conquest of the Seychelles Islands 1742-1818*. Lewes, Sussex: Book Guild.

Masefield, John. 1971. *Sea life in Nelson's time*. Greenwich: Conway Maritime Press.

Mason, Michael, Basil Greenhill and Robin Craig. 1980. *The British seafarer*. London: Hutchinson with National Maritime Museum.

Nicol, John. 1997. *The life and adventures of John Nicol, mariner*. Edited and introduced by Tim Flannery. Melbourne: Text Publishing.

O'Brian, Patrick. 1970. *Master and commander*. London: Fontana. (First of a series of twenty historical novels set in the Napoleonic Wars.)

Parkinson, C. Northcote. 1977. *Britannia rules: The classic age of naval history 1793-1815*. London: Weidenfeld & Nicolson.

Pengelly, C.A. 1966. *The first Bellerephon*. London: John Barker.

Rawson, Geoffrey. 1963. *Pandora's last voyage*. London: Longmans.

Watt, J., E.J. Freeman and W.F. Bynum (eds). 1981. *Starving sailors: The influence of nutrition upon naval and maritime history*. Greenwich: National Maritime Museum.

Miscellaneous

Edwards, Philip. 1994. *The story of the voyage: Sea-narratives in eighteenth century England*. Cambridge: Cambridge University Press.

Flemyng, Francis P. 1862. *Mauritius: or, the Isle of France*. London: SPCG.

Foulke, Robert. 1997. *The sea voyage narrative*. New York: Twayne Publishing.

Herotodus. 1996. *The histories*. Translated by Aubrey de Sélincourt. Ringwood, Vic.: Penguin Books.

Warner, Oliver. 1958. *English maritime writing: From Hakluyt to Cook*. London: Longmans.

Index